Raphael Scl

Technological Knowledge Transfer

Raphael Schneeberger

Technological Knowledge Transfer

A Study on ICT and Development by an Austrian in Uganda with a View to Intercultural Awareness

VDM Verlag Dr. Müller

Imprint

Bibliographic information by the German National Library: The German National Library lists this publication at the German National Bibliography; detailed bibliographic information is available on the Internet at http://dnb.d-nb.de.
  Any brand names and product names mentioned in this book are subject to trademark, brand or patent protection and are trademarks or registered trademarks of their respective holders. The use of brand names, product names, common names, trade names, product descriptions etc. even without a particular marking in this works is in no way to be construed to mean that such names may be regarded as unrestricted in respect of trademark and brand protection legislation and could thus be used by anyone.

Cover image: www.purestockx.com

Publisher:
VDM Verlag Dr. Müller Aktiengesellschaft & Co. KG , Dudweiler Landstr. 125 a, 66123 Saarbrücken, Germany,
Phone +49 681 9100-698, Fax +49 681 9100-988,
Email: info@vdm-verlag.de

Zugl.: Wien, TU, Diss., 2007

Produced in USA and UK by:
Lightning Source Inc., La Vergne, Tennessee, USA
Lightning Source UK Ltd., Milton Keynes, UK
BookSurge LLC, 5341 Dorchester Road, Suite 16, North Charleston, SC 29418, USA

ISBN: 978-3-639-06926-6

Kurzfassung

Im Zeitalter der Neuen Technologien liegt im Einsatz von Informations- und Kommunikationstechnologien (IKT) viel Hoffnung zur Reduktion der Armut. Bereits 2001 dotierte das United Nations Development Programme (UNDP) einen eigenen Fonds für IKT-Projekte in den (so genannten) Entwicklungsländern in der Höhe von 30 Millionen US-Dollar. Gedanken über die Schwierigkeiten, die beim Einsatz des ("westlichen") Computers in anderen Kulturen auftreten, bleiben unbemerkt oder werden totgeschwiegen. Veröffentlichte Anwenderberichte und Projektstudien sind Mangelware und es gab - bis dato - keine Untersuchung die von einem Computerlehrer mit Praxiserfahrung durchgeführt wurde.

Diese Fallstudie untersucht ein Computerprojekt an einer Grundschule in Südwest - Uganda. Während des Projektes im Jahr 2002 nahm ich mehrere kulturell bedingte Missverständnisse wahr, auf die unter der Anwendung von qualitativen Methoden im Jahr 2003 untersucht wurden. Durch den Einsatz der neuen Methodenkombination des "problemzentrierten Interviews" [Witzel, 1985] und der "objektiven Hermeneutik" [Oevermann, 2002] wurde die Integration des Computers in den Alltag im Jahr 2003 beleuchtet. Diese Untersuchung präsentiert die soziale Funktion des Computers innerhalb der regionalen Bevölkerung und diskutiert die sozialen Beziehungen der involvierten Akteure wie Lehrer, Schüler, Computerlehrer, Evaluator und Geldgeber. Dabei werden die (zu) hoch gestellten Erwartungshaltungen (gegenüber Computer) aufgezeigt und der latente und erwartete Zugewinn an Macht durch den Zugang zum Computer präsentiert. Darüber hinaus zeigt diese Studie auch die kulturelle Erwartungshaltung der Technologiekontrolle auf: Die praktische Auseinandersetzung mit der Technologie (z.b. mit dem Computer) erfolgt erst nach einer theoretischen Aneignung von Fachkenntnissen. Diese Vorgehensweise steht im Kontrast zu der von den externen Computerlehrern durchgeführte und geförderte explorierende Aneignung mithilfe schülerzentrierter Methoden. Diese Studie liefert Hinweise, dass die stark hierarchische Lehrer-Schüler-Beziehung den Drang zur selbstständigen Technologieauseinandersetzung einschränkt, da finanzielle und körperliche Strafen bei Fehlern zu erwarten sind. Diese Studie präsentiert als Lösung das Konzept des "Cultural Brokers" [Aikenhead, 2002]. Ein "Cultural Broker" erkennt unterschiedliche Weltanschauungen zur Wissensaneignung und thematisiert diese im Prozess des Technologietransfers, um zu verhindern, dass der Computer kulturimperialisitisch transferiert wird. Der Lehrer nimmt so die Rolle eines "Cultural Brokers" ein, dessen Aufgabe in der Wahrnehmung und Vermittlung kultureller Differenzen liegt. Ziel ist eine relativ "wertungsarme" Kommunikation bzw. der Aufbau eines gemeinsamen "Kommunikationsraums". Er/Sie handelt gemeinsam mit den Studenten geeignete Vorgangsweisen aus und ermöglicht so einen erfolgreichen technologischen Wissenstransfer. Aus dieser Fallstudie geht hervor, dass ein "Cultural Broker" idealerweise über lokales kulturelles, historisches Wissen, den Gebrauch der lokalen Sprache, Bewusstsein über die soziale Rollen involvierter sozialer Akteure sowie Kenntnis von Sanktionierungsmaßnahmen verfügt. Die Annahme dieses neuen Selbstverständnisses bedingt jedoch die Bereitschaft der Computerlehrer einen Machtverlust in Kauf zu nehmen, welcher für einen erfolgreichen Technologietransfer unerlässlich zu sein scheint.

Abstract

In the last few years ICT projects in *developing countries*[1] have gained tremendously in importance. Nevertheless there has been a lack of case studies dealing with teaching computing in LDCs. So far there has been no case study performed by a scientist with practical experiences in computer related interventions.

This case study describes a computer project, which was undertaken at a primary school in South-West Uganda, in the year 2002. During the intervention the author encountered - in parallel to technical problems - several misunderstandings related to culture. It became apparent that the participating people from that region approach new technology differently than we do (in the mind of members of different cultural backgrounds, namely teachers from Austria) and therefore integrate knowledge in a different way. This led to a cultural perspective, and a qualitative research was conducted. The collected empirical data enabled a systematic hermeneutical analysis with regard to Witzel's programme of problem centred analysis.

This research outlines the social function of computers among the regions' population and discusses the socially reflected interrelationship between involved social actors and the reigning hierarchic positions. It shows that too high expectations (towards computers) are set by participants and that computers are seen as an empowering status symbol. This study shows that the cultural expectation towards gaining control over technology, such as computers, were seen as to be represented in a non explorative way of teaching, whereas the external instructors were causing a cultural break by applying explorative student-centred teaching methods. The strong hierarchic position of teachers and parents hinder the pupils eagerness to explore new technology on their own, as they have to expect different forms of (negative) sanctions, when mistakes are made. A solution might be that a computer teacher becomes a "cultural broker" [Aikenhead, 2002] who is aware of the student's underlying world views, "*regionally used teaching methods/habits*", "*ways of sanctions*", use and status of "*local language*", the social role of "*involved institutions*", and draws upon knowledge on regional historical and cultural specifics. This is also related with the readiness to take into account a loss of power which a successful technological knowledge transfer implies.

---

[1]For "*developing countries*" there is a degree of self-selection and no definition. Currently all *least developed countries* (LDCs) have declared themselves as DCs. [WTO, 2004].

Preface and Acknowledgements

The study presented is the result of my Ugandan encounters in the years 2002 and 2003. It is part of my own intercultural reflective period and shall help other people to enrich their understanding of living and working in a new cultural context. This work also aims at giving insights to all development oriented actors *how* an ICT intervention can take place and what kind of implications arise. Though my own Ugandan experiences are far too limited to generalize facts, I hope to contribute valuable inputs. Furthermore, I would love to see further case studies related to this topic in the near future to discuss and broaden my results. Even though most work has been carried out by myself, the research and its results were shaped based on uncountable discussions and critical inputs by my advisors and colleagues. It is important to stress that whenever I consider the statement as widely accepted or as a result of severe social interaction I have decided to use - if necessary - the pronoun "we" rather than "I". Typically, the latter one will be used throughout the text. Thereby, the way I have perceived and interpreted certain literature is emphasized, too. It is also important to keep in mind that "I" can mean one of my inevitable multiple-roles as a "donor, evaluator, or teacher", which sometimes has been quite difficult to keep separated.

Many people contributed to this thesis directly or indirectly and I would like to thank them very much for their support and patience. First of all, my advisor Gerald Steinhardt deserves many thanks for supporting me so well and encouraging, helping, and motivating me throughout the work on this thesis. Likewise, I would like to thank my second advisor Michael Zach for his valuable comments and input to this work. Thanks to my numerous discussion partners Wolfgang Aigner, Ines Anzengruber, Karin Dauda, Sebastian Fasthuber, Edit Fasthuber, Simon Gampenrieder, Katrin Gatterbauer, Ingrid Hartl, Susanne Höfler, Jan Lengelsen, Harald Nowak, Manfred Perthold, Miriam Pilters, Elke Pürgstaller, Janine Werneck-Reich, Martin Jan Stepanek, Evi Tscholl, and Himal Trikha. Furthermore, I would like to thank Ilse Kroisenbacher for supporting the Ugandan period and Oswald Sedlacek for the shared experience in Uganda. Special thanks to Oliver Vettori for reviewing my work in depth and Harald Pichlhöfer for his help on the research development and concept. Moreover, I would like to thank Richard Crusader, Agnes Kitzler, Roland Krebs, Selina Christina Morrison, and Marion Pachernegg for proofreading my thesis. Thanks also to my Ugandan supporters Doreen Atim, Helen Wangusa, and the UBOS. Last but definitely not least, many thanks go to my grandparents, my parents, brothers and sisters for their love and great support during all the years. Special thanks to all my friends for their understanding, motivation and support during this research and especially during the last few months.

Contents

Chapter 1

Introduction

1.1 Motivation

Information and Communication Technologies (ICTs)[1] are widely promoted as the panacea for LDCs and millions of US-Dollars ($) are spent to "introduce" ICTs in LDCs. Nevertheless, over and over one receives messages about the failing of ICT projects [Avgerou & Walsham, 2000] and studies claim failure rates up to 80% [Heeks, 2000]. But why? How and when does a project become a "failed" one? Who had fixed the projects' goals? Why were these goals not reached and how were these measured/evaluated?

This thesis attempts to deduce relevant factors for a "successful" (see Page 15) implementation of ICT projects in LDCs by trying to investigate a computer project which took place in a primary school in Summer 2002 in South-West Uganda.

The point of departure: In the year 2002 the *non-governmental organization* (NGO[2]) "*Projekt=Uganda*"[3] sent fourteen computers and two computer scientists to a primary school in South-West Uganda to install a computer room and establish a functioning computer network. During the period of three months, 300 people were

---

[1] ICTs do not only mean computer, but encompass also fax machines, state lights, television sets, radios, Internet, e-mail, telephone, scanner, plus procedures and processes that together support the processing, storage and delivery of information [Mugisha, 2002].

[2] I preserve the term NGO, though most NGOs are financially supported by their home country government. Of course, it is the NGOs choice to accept this aid; e.g. in Austria an NGO can get up to 80% of an intervention's budget, if the NGOs activities are placed within the politically defined "Schwerpunktland" (focus country). Currently, Austria supports thirteen countries: Uganda, Ethiopia, Nicaragua, Bhutan, Mozambique, Palestine, Burkina Faso, Cap Verde, Albania, Bosnia-Herzegovina, Croatia, Macedonia, Serbia and Montenegro incl. Kosovo. To my knowledge almost all NGO activities are placed within these countries. Another negative effect of this politics is that if a LDC "loses" its status (e.g Ecuador in 2002) because as big "national" projects have been finished, smaller NGO activities are prematurely interrupted and might not be able to finish their plans - cp. ([Gomes, 2003], p.24).

[3] The "Projekt=Uganda" builds upon the historical relationship of Ilse Kroisenbacher and Father Albert Byaruhanga, Yoveri Museveni's, personal consultant. When the constitution of Uganda was proclaimed in the small village "Unterolberndorf" in Lower Austria in 1986, Ilse Kroisenbacher got to know Father Albert Byaruhanga. Since that point of time Ilse Kroisenbacher collects money by several fund-raisers per year to support interventions like the discussed computer intervention in south-western Uganda.

taught and instructed in the basics of computing. In this period it became apparent that the people from this region approach technology differently than we do (in the mind of members of different cultural backgrounds) and therefore integrate knowledge in a different way. Throughout this stay misunderstandings, as well as smaller discussions about the access to computers among the actors, led to an unsystematic documentation of relevant data, in terms of cultural, ethnological and educational.

In a further research period in 2003 interviews were conducted and the sociocultural context explored. Thus, the collected data has enabled systematic analysis in order to focus on computer teaching within a social environment and answers to the following research questions are aimed to clarify:

- What are the social functions of computers within the geographical area of Kabale (Uganda) and (how) is a computer integrated into a participant's everyday life? Does the fact of "having computer access" influence power relationships among pupils, staff, non-teaching staff and other schools as well as between them? How is computer knowledge represented and posed?

- In which way does power manifest itself in computer teaching? Which signs of social function can be spotted among various social actors? I.e. a computer teacher or a teacher at a primary school.

- What kind of intercultural phenomena can be observed? When and how is referred to different (cultural) systems?

- In what way does the multifunctional role as an interviewer, teacher, colleague and donor influence the progression of interviews? Should and can donors evaluate[4] such interventions[5] on their own?

The overall goal of this research is to systematically analyze practical experiences of computer teaching. Therefore, I shall make use of appropriate methods to analyze the social impact of technology transfer. This thesis goes beyond the mere statement about evaluating an intervention's "failure" or "success". It depicts both its process and possible ways how arising intercultural interactions can be understood and interpreted.

1.2   Shortcomings in on hand Literature

Reviewing current[6] research approaches of ICT related interventions in LDCs and their underlying theoretical and methodological tenets is not neglectable when one wants to analyze and evaluate an intervention on his/her own. Case studies on teaching ICTs in LDCs can be utilized to compare results and methodological approaches. A comparison of methodological approaches and the presentation of prevalent theoretical

---

[4]Please see Section 2.2 for further thoughts on the concept of evaluation.

[5]Please see Section 2.2 for further elaborations why I prefer to use the term "intervention" instead of "project".

[6]Of course this list does not demand for completeness, as especially in the field of empirical studies, many projects are not documented and not reflected and therefore not accessible to the researcher.

and methodological tenets in literature approves that the interpretative paradigm is predominant. This is also the theoretical perspective I am building upon, though different, as I am uniquely drawing upon the interlinking of Witzel's "*problem centred analysis*" [*orig.* "Das problemzentrierte Interview*"] [Witzel, 1985] and "*objective hermeneutics*" by Oevermann [Oevermann, 2002].

Furthermore, in literature there are hardly academic studies at all and there is a lack of case studies

... *on social implications at an individual's level*

When examining the organization's publications it seems that the argumentation is mainly focused on the organizational level and not on the level of individual experiences. Most research on ICTs in LDCs is limited to a macro and partly meso view and very rarely examines the micro level of the individual as to be read in Helana Scheepers' documented experiences about conducting IT training at the SEIDET Centre in Siyabuswa, South Africa [Scheepers & de Villiers, 2000]. Though, I note that if there is a study on an individual's level it considers both micro and macro levels of research. Latter one is used to investigate on social frames, in which the action of the individual is placed within. Even well known Chrisanthi Avgerou focuses in her "Information Technology in Context" on the macro level of analysis [Avgerou & Walsham, 2001]. This emphasizes the disregard of case studies on social implications at an individual's level. The situation seems not to have changed much, when Geoff Walsham and Sundeep Sahay claim: "*more work on individuals, and issues such as identity, would be a valuable direction for future IS research*" ([Walsham & Sahay, 2005], p.14).

... *on teaching computers in LDCs*

Among found publications the theme "learning & education in DCs" is rarely treated. Scheepers' and Villier's researches [Scheepers, 1999] [Scheepers & de Villiers, 2000] and Sandy Turner's "The Use of ICT in Secondary Schools in Ghana" [Turner, 2005] are exceptions. Together with Roger Harris's paper on attitudes towards new technology and computer anxiety among secondary school teachers at a school in Borneo [Harris et al., 2001] these case studies are almost predominantly lonesome in the fields of qualitative research.

When Scheepers emphasizes in "The Computer-Ndaba experience: Introducing IT in a rural community in South Africa" [Scheepers, 1999] the importance of the relationship between the "change agent" and participants, she does not consider the issue of interculturality during the teaching of computers as Jegede et. al. do [Aikenhead, Glen, 2001], [Cobern, 1991], [Jegede, 1994], [Slay, 2002]. The same situation can be found at Roger Harris's Kelabit study which does not discuss the knowledge transfer process itself [Harris et al., 2001]. To my knowledge the initial computer teaching in ICT projects in LDCs is mainly done by members of technologically developed cultures. Taking into account the complexity of cultural implications it seems obvious that a solid study can offer new perspectives in teaching computers in LDCs.

... *being performed by a scientist with practical experiences.*

As far as my literature research is concerned, further case studies, are based on reports made by administrators to academics. When Odedra-Straub complains about "*the little participation from practitioners*" ([Odedra-Straub, 2002], p.1), it becomes apparent that there is a real lack of case studies being carried out by a scientist with practical experiences. This allows a skilled and ex post inspection of the (technological) knowledge transfer itself. Starting from this particular point of departure I would like to contribute to the provision of further evidence.

1.3   Aim of Research and its implied Research Questions

The aim of this interdisciplinary study is to concentrate on the relationship between computers, teachers, donors and pupils, in other words: to investigate "*symbolic instruments as both structuring and structured structurals and as instruments of domination*", as Bourdieu puts it [Bourdieu, 1991].

Firstly, I identify involved actors and analyze the hierarchic position of a teacher in particular, the Ugandan education system and meaning of education in Uganda. General prevalent but latent attitudes towards learning and teaching shall be derived and explored by the means of hermeneutics. Thereby this work aims to make an important contribution to understand how ICTs are integrated into everyday life, which is needed to realize and evaluate a "successful" ICT intervention in LDCs. Thus, I will debate the social function of computers within this geographical area and among its main ethnical group, the Bakiga[7]. The debate will both focus on the cultural context of computer lessons as e.g. the status and the meaning of a computer itself. It will examine to which extent a computer influences a Mukiga in his/her daily life and if other technologies are gaining a normative function in social life.

Secondly, I investigate on experiences with computers witnessed by the Bakiga and their confrontation with computers in their daily life. My own experiences are consulted and analyzed, too. A systematic analysis gives several examples of encountered misunderstandings related to culture and attempts to relate them to the context of culture. This intercultural social interactions are pointed out to sharpen the mind of those intending to finance and perform computer interventions in LDCs.

1.4   Theoretical Approach

This thesis is philosophically grounded in an actor-oriented social constructionist view of change and continuity - (cp. [Long, 2001]). It assumes that both actors and social structures mutually reconstruct each other in everyday life. Putting (inter-)action, which is social in any case, into a context means that it is cultural as well [Geertz, 1973]. Thereby, I take up an intercultural[8] perspective. From a methodological point of view this theoretical stance asks to perform ethnographical research with the regard to three different perspectives (problem perspective, process perspective and substantive perspective) [Witzel, 1985] to allow a solid investigation of the everyday life of the involved participants of this intervention. This investigation goes also along with the goal to understand people's everyday life, which I will explore with regard to hermeneutics [Oevermann, 2002].

---

[7]Rukiga term (language of the Bakiga): Singular is built by the prefix "Mu-", i.e. "Mukiga" and plural by the prefix "Ba-", i.e. "Bakiga".

[8]Please see Section 2.2 for further elaborations on "culture" and its implications to this research.

1.5 Outline of Applied Methodology

Since this research focuses on the subjective relevance of computer and computer teaching and its integration into the participants' everyday life, the selected methods are qualitative ones. Quantitative methods would also be unusual to the local communication structures and the local tradition in transferring knowledge of the Bakiga [Vansina, 1973]. Thus, only by establishing a daily life situation as in talks, good results can be realized. Whereas Scheepers' [Scheepers, 1999] and Roger Harris's [Harris et al., 2001], studies are based on the method of action research [Rapoport, 1970] implying the analysis during intervention, this upcoming research applies a different approach: the whole intervention is mainly analyzed one year after its main introductory phase.

The following paragraphs will briefly depict the circular processes of data collection, interpretation and analysis:

1.5.1 Data Collection, Interpretation & Analysis

All in all the gathered empirical data represents the backbone of this research project. The collection has been conducted with a view to qualitative research in the field of intercultural studies and socio-scientific hermeneutics – (cp. [Geertz, 1983], [Oevermann, 2002], [Witzel, 1985]). The ethnographic methodological instruments include qualitative interviews and participant observation. To do justice to my socio-constructionistic perspective and to enable a problem centred interview appropriate to the region I explored the regional sociocultural context as much as possible and collected relevant meta-documents.

When I was teaching in 2002 I realized that there obviously were problems in applying and conveying the technology. As Witzel's programme of the problem centred interviews [Witzel, 1985] enables the exploration of "problem areas" as computers seemed to be, I have decided to apply a nested variant of Witzel's programme [ibi.] and objective hermeneutics [Oevermann, 2002] for the subsequent research period in 2003. Whereas the first one encourages to explicitly focus on certain problematic situations, latter one aims to reveal and reconstruct the context of creation and perception of texts[9]. Thus, hermeneutics enables the researcher to uncover (latent) meanings and thereby the social function of computers within a certain sociocultural environment. The transcribed interviews are analyzed by Oevermann's "*sequence analysis*" [*orig.* "Sequenzanalyse"] with a regard to Witzel's problem, process and substantive perspective [Witzel, 1985]. This means that Oevermann and Witzel emphasize on reflection throughout the research process. Doing so it incorporates more "*interactive moments of text production*" [Witzel, 1985] than other qualitative methods.

The results of the text analysis and my text about the period in 2002 – similar to Geertzs' suggested *thick description* [Geertz, 1983] – is supposed to give more information on experienced intercultural phenomena.

---

[9]For further information on the term text please see Chapter 6.

1.6   Results of this Work

This study discusses roles of the involved actors, their relationships and hierarchic positions. It will show the constant change of roles (both the interviewees' and the interviewer's) and indicates that a trustful personal relationship enables to point out other crucial issues in the transfer of technological knowledge rather than the evaluation by an outsider might reveal. I will also present that the access to computers is correlated with a gain of status. This becomes not only evident by the *how* the access to computers is constructed, but also by the fact that the prevalent expectation towards a computer is to improve ones financial situation.

Consequently, the desire to acquire computer knowledge is high, but there are indications that this wish refers to a distorted image of a computer.

Apparently, the benefit of an achieved certificate is more real than benefits through the application of a computer. Thereby, I will show the deep nestedness of formal education, which is evidently reconstructed by a computer related intervention.

I will also discuss that computer knowledge is obviously instrumentalized as a sign of power. Thereby, I will reveal the possible deadlock situation that a local computer teacher might not be motivated to transfer computer knowledge in an adequate way, as the fear of loosing power might prevail to him/her.

This correlates with another finding of this thesis I will deduce: The applied explorative way of teaching computers was untypical, regarding the cultural expectations towards technology control and the manifested hierarchic positions. This causes a cultural break, which was welcomed on the one hand, but not internalized on the other hand. Possible factors like lacking of teaching methods or of technical skills among local computer teachers are discussed and presented.

Further, I will show that the best case scenario to transfer ICTs seems to be the concept of a cultural broker. Such a teacher applies his/her mother tongue, draws upon a variety of teaching methods and has deep computer relevant skills. Another result I will indicate is that it seems as if the Internet's integration into the participants' everyday life might be hindered by the mobile's phone of gaining ground. Access to mobile phones is easier managed than to the Internet. Variable costs of both are comparable and in terms of fixed costs a mobile phone is much cheaper. This is accompanied by the fact that the physical access to a mobile phone (for the participants of this study) is easier than to an Internet-ready computer.

A limiting factor to the general integration of computers into a teacher's life is their high workload during the terms. It turns out that if the knowledge transfer takes place during the vacations, there are higher chances that the participants dedicate themselves to a higher extent than during the year. However, such a course competes with other advanced courses for teachers. This asks to promote future courses with standardized certificates to become comparable to other institutions' courses.

1.7   Overview

As this Chapter 1 tries to expound the structure of the subsequent work the next Chapters deal with both the *"Theoretical Framework" (Part I)* and the *"Empirical Study"*

(Part II+III) itself. Within the "Theoretical Framework" I present the "Motivation and Background" of this study (Chapter 2) and thereby arising questions. Chapter 3, "Theoretical Stances", enables the reader to get a general idea of the current situation of available literature concerning the application of ICT in developing countries. It presents (academic) empirical case studies (3.2), their current theoretical approaches (3.3), and theory of methods (3.4). Thereby, I deduce the shortcomings in current literature I aim at remedying. Afterwards I refine my "Aims of Research" (Chapter 4) and present the underlying theoretical concept of actor-oriented socio-constructionism (Chapter 5). In the subsequent methodological part (Chapter 6) I describe the characteristics of objective hermeneutics [Oevermann, 2002] in combination to Witzel's programme of problem centred analysis [Witzel, 1985] (Section 6.3) and depict the applied research methods (Section 6.4). I underline the necessity and correctness of the methods' application and depict its utilization on the circular research activities of "data collection", "analysis", and "interpretation". To round off the theoretical analysis of technological knowledge transfer in LDCs I investigate on the political prevalent situation of ICTs in LDCs and especially within Uganda - Chapter 7. The theoretical framework will be concluded by a Chapter "Education" (Chapter 8) and discusses ICT & Education related stances (Section 8.1) and Education in Uganda (Section 8.2).

Building upon this theoretical framework, Part *II* deals with the preconditions of the Ugandan case study. I explore the sociocultural context of the Ugandan case study (Chapter 9) and sketch the settings of the intervention (Chapter 10). This is followed by a description of the selected and applied methods of objective hermeneutics and problem centered interview. Thus, the reader shall be enabled to track the study's process to reach a highest degree of understanding of the analysis as possible. In the third Part (Page 141) I list the analysis's results. I present the "Actors and their Roles" (Chapter 13), the way how the object computer is represented (Chapter 14), and the findings about the prevalent usage of computers in the participants' daily lives (Chapter 15.1). In Chapter 16 I put the results of the previous analysis into context to literature and answer the prevalent research questions. The last chapter (Chapter 17) points out the findings and conclusions of this thesis. It will critically reflect on the applied methodology and discuss to what extent further case studies can learn from this investigation. In the Appendix the reader can find a list of figures, tables, and presented text passages to have quick access to the underlying categories.

1.8 Conventions

Direct quotations are denoted by "*italicized text*" [Reference, 2006] and the according reference in square brackets right after the quotation. Whenever ideas or concepts of others are referred to, the respective reference is put at the corresponding position within a sentence, or after the end of a sentence (after the "full stop") if the whole paragraph is meant to be related to the given reference.

Comments within direct quotations are put into square brackets and concern mainly for the terms "*African*" and "*Western*". As there is nothing like one African or one Western culture I avoid this term and name corresponding substitutional, content related phrases. This is indicated by the "in the meaning of" which is abbreviated by the "*itmo*". If a quote is a translation it is put into square brackets and whenever it is done by myself it is annotated as "[my translation]". Whenever needed I refer to the original term by "*orig.*" in square bracket or in the footnote. Following definitions of concepts and terms are used throughout this text, but are discussed for clarification, if mandatory, once more inside the thesis in detail:

Concept	Definition
Culture	is an ordered system of meaning and symbols, in terms of which social interaction takes place [Geertz, 1983]
Development	is a process of enabling people to accomplish things that they could not do before [Todaro, 1989].
Intercultural awareness	is the recognition that I carry a particular mental software because of the way I was brought up, and that others brought up in a different environment carry a different mental software for equally good reasons ([Hofstede, 1991], p.230).
Intervention	is a transformational process that is constantly reshaped by its own internal organizational, cultural and political dynamic and by the specific conditions it encounters or itself creates, including the responses and strategies of local groups who may struggle to define and defend their own social spaces, cultural boundaries and positions within the wider power field and means change in culture.
LDCs	According to the definition of UNCTAD. Internet: http://esa.un.org/unpp/definition.html
Lifeworld	are "lived-in" and largely "taken-for-granted" social worlds centring on particular individuals. Such worlds should not be viewed as "cultural backcloths" that frame how individuals act, but instead as the product of an individual's own constant self-assembling and re-evaluating of relationships and experiences. Lifeworlds embrace actions, interactions and meanings, and are identified with specific socio-geographical spaces and life histories ([Long, 2001], p.241).
Power	in the meaning of a relation, and not something stable that automatically goes along with specific characteristics. Power is negotiated in discourses the way cultures or social relationships are: by representation. The design of power can also be observed in such text as transcribed interviews are (cp. [Long, 2002]).
Social actors	are all those social entities that can be said to have agency in that they possess the knowledgeability and capability to assess problematic situations and organise "apppropriate" responses. Social actors appear in a variety of forms: indiviual persons, informal groups or interpersonal networks, organisations, collective groupings ([Long, 2001], p.241).
Social meaning	The social meaning of computers is attributed by assumptions, knowledge and expectations of its individual, group or organization. It emerges both from interactions involving actors in context and shape those contexts in turn [Heaton L. and G. Nkunzimana, 2006]. These contexts manifest in cultural practices.

Table 1.1: Key concepts & definitions of this research

Part I

Theoretical Framework

Chapter 2

Motivation and Background

2.1 Background

"Ultimately, our aim must be to ensure that people everywhere have access to information technology, and can use it to build better lives, for themselves and for their children" [Annan, 2003].

In this press release the United Nations Secretary General Kofi Annan emphasizes the importance of ICTs for LDCs, and was looking forward to yield new results at the *World Summit on the Information Society (WSIS)* in December 2003 in Geneva. This event was supposed to mirror an ongoing discussion about technology transfer from developed to less developed countries, existing since the late seventies [Golden, 1978].

Governments all over the world and more and more international donor organizations demand ICT for LDCs, because they consider ICT as a panacea for Developing Countries. Starting from their own position, they are convinced that reaching the status of an information society will result in economical and social benefits. They believe that ICTs can eradicate poverty and *"ICT related "initiatives" on behalf of international and donor institutions became increasingly colored by the rhetoric of an emerging global information society"* [Audenhove, 2001]. International development projects tend to be very costly and huge amounts of money are usually spent on them; e.g. the " UNDP THEMATIC Trust Fund Information and Communication Technology for Development" announced an "Indicative Contribution Target 2001-2003" with the total volume of $ 30 mio. [UNDP, Thematic Trust Fund, 2001] or the ACACIA Initiative provided about $ 7.2 mio. in the first year [Mansell & Wehn, 1998]. To stimulate economical and social growth in Africa the *International Telecommunication Union (ITU)*, the *United Nations Educational, Scientific and Cultural Organization (UNESCO)*, the *International Development Research Centre (IDRC)* and the *World Bank* formed the *African Information Society Initiative (AISI)*, a programme for the utilization of ICTs in Africa. Its main document called "An Action Framework to Build Africa's Information and Communication Infrastructure" *"is regarded as a guiding framework on which to base other information and communication activities in Africa"* [Audenhove, 2001]. Together with a ITU Green Paper on Africa's ICT infrastructure these documents pose the framework of Africa's ICT policy. Though this framework is not compulsory, many major donor agencies consider these documents as a guiding framework for the accomplishment of their initiatives. This is important as well as all ICT projects are donor driven [SchoolNet Africa et al., 2003]. ICTs are deeply interweaved in any sector of a technologically developed society and as those societies are considered to be more advanced the AISI concludes that ICTs accelerate development in any African economic and social activity. The AISI tries to adopt a concept of an information society in which ICTs replace communication[1] means as the great hope for economic and social development. AISI's approach is based on a consequent technology variant of modernization [Audenhove, 2001] and relies on a threefold latent assumption: technology is neutral, technology is easily transferable and technology shows the same effects in any society. From a sociocultural view, technology is produced by interconnecting activities of certain actors (inventors, producers, and users), coincidence and socio-technological constraints. Therefore it is definitely not neutral (although some practitioners wrongly think so) and depends drastically on its "original" sociocultural context [Audenhove, 2001]. Many studies have shown that a technology transfer can only happen if it is adopted and enhanced by the implementing country and therefore is the opinion of several authors [Olivier, 1994], [Mansell & Wehn, 1998], [Audenhove, 2001], [Latu, 2006].

As Ess has argued, "*if IT [itmo ICT] is to be implemented in ways that will prevent its spread around the globe from becoming yet another exercise in colonization and imperialism* (cp. [Postma, 2001]) – *the users of ICT "must become cultural hybrids and polybrids who are fluent in the worldviews[2] and communication preferences of more than one culture"* ([Ess & Sudweeks, 2001], p.259f) This aims that inter cultural communication does not occur simply through the imposition of a single worldview and communication style. *"To become such cultural hybrids requires, in part precisely our becoming aware of the diverse cultural values and communicative preferences"* ([ibi.], p.260). This claims to reveal and gain such diverse and different values and communicative preferences as to contribute new perspectives on ICTs in LDCs.

Therefore this thesis attempts to deduce relevant factors and preferences for a successful ICT transfer in LDCs by trying to investigate a computer project which took place in a primary school in Summer 2002 in South-West Uganda. In this year the ADC Austria sent fourteen computers and two Austrian computer scientists to a primary school in South-West Uganda to install a computer room and establish a functioning network. During a period of three months, 300 people were taught and instructed in the basics of computing. In this period it became apparent that the people from this region approach new technology differently than we do (in the mind of members of different cultural backgrounds) and therefore integrate knowledge in a different way. As one of the teachers, I became interested in identifying relevant factors, which

---

[1]In the 60's UNESCO "media minima" was established which recommended that every nation should aim to provide, for every 100 of inhabitants: 10 copies of daily newspapers; five radio receivers; two television receivers; two cinema seats ([Felician1976, p.92] cited in [Burton, 2001]). By the mid 70's NO correlation between economical growth and quantitative increase of mass media was found in those African countries.

[2]According to Cobern, 1991 the term "world view"has two different connotations in English. The first has a philosophical meaning and involves a person's concepts of human existence and reality; the second is an individual's picture of the world, which is the one applied in this context. The term "*world view*" as used in anthropology refers to the "*culturally-dependent, implicit, fundamental organisations of the mind*" ([Cobern, 1991] cited in [Slay, 2002]).

facilitate a **successful** (see Section 2.2) ICT project in LDCs and what kind of intercultural phenomena occurred during the process of technological knowledge transfer. In July 2003, after a solid preparation in socio-scientific research, I scrutinized the Ugandan project by interviewing participants and exploring the sociocultural context in their everyday life. So starting from the former position as a practitioner allowed me to focus, refer and reflect on key problems I encountered during the projects first phase of implementation, which poses evidently a crucial advantage in opposition to a researcher who is a technological layman. Besides, although more and more computer projects in LDCs are setup, there is a lack of academic case studies in general and especially in the fields of teaching computers in LDCs and those being performed by a former practitioner. [Odedra-Straub, 2002]. [Walsham & Sahay, 2005] identified a dearth of case studies focusing on the social implications of technologies in LDCs on an individuals level emphasizes our focus of investigation, too. The study "Technological Knowledge Transfer in Least Developed Countries with a view to Intercultural Awareness. Depicted by a Case Study of a Computer Project in Uganda" tries to contribute valuable inputs on an empirical technological knowledge transfer. The following topics are investigated: What are the cornerstones of a successful computer project in a LDC? When does a project become a failure? What are the project's goals and who fixed the projects' goals by what measures? Why were these goals (not) reached and how can these be measured/evaluated? What are the "real benefits of an ICT project and who gains the (most) benefit from them? Is there a demand for a more homogenous/heterogenous technology? Have the economical and sociological conditions for ICT in LDCs been set yet?

2.2 Underlying Concepts

Before we are able to discuss the interventions' process and substantiate our research goals, I have to clarify certain concepts and meanings associated to the context at hand. Although I go along with [Walsham & Sahay, 2005] who say: "*If researchers wish to address topics such as globalization, culture, power or the meaning of development itself, there are vast literatures to draw on and it would surely be foolish to ignore them. Such theories should, however not be drawn on uncritically, and it is incumbent on the IS[3] researcher to select what is relevant to his or her study, and in some cases to develop the ideas within the specific IS domain*", but I do not elaborate on concepts' historical and developing background in an exhaustive way. Nevertheless, I am trying to depict my stance on key concepts in a pursuable manner to allow a solid auditability of my underlying stances. The following concepts are bound to be clarified:

2.2.1 Categorizing Success and Evaluation

Reconsidering above mentioned research topics points out an evaluative character of this study. Evaluation (especially in the fields of developmental intervention) proves to be complex and several crucial factors have to be obeyed. Evaluating developmental activities means to think ex-post about "success" and "failure" of it. Any attempt

---

[3] Actually, [Walsham & Sahay, 2005] refer to Information Systems, which are abbreviated with IS.

to decide on "success" and "failure" must start by categorizing success and failure. As Heeks noted this cannot completely be resolved, because this categorization is subjected to two main difficulties:

1. "[...] *Subjectivity of evaluation - one person's failure may be another's success* ([Lyytinen & Hirschheim, 1987], [Sauer, 1993] cited in ([Heeks, 2002], p.101)). *The categorization does try to address this within the limits imposed by the subjectivity of the case study writers themselves* [...]

2. *[...] Timing of evaluation - today's IS success may be tomorrow's IS failure, and vice versa [...]"* ([Heeks, 2002], p.101).

So deriving from the qualified terms *"total failure"* and *"partial failure"* Heeks attempts to define "success" for each initiative, in which *"most stakeholder groups attain their major goals and do not experience significant undesirable outcomes."* Whereby this sounds reasonable at a first glance it requires once more the relatively sophisticated approach of evaluative outcomes [Heeks, 2002]. *"Summative evaluation[4]"* is difficult to perform because profound understanding of local cultural norms and values are necessary, which implies dedication with indigenous culture through reading, discussion and engagement [Walsham, 2000]. Assuming an intense dedication to local structures, the above mentioned statement by Heeks sounds solid at first, but appears to be quite "hazardous". Quite often donor agencies set-up major goals of interventions and not all participants may contribute to intervention's plans in an egalitarian way. Well, several times donor agencies claim to discuss interventions goals openly/freely, but still setup basic rules like equalizing strategies. In fact, this means that different intervention strategies are negotiated, but are subjected to cornerstones of eurocentristic ideologies (e.g. "empowerment", "participation", "social accuracy", "gender", "sustainability"), which pose a framework where other stakeholders are allowed to move within ([Obrecht, 2005], p.240, [my translation]). As stated above, political guidelines (like e.g. AISI's documents) result in more donor oriented *top-down* approaches instead of a combination of *top-down*[5] and *bottom-up*[6] invention strategies. Of course, a careful reader might claim that former statements underlie again an eurocentristic democratic principle, but to my view it enables all stakeholder groups to explore further possibilities of how the intervention can be designed. According to my understanding the more different perspectives and information are available to enlighten an interventional situation and its context, the higher is the probability of a more "successful" intervention for all participants. Concluding from above, and in order to avoid applying eurocentristic ideologies to describe *sensible interventions*, I go along with ([Obrecht, 2005], p.246f) who defines two criteria:

- The intervention is sensible if it is received and experienced as something which is "Owned" [*orig.* "Eigenes"], "Communal" [*orig.* "Kommunales"], "Reflexive

---

[4]Scriven distinguishes between summative evaluation and formative evaluation. Whereas the first one is concerned with assessing the worth of programme outcomes after the initiative, the latter one is used to provide systematic feedback to programme designers and implementers during the implementation process of information systems. [Scriven, 1967].

[5]Donor-oriented intervention strategies are called "top-down approaches".

[6]Approaches developed by participants of interventions are called "bottom-up".

belonged" [*orig.* "selbstbezügliches Dazugehörendes"], and "real" [*orig.* "Wirkliches"] and thereby is integrated into everyday life, can be sustained and modified (adapted) in this everyday life.

• The intervention is sensible if most (financial) means are used for the *Grass-Root-Level*[7] instead of maintaining administrative and operational structures to keep the initiative running and alive.

To find out if something has been integrated into everyday life means to explore one's lifeworld[8]. This calls for exploring e.g. attitudes, fears, and practices of all involved actors. A solid investigation of lifeworlds can be reached by the means of qualitative research (see Chapter 6). This implies "on the spot research" by at least one researcher, which has - in addition to the prior intervention - again an impact on its social environment. This happens solely through his/her presence. Therefore, when a researcher attempts to explore other people's everyday life, and e.g. talks to people, it inevitably becomes a matter of social interaction, as any face-to-face encounter becomes necessarily a social interaction [Berger & Luckmann, 1966]. So the way "how" actors interact depends on how their interpersonal relations are managed. ICT cases from developing countries therefore provide fertile ground to help understand the complex interplay of action and context that underlies all organizational change [Heeks, 2002]. Social interaction can involve negotiation[9]; this implies analyzing "*how differential conceptions of power, influence, knowledge and efficacy may shape the responses and strategies of the different actors*"([Long, 2001], p.19). This reveals the fact that evaluation has a lot to do with power[10] and asks me to put a focus on "how" power manifests itself during technological knowledge transfers. Referring to the second cri-

---

[7]Grass-Root-Level stands for (common) people or society at a local level rather than at the center of major political activity.

[8]*Lifeworlds are 'lived-in' and largely 'taken-for-granted' social worlds centering on particular individuals. Such worlds should not be viewed as 'cultural backcloths' that frame how individuals act, but instead as the product of an individual's own constant self-assembling and re-evaluating of relationships and experiences. Lifeworlds embrace actions, interactions and meanings, and are identified with specific socio-geographical spaces and life histories* ([Long, 2001], p.241). Long draws upon Schütz's term "*Lebenswelt*" who applied this concept in the context of sociological analysis first [Schütz & Luckmann, 1973].

[9]Interaction can alter views, temper emotions, modify intentions, and change actions - all without negotiation. The strategic quality of negotiation may be limited or absent during much sociability. People can be persuasive without attempting to negotiate. Negotiation assumes actors who are explicitly aware of the content and structure of the ensuing interaction. Negotiation also assumes that participants' interactional goals conflict or need realignment if future mutual endeavors are to occur. For that matter, the term assumes that all participants have sufficient power to make their voices heard, if not also to affect outcomes ([Denzin & Lincoln, 2005], p.525).

[10]Power in the meaning of a relation, and not something stable that automatically goes along with specific characteristics. Power is negotiated in discourses the way cultures or social relationships are: by representation. The design of power can also be observed in such text as transcribed interviews are. ([Long, 2002], [Pichlhöfer, 2000]).
Power is, just like knowledge, not simply possessed, accumulated and unproblematically exercised. "*Power implies much more than how hierarchies and hegemonic control demarcate social positions and opportunities, and enable or restrict access to resources. It is the outcome of complex struggles and negotiations over authority, status, reputation and resources, and necessitates the enrolment of networks of actors and constituencies*" ([ibi.], p.5).

terion I stress that profound structural knowledge about the intervention's means is necessary. For further aspects please see Chapter 5

2.2.2 Concept of Culture and Intercultural Awareness

Culture and Nation

A term commonly used in the fields of technology transfer is *"culture"* and its implications on technology transfer. There are manifold definitions of culture and most of them cover one special aspect of culture in its applied context. Kroeber and Kluckhohn identified more than 150 definitions already in the early 50's [Kroeber & Kluckhohn, 1952]. In advance I note that I draw upon the anthropologist Clifford Geertz where culture means *"an ordered system of meaning and symbols, in terms of which social interaction takes place"* ([Geertz, 1973], p.19). Often, culture is represented uniformly distributed across a nation. There are some models in literature which try to capture this concept of national culture. Hall's and Hofstede's dimensional models are the most popular ones. Edward T. Hall identifies dimensions of a human living together, which are applicable to all cultures over the world. In his work he has figured out three dimensions to differentiate between national cultures: time, space, and communication.

The dimension space describes the physical distance one assumes freely in a talk with friends, business partners, or foreigners. The dimension "time" is split up into two different attitudes to time: monochronous and polychronous. A monochronous time attitude means that the running time represents a linear axis on which all activities have to be placed singularly and therefore one after the other. This requires the capability of planning and reliability not only of single persons, but also of the whole society, e.g. public transport. Consequently, such monochronous oriented societies quarrel with overlapping appointments, interruptions and problems.

Polychronous oriented groups track their intended activities at multiple levels at the same time, which claims for a high flexibility in terms of time. Overlappings related to time and appointments are tolerated as well as interruptions and problems are.

Within the dimension of communication he distincts between high context versus low context. This dimension draws upon the degree of communication between individuals in a particular national culture.

Low-context countries show a preference for literate (i.e., textual), high content information transfer. As much information as possible is expressed explicitly in order to avoid (mis-)interpretations. High context communication conveys less content, but relies more on the context of a dialog situation, non-verbal signs. Metaphorical phrases influence the progress of conversation, too. [Hall, Edward T., 1976]. To compare the cultures of two nations Hall uses the categories time, space, behavior & communication style, high/low context, work, belonging, fellowship, which are found again in Cushner and Brislin's general phenomena of intercultural interactions - see Section 3.1.3.

Hofstede, who takes up a sociological view, proposes a multidimensional model and suggests, on the basis of a survey conducted amongst IBM employees[11], that national culture and values, can be categorized on the basis of five dimensions, namely: power

---

[11] 116.000 people participated in this study.

distance, uncertainty avoidance, individualism–collectivism, masculinity-femininity, and long-time orientation [Hofstede, 1980]. In 1998 Trompenaars and Hampden-Turner draws upon the work of Hofstede and others e.g. on Hall. Basically, they retained the dimension of individualism-collectivism and named six further dimensions, which are obviously based on the pattern variables by Parsons [Parsons, 1951]: universalisms-particularism, neutral-affectiv, specific-diffuse, acquired-attributed status, and level of orientation towards nature [Trompenaars & Hampden-Turner, 1998]. As such a categorizing and inflexible concept of culture is not applicable in the context of this research it is not within the scope of this work to discuss these approaches in detail. These models might be helpful and easy[12] to understand for decision makers of MNCs - (cp. [Ess, 2004]), but for my research following critics arise:

1. *One dimensionality.* As for a nation's representation only one variable is used and a 1-dimensional and thereby hierarchical structure is presented; e.g. when Hofstede presents a power index of 19 for Denmark, 52 for Japan and 70 for France these countries are put in a hierarchical relation which presents culture oversimplified and reductionistically. Hofstede talks about statistical data and is fully aware that individuals behave differently and points out that behavior of individuals may not be the same. Nevertheless, his structure remains to be hierarchical and so it is a relational representation and can easily be interpreted in oversimplified and reductionistical ways. [Thomas et al., 2003a].

2. *Problem of perspective.* Any description of a different culture can only be made from an own specific cultural perspective. The attribution of dimensions to cultures (nations) are difficult to define, because e.g. a Japanese might be perceived by an Austrian as aggressive, whereas a Ugandan won't describe the Japanese like that. Thus, ethnocentrism gets inevitable [Thomas et al., 2003a] and one has to be aware of his/her own perspective.

3. *Culture as Nation.* The simplification or aggregation of culture on a nation's level is difficult to understand, as many contemporary countries were artificially defined by hegemonic governments; e.g. Uganda comprises at least 63 different ethnic groups ([Okoth, 1995], p.xiii). Similarly, Wolfgang Welsh proposes the term "*transculture*" to describe such a reality of cultural compounds [Welsch, 1995].

Conclusion: intercultural phenomena are manifestations of culture.

By above statements it becomes evident that any study which draws upon a cultural perspective has to become detached from such existing models and has to look beyond those. "*We should give up the idea that cultures are isolated wholes or that we could draw strict borders between them*" ([Kamppuri & Tukianinen, 2004], p.12).

Culture and Classification

As a matter of fact, from a sociologically point of view, life consists from nothing else than from social encounters and as a human being one constantly registers informa-

---

[12] Alexander Thomas notes that these models are easy to understand, which might have been a reason for their success [Thomas et al., 2003a].

tion flows, which areinterpreted and learnt incessantly. "*To enable the absorption and processing of multi-layered content of learnings, like culture, complexity has to be reduced - a process, which permanently takes place in human perception and information processing. That is what stereotyping is exactly for. On the one hand stereotyping implies distortions, but on the other hand it facilitates the orientation in new situations. It is determining how consciously this process is performed and how close these stereotypes are constructed to reality and to what extent they are amendable to further differentiations*" ([Thomas et al., 2003b], p.21, [my translation]). In other words, on an individual's level, "*culture can be seen as a classification system to master ones social environment*" ([Thomas et al., 2003a], p.22, [my translation]) "

This goes along with Hofstede, who says: "*Culture (two)*[13] *is always a collective phenomenon, because it is at least partly shared with people who live or lived within the same social environment, which is where it was learned. It is the collective programming of the mind which distinguishes the members of one group or category of people from another*" ([Hofstede, 1991], p.5). As almost every human being belongs to a number of different groups and categories of people at the same time, people unavoidably carry several layers of mental programming within themselves, corresponding to different levels of culture (e.g. generation level, social class level, national level). Hofstede identifies four terms to describe manifestations of culture: symbols, heroes, rituals, and values [Hofstede, 1991]. By practicing these terms culture becomes evident. So culture is reproduced and reconstructed by its practices everyday. Over time e.g. new symbols and heroes arise which shows us that culture is constantly changing, too. Therefore, culture can be seen as dynamic and processual (cp. [Bourdieu, 1992]). As mentioned above, Hofstede has been criticized for being too fixed on national and ethnic levels [Kamppuri & Tukianinen, 2004]. Consequently, I do not apply his dimensional perspective to perform this case study, but go along with Clifford Geertz where culture means "*an ordered system of meaning and symbols, in terms of which social interaction takes place*" ([Geertz, 1973], p.19). So culture is both changeable, but can also stabilize certain aspects of its environment ([Kamppuri & Tukianinen, 2004], p.12). Culture applies changes that may have a stabilizing effect, because it is something constantly being negotiated by the social actors involved. This process makes them part (*itmo* stabilizing) of the social system they are living in.

Cultural Differences

The above mentioned "distortion" of perceived culture is crucial to this thesis. By the application of ones inherent classification system the distinction between an own[14] viewpoint and an foreign[15] viewpoint becomes evident and thus these distortions are perceived as differences. "*Cultural differences are the result of differing viewpoints, which are always biased and just partly observations*" ([Demorgon, 1999], p. 73, [my translation]). Demorgon concludes further, "*it seems is if an observer perceives a*

---

[13]Hofstede recognizes another usage of the term "culture": It is the meaning of "*civilization*" or "*refinement of the mind*" and results in disciplines as education, art, and literature are ([Hofstede, 1991], p.5).

[14]Das Eigene [my translation]

[15]Das Fremde [my translation]

sign of behavior as different, if it is different to his own" ([ibi.], p.73, [my translation]. Since culture manifests in practices[16] the term "culture" becomes "differenzstiftend"[17] per se as Hakan Gürses [Gürses, 1998] has stressed. Discovering a disparity necessarily includes "something" which is "inbetween": interculturality. In fact there is a constant desire to distinct oneself from others, which is inseparable to the desire for identification ([Demorgon, 1999], p.77). According to Stuart Hall everyone has several different identities he/she is looking for[18]. One is member of different societal groups (e.g. sports club, employee, family member) at the same time which leads to a situation which is identified by Hall as "the biggest problem": *"The proliferation of identities provides within the field of identities and antagonism that those mutually expel each other"* ([Hall, Stuart, 2002], p.109, [my translation]). Putting this into relation to culture Hofstede notes wisely that identities and individuals are in the same relationship like cultures and societies.

Discussing the process of technology transfer from one social-environment (level of culture) to a *different* one, demands identifying those differences, which manifest in practice. From a historical point of view, the term *"difference"* is connotated with deficiency, especially in the field of development cooperation ([Wimmer, 2004], p.11f). Therefore the disentanglement of a "cultural difference"-grounded thinking would supersede interculturality. As a human being, I cannot decontextualize and detach myself from my culture to resolve those differences and so it becomes necessary to accept those. Wulf depicts the problem of cultural difference as follows: *"Only by an underlying acceptance of differences to other cultures and to other (foreign) people it becomes possible to reveal transnational common grounds and to bring forward the grounds' development [...] The preconditions for intercultural competence are based on a mutual acceptance of difference. Knowledge and acceptance of the other's disparity paves the way for communication, cooperation and sympathy. The other one has to be explored.* ([Wulf, 1998], p.45, [my translation]).

The following definition of interculturality derives rather from the competence of individuals to act interculturally than from a disparity between categories and classifications, but it allows us to gives us a further insight into the meaning of interculturality: *"Nowadays interculturality is understood in a way which aims to develop adequate capabilities of acting in the meaning of a fundamental law. This is about the acceptance of differences, development of appropriate attitudes and an adoption of capabilities to change perspectives, to reconcile interests and to evolve empathy. Thus, mechanism shall be developed, which allow to perceive "own" or "foreign". Everyone has to be capable to experience, understand, interprete, and appreciate diversity and potentially to cope with it. "Intercultural" indicates the capability of every single one to accept*

---

[16] Edward T. Hall describes culture as a man's medium: *"there is not one aspect of human life that is not touched and altered by culture. This means personality, how people express themselves (including shows of emotion), the way they think, how they move, how problems are solved, [...]"* ([Hall, Edward T., 1976], p.16f).

[17] To bring about differences [my translation].

[18] Though this concept of identity is critizised by Lenzen who notices that the concept of ego-identification ("Ich-Identität"), which is the underlying concept, has emerged within the Age of Enlightenment, is therefore ethnocentristic and not valid across all cultures ([Lenzen, 1991], p.155f). For this case study at least one actor, me, draws upon this concept and therefore remains valid, at least partly.

and participate in a society, which is heterogenous in ethnical, cultural, linguistical, religious and social terms" ([Comenius-Institut, 2004], [my translation]). Accordingly, this asks a researcher to be open and reflexive as much as possible and to interlink the difference instead of opposing it dualistically ([Hall, Stuart, 2002], p.107).

Intercultural awareness

Although, the above discussed dimensional concepts of culture are too inflexible and categorizing and consequently unappropriable for this study, Hofstede elaborates three stages of quality of intercultural encounters which are applicable in the context of this study. At the beginning of an encounter *"awareness"* is needed: *"the recognition that I carry a particular mental software because of the way I was brought up, and that others brought up in a different environment carry a different mental software for equally good reasons"* ([Hofstede, 1991], p.230). Ideally awareness is transformed into *"knowledge"* which is obtained by learning about other's symbols, heroes, rituals, and values. Putting awareness and knowledge into practice leads to *"skills"*. For this study I want to point out those differences I became aware of and to create knowledge by describing intercultural phenomena to local symbols and values to enable intercultural skills for further interventions.

Describing differences is always and entirely subjective and cannot aver completeness. Studying cultural differences among groups and societies pre-supposes a position of cultural relativism, which is defined by Claude Levi Strauss as: *"Cultural relativism affirms that one culture has no absolute criteria for judging the activities of another culture as "low" or "noble". However, every culture can and should apply such judgment to its own activities, because its members are actors as well as observers"* ([Lévi-Strauss & Eribon, 1988], p.229). It is important to stress that according to Hofstede, G. & Hofstede.G.J *"cultural relativism does not imply normlessness for oneself, nor for one's society. It does call for suspending judgement when dealing with groups or societies different from one's own. One should think twice before applying the norms of one person, group, or society to another"* [Hofstede & Hofstede, 2005].

For a researcher or teacher who perceives his culture different to the one he is interacting with and becomes part of another society for a limited period of time this means that he/she will have to negotiate interventions. Stuart Hall underlines this when he claims that all cultural differences are negotiated, be it gender, sexuality, etc. [Hall, Stuart, 2002]. Again, negotiation is more likely to succeed when the parties concerned understand the reasons for the differences in view points [Hofstede, 1991]. So this has to do a lot with interaction and communication and leads us to the concept of cultural brokerage (see Section 3.1.3). *"[...] The quality of development cooperation depends on the effectiveness of the intercultural encounter of members of two [...] different societies. Nobody can develop a country but its own inhabitants; so foreign experts are only effective to the extent that they can transfer their know-how in the local context [...] This demands intercultural understanding, communication, and training skills"* ([Hofstede, 1991], p.219).

2.2.3   Concept of Development

The term "*development*" was originally used by Auguste Comte and up to the 1960's and 1970's "development" was actually equal to economical development. Gross National Product (GNP), Per Capita Income were indices to present levels of development. Developmental experts were convinced that by an increasing of GNP the levels of living would increase as well[19] [Burton, 2001]. When living circumstances of masses of people remained for the most part unchanged although LDCs' GNP grew, it's deficient definition became apparent and had to be adapted. Hence, development was not only defined on economic terms, but also in terms of social well-being and political structures, as well as in terms of the physical environment [UNDP, 1991]. Since 1992 the UNDP has taken into account alternative dimensions of development such as human autonomy, equity, sustainability, empowerment, and cultural identity [Madon, 2000]. Up to now development is redefined over and over again, factors are added and de-emphasized. This leads to my perception that development is difficult to define and constantly subjected to changes and manifests thereby in cultural changes and practices. So it makes sense to abandon mathematical definitions and look beyond this cultural-specific view as it relies on measuring, weighing, and counting. This is underlined by Sen's "*capability approach*", who criticizes: "*Development can be seen [...] as a process of expanding the real freedoms that people enjoy. Focusing on human freedoms* [or capabilities] *contrasts with the narrower views of development, such as identifying development with the growth of gross national product, or with the rise in personal incomes, or with industrialization, or with technological advance, or with social modernization*" ([Sen, 1999], p.36). One possibility can be found in Todaro who describes three major values which are supposed to represent the inner meaning of development:

- Life-sustenance: The Ability to Provide Basic Needs

  To increase the availability and widen the distribution of basic life-sustaining goods such as food, shelter, health and protection;

- Self-esteem: To Be a Person

  To raise levels of living including, in addition to higher incomes, the provision of more jobs, better education, and greater attention to cultural and humanistic values, all of which will serve not only to enhance material well-being but also to generate greater individual and national self-esteem; and

- Freedom from Servitude: To be Able to Choose

  To expand the range of economical and social choices available to individuals and nations by freeing them from servitude and dependence not only in relation to other people and nations, but also on the forces of ignorance and human misery [Todaro, 1989].

Whereas Todaro's concept is partly focused on a macro level, Thompson defines development more individualistically by saying: "*Development is a process of enabling*

---

[19]This implies tacitly that development is improvement in human well-being.

people to accomplish things that they could not do before. That is attained when people learn to apply information, attitudes, values and skills previously not available to them. Learning is not usually enough by itself. Most aspects of development require capital investment and technical process. But capital and technology are not without human knowledge and effort. In this sense, human learning is centred on development" [Thompson, 1981]. Through this, knowledge transfer becomes pivotal to development processes.

2.2.4   Concept of Knowledge

Nath has linked knowledge to improved quality of life, which is the ultimate goal of development activities:

"Knowledge is empowering[20]. Lack of knowledge is debilitating. Knowledge empowers an individual to form his or her own opinion, to act and transform conditions to lead to a better quality of life. Approaching development from a knowledge perspective can vastly improve the quality of people's lives" [Nath, 2000]. The more people know about certain possibilities and the more "right information" they have at the "right time" (e.g. about nutritional values of food grains or how a computer can used in a certain context) the greater is their control over their own destiny. Knowledge is manifold and discontinuous, constructed through the practices of social actors in their interactions and negotiations within certain political, economic and cultural contexts and constraints. These contexts are formed by the intersection of local and global factors, which influence and are in turn influenced by the actors' behavior. In other words: Knowledge is a cognitive and social construction that results from and is constantly shaped by the experiences, encounters and discontinuities that emerge at the points of intersection between different actors' lifeworlds. Knowledge can be seen as a *product* of dialogue and negotiation. [Long, 2002]. According to Giddens knowledge is provision and mutable in contemporary society, and is *constantly* being assessed and revised. Changing over time signifies that a concept of knowledge process is more suitable than solely a concept of knowledge. *"Knowledge processes are embedded in social processes that imply aspects of power, authority and legitimation [...]"* ([Long, 1992], p.27). An intervention which transfers technological knowledge in LDCs induces several social processes. Emerging social interactions are likely to cause new forms of experiences, encounters and discontinuities. In the context of this case study, the transferred technology brings along differing cultural values and if so - as in this intervention - knowledge is conveyed by actors with a different cultural background. In this case less tacit conventions are available and an increased (intercultural) awareness is essential. *"The appearance of knowledge transfer builds upon a set of communicatively-formed conditions: agents share a sufficient amount of background or contextual knowledge; they agree about the meanings (e.g. values, believes, expectations) of activity, and they understand the requirements and resources available. If these conditions are present, then actors do not need to negotiate the preconditions for*

---

[20]There are manifold definitions of "empowerment", but mainly it is understood as follows: *"Disadvantaged people are strengthened to gain control of their own businesses. By this they get aware of their own capabilites and develop ideas and strategies for forming and improving their life situation"* ((cp. Crawley 1998, p.25f; Theunissen und Plaute 1995, p.12) cited in [Krummacher, 2004]).

knowledge and, knowledge can be constructed as information and easily transferred" ([Heaton L. and G. Nkunzimana, 2006], p.7).

Consequently, investigating the success of knowledge transfer involves elucidating aspects of power, authority and legitimation during its implicit social processes; basically *"they* [social processes] *are just as likely to reflect and contribute to conflict between social groups as they are to lead to the establishment of common perceptions and interest"* ([Long, 1992], p.27). Power emerges out of social processes and is better considered a relation and not something stable that automatically goes along with specific characteristics. Power is negotiated in discourses the way cultures or social relationships are: by forms of representation. The design of power can also be observed in such text as transcribed interviews are.

2.2.5   Concept of Intervention

When reading literature on development one stumbles across the term *"development project"* again and again. Basically it should be stressed that a "project" is defined by a certain launch and end date; e.g. in business, special appointed meetings mark the beginning of projects ("Kick–Off"-Meetings) and "final evaluation reports" or "product launches" determine its end date. In fields of development co-operation this boxing-in of space and time is also present and various kinds of interventionist discourse promote the idea that problems are best tackled by dividing up empirical complexity into *"a series of independently given realities"* based on "sectoral" criteria (i.e. by designing policies specifically focused upon agriculture, health, housing etc.). *"Decision makers, before they act, identify goals, specify alternative ways of getting there,* [...] *and then select the best alternative"* ([Long, 2001], p.32). Consequently, a project *"is still often visualized as a discrete set of activities that takes place within a defined time-space setting involving the interaction between "intervening" parties and "target" or "recipient" groups. Such an image isolates interventions from the continuous flow of social life and ongoing relations that evolve between the various social actors, including of course (though not exclusively) the manifold ways in which local actors (both on- and off-stage) interact with implementing officials and organizations"* [ibi.]. So, as in fields of development a project is not continuous, I rather prefer to use the more appropriate term *"intervention"*. According to [Long, 2001] *"Development interventions are always part of a chain or flow of events located within the broader framework of the activities of the state, international bodies and the actions of the different interest groups operating in civil society. They are also linked to previous interventions, have consequences for future ones and more often than not are a focus for intra and inter-institutional struggles over perceived goals, administrative competencies, resource allocation, and institutional boundaries"* ([Long, 2001], p.32).

Practically and regretfully, just the opposite image is present and often expanded with a *"cargo"* image of intervention which implies that development results mainly from bringing outside inputs into a specific locality. Thereby, previous identified problems are supposed to be solved, which is colored by the notion that these presents or gifts have supreme quality which cannot be produced within the local situation itself. The local (perceiving) environment is not able to solve those problems on its own and depends on external inputs. This is regardless whether a *"cargo"* is a tangible product e.g. computers, or intangible as technological knowledge is. "*The situation chosen for intervention is deemed inadequate or needing change; thus local bodies of knowledge, organizational forms and resources are implicitly*" (and sometimes quite explicitly) de-legitimised; and consequently external inputs are assessed as necessary and indispensable ([Long, 2001], p.35). Thereby, "*intervention becomes a way of reshaping existing social practice and knowledge and of introducing new elements [...] that either replace or accord new meanings to already established ways of doing things*" ([Ploeg, 1989], p.154, 161). External factors become "internalized" and come to mean different things to each actor. Therefore, an intervention is not simply a top-down process, as it is often implied, since initiatives may come as much from below as from above [Ploeg, 1989]. Hence it is important to focus upon intervention practices as shaped by the interactions among the various participants, rather than on ideal-type implementation processes. Any intervention promotes new social values and ways of organising society.

Hence "planned intervention" is considered as "*a transformational process that is constantly re-shaped by its own internal organizational, cultural and political dynamic and by the specific conditions it encounters or itself creates, including the responses and strategies of local groups who may struggle to define and defend their own social spaces, cultural boundaries and positions within the wider power field*" ([Long, 2001], p.72). Intervention, thus, means a change of cultural practices by which culture can be manifested.

2.3   Refining Aims of Research

So reconsidering how success of a development intervention with technology can be defined; and integrating the above concepts, the following claims for an "ideal" Ugandan case study arise:

- The better the intervention is integrated in the participants everyday life, the more successful the intervention can be considered.

- The lab can be self-maintained by the community/institution and, ideally, will be up and running for ever.

- People can use the new technology in the way they want to use it. They can adapt and enhance it. This implies a certain knowledge of new technologies' applications.

On a meta-level this leads to more abstract questions:

- Who can evaluate above mentioned goals and how can these be approached?

- Is given information by participants true, as a researcher is part of the environment and involved within the field of power?

To go beyond mere statements our identified "research questions" can be (although so far still rather simplified) summarized as:

- How is access to computers constructed?

- To what extent is the new technology integrated into a participant's everyday life?

- What attitude[21] and expectations towards the "receiving" of computers can be found?

- What kind of cultural differences occurred and how can they be explained/categorized?

- Did relevant hierarchical structures change with the introduction of computers?

- In what way does the multiple role as an interviewer, teacher, colleague and donor influence the intervention's implementation?

- Should donors evaluate such interventions on their own?

Let us keep in mind these statements and questions during the following chapters to have an informal and "blunt" idea what this research aims to investigate. These questions are refined to explicit research questions in Chapter 4 and their answers will be given in the Chapters 13, 14, 15. These chapters deal with forms and ways of daily representation and usage of computers which are discussed in Chapter 16.

2.4 Summary

Beginning from a short statement on ICT policies in LDCs and Africa, I briefly presented the background of this research and explained its underlying concepts. Pillars of technological knowledge transfer in LDCs were elucidated and I tried (in order to get closer to the core issues) to present my view points on certain concepts. Through the depiction of underlying concepts I made it evident that any intervention and any developmental study is inevitably closely interlinked to culture and its inherent constant changes. Consequently, the investigation of cultural practices (its values, believes, etc.) enables to understand these changes and "how" computers are integrated into a participant's everyday life. Together with refined research questions I evolved my research aims to allow a better clarified presentation of available literature, theoretical approaches and empirical studies.

In order to get even closer to my subject I will discuss available (academic) studies on technological knowledge transfer within the domain of computers in LDCs in the following sections. Of course, this presentation does not aver completeness, but tries to give a solid overview on relevant (academic) literature on teaching ICTs in LDCs. This overview is relevant for my purpose to that effect, as important stakeholders and scientific communities are identified within the framework of this thesis and their outcomes of previous works are scrutinized. Case studies of other scientists give the opportunity to reveal similarities and differences and enables me to put the case study

---

[21]Attitude refers to an individual's feeling towards the personal and societal use of computers. A positive attitude includes an anxiety-free willingness or desire to use the computer, confidence in one's ability to use the computer and a sense of responsibility when using computers. Computer anxiety is the fear or apprehension felt by individuals when they use computers, or when they consider the possibility of using the computer [Harris et al., 2001].

at hand in a broader context. This chapter will also discuss other studies' methodological assumptions and approaches to point out different scientific contexts and tenets how ICT related interventions are currently tackled. Thereby, I pave the way why this Ugandan case study offers new possibilities how ICT related interventions can be investigated and how I aim to fill the identified gap of lacking case studies in prevalent literature.

Chapter 3

Theoretical Stances

To be able to investigate a technological knowledge transfer intervention in LDCs a researcher asks to examine available literature within the field of interest. In order to get an overview of academic literature and approaches to this domain means to identify relevant implied scientific disciplines. In this case "social implications of computer teaching as an Austrian in LDCs" can be approached in multiple ways: it can be seen e.g. as a matter of education, as a technological-oriented intervention, as an intercultural encounter, and also as a development sociological issue dealing with planned intervention. In order to receive less quantitative but more detailed and relevant information, a researcher has to clarify on a certain perspective from which he/she wants to concentrate his/her field of research. It asks to examine the intersections of the involved disciplines to identify relevant literature. As deduced in Chapter 2, all previously mentioned perspectives become evident through cultural practices and therefore the following (colloquial) statement can be considered as the "real" area of interest in this study: "introducing computers in LDCs with a view to intercultural awareness[1]". As already explained in Chapter 2 this research concentrates on the social meaning[2] of computers in South-West Uganda and their integration in the participants everyday life. Therefore, I limit the latter presentation of empirical studies to those examining implications of ICT related interventions on a participants everyday live in LDCs. Of course, outstanding studies and achievements of studies which are not directly related to the tenor of my stance, must not be disregarded and several of their allocated results are incorporated in this research, as they contribute valuable insights to the case study's investigation. Nevertheless, I omit a detailed description of those empirical studies due to their irrelevance for this research.

---

[1] See Section 2.2.2

[2] The social meaning of computers is attributed by assumptions, knowledge and expectations of its individual, group or organization. It emerges both from interactions involving actors in context and shape those contexts in turn [Heaton L. and G. Nkunzimana, 2006]. These contexts manifest in cultural practices.

3.1 Literature

To my knowledge there is no institution explicitly focusing on social implications of teaching computers in LDCs with a view to intercultural awareness, which I am interested in. Therefore, disregarding any intercultural perspective - in order not to miss any relevant literature - I firstly concentrate on literature on ICTs in LDCs in general. The whole literature research was mainly desktop-performed, because there is a general restriction in searching for sources on ICTs in LDCs. As the aim of literature on certain topics, in our case ICTs in LDCs is to enlighten disciplines (and researchers and institutions) on their research domains, most literature on ICTs in LDCs is published via Internet, because library access in LDCs is even more difficult than access to the Internet. So this is the only way researchers from sometimes called "disadvantaged" countries can gain access to other researcher's outcomes. Unfortunately, tracking and validating certain information sources becomes difficult. Studies by (non-)governmental organizations in particular have to be considered carefully, because these studies often lack references and reasonable academic differentiations and backgrounds. Though, to my knowledge quite often academics are deeply involved in the studyies's preparation and creation - e.g. S.C. Bathnagar, an academic researcher within the IFIP has become a consultant of the World Bank. A new way how this kind of literature can be integrated in academic researches has to be found. For this study I cannot incorporate available literature in sections of methodology and theoretical approaches, but general results are taken into account in discussional and historical sections. This helps to underline or de-emphasize certain discourse results. As a next step I am going to illuminate separate academic threads, namely, after a short introduction on general literature on ICTs in LDCs and a short depiction of the Working Group 9.4 of the IFIP, relevant literature on culture and ICTs in LDCs is followed by literature on science teaching in LDCs regarding intercultural awareness.

3.1.1 Literature on ICTs in LDCs

Examining the emergence of publications for half a decade now shows a rapid increase of published studies about ICTs in LDCs since 2003. This happened most likely as a consequence of the World Summits of Information Society (WSIS) in Tunisia 2003 and Geneva 2005. For these conferences many NGOs which are engaged in the fields of development co-operation put high efforts into ICTs in LDCs and published literature on ICTs in LDCs. In general, as mentioned above, it is quite difficult to take these (case) studies into account, as concealed statements about applied methods and underlying theoretical frameworks do not allow a solid reconstruction of their contributions. Furthermore, looking on the list of publishing authors political and economic motivations seem to be widespread as well. Work by authors from big firms like Intel, Cable & Wireless or Microsoft raise concerns about "purposeless/non-intentional" and free studies. Of course, these studies offer valuable insights, but might be considered as inappropriate for the realistic and true assessment of a state of the art. Valuable and contributing literature and its policies on ICTs in LDCs will be put in context in Chapter 7.

A closer look at the (more or less) academic discussion about ICT in LDCs, reveals that there are a lot of actors involved coming from different origins and (political/-scientific) backgrounds. Both members of the World Bank group, project managers of multinational development organizations on governmental level such as the *International Institute for Communication and Development* (IICD) or the *International Development Research Center* (IRDC) and researchers of the *International Federation of Information Processing* (IFIP) publish academic articles, edit books and organize conferences on this topic and related issues. All in all there are many organizations and institutions examining ICT in developing countries in an academic manner, but up to now premier journals for example MIS Quarterly or The Information Society, did not care much about ICTs in LDCs [Walsham & Sahay, 2005]. Up to 2002 the situation was the following one as Sahay and Avgerou said: "*If one picks up back issues for the last five years of journals like MIS Quarterly, Information Systems Research, Organization Science, The Information Society, Journal of MIS, and so on, in all probability not even five articles could be be obtained that deal explicitly with developing countries issues*" ([Sahay & Avgerou, 2002], p.2). Well, the situation changed recently when "The Information Society" called for a special issue on IS in LDCs in 2002 and the MIS Quarterly has announced a special issue for August 2006 [Walsham & Sahay, 2005].

The amount of research decreases drastically, when one focuses on studies dealing with the social implications of ICTs in LDCs, as e.g. the Working Group 9.4 of the IFIP does. For what reason? There is either a problem in tackling the social and cultural consequences, or sociocultural aspects are considered to be not important at all. Latter assumption goes along with Nulens study "Information Technology in Africa: The Policy of the World Bank", who examined relevant papers and documents of the World Bank from 1986-1996 to analyze the World Bank's IT policy. Nulens follows Okot-Uma's classification of IT problems into operational, contextual (sociocultural) and strategy problems [Okot-Uma, 1992] and claims: "*The World Bank focuses on operational and strategy problems*" ([Nulens, Gert, 1997], p.21). He concludes, according to his perceived stance of the World Bank on ICT in LDCs, that omitting sociocultural issues is no surprise as, "*when one is convinced that technology will overcome all sorts of problems, it is indeed unnecessary to care much about culture*" ([Nulens, Gert, 1997], p.21). Although Nulens might be considered as a bit outdated it brings up the heated discussion on ideology of technology and on a predominance of technological concentration in multinational organizations. In Chapter 7 I will discuss this topic in more detail. Recent trends just partly live up to Nulens expectations, as the World Bank's financed *info*Dev (The Information for Development Program) has developed 50 research questions related to ICTs in education (in LDCs) that merit further (future!) research. Among these is: "*Research question #15: * What are the emotional, psychological and cultural impacts of ICT use on learners from disadvantaged, marginalized and/or minority communities? The impact of ICT use on learners may be most pronounced not on student achievement, but rather on a learner's sense of self and cultural identity*" [infoDev, 2005a]. As these questions are part of work which is supposed to be done, it underlines the reasonableness and worthiness of our study.

Let us now bring the IFIP into focus. The IFIP was founded by the UNESCO at the first World Computer Summit as early as 1960. In 1989 the IFIP founded the Workgroup 9.4 as a subgroup of the Technical Committee Nr. 9 (TC) named: "**WG**

9.4 - Social Implications of Computers in Developing Countries". The WG 9.4, which is currently the leading body in dealing with social implications of ICT in developing countries [Odedra-Straub, 2003], defines one of its aims with the phrase "*To collect, exchange and disseminate experiences of developing countries"*. The conference in July 2003 was organized together with the WG 8.2., which also "*seeks to generate and disseminate knowledge about [information technology] and improve understanding in the role and impact of information technology across a range of social levels (society, organization, individual) and across a diversity of spheres (marketplace, workplace, home, community)"* [IFIP WG 8.2., 2003]. So most relevant literature for the presented research can be found in publications of and in correlation to this institutional body[3]. Studies by Helen Scheepers and Sandra Turner will be depicted in Section 3.2. As a matter of fact scientific communities tend not to see further than the end of their nose and this is what happened to the WG 9.4. Sahay and Avgerou criticize the isolation of 9.4. as well when they claim that "*Most researchers working within the 9.4. field also tend to function as an isolated community doing pockets of research and writing for and presenting their research to each other in the safe haven of their group meetings"* ([Sahay & Avgerou, 2002], p.74).

3.1.2 Literature on Culture and ICTs (in LDCs)

Looking beyond the WG 9.4 with an eye on culture leads to the biennial "Cultural Attitudes towards Technology and Communication (CAtTaC)" conference - organized by the School of Information Technology Murdoch University and the Drury University in Australia. CAtTaC does not explicitly deal with LDCs, but investigates the impact of ICTs on local and indigenous languages and cultures.

Theoretical Stances

In literature one can find two major perspectives on ICTs and its cultural impact, whereas both perceive ICTs as not value-neutral. On the one hand ICTs are perceived as new means of colonization. Postman, a technological determinist, forecasts a loss of the "*psychic, emotional and moral dimensions"* of human thought, the loss of subjectivity and traditional values ([Postman, 1992], p.118). According to this view the adoption of the technology by Indigenous people implies "*struggling to maintain the integrity of their culture in a world dominated by Western [itmo technological orientated] ideologies and lifestyles"* ([Dyson, 2004], p.60). The science of technologically developed societies is to be seen as a hegemonic icon of cultural imperialism ([Ermine, 1995], [Maddock, 1981], [Simonelli, 1994] cited in [Cobern & Aikenhead, 1997]).

The other view derives from the image that cultural-specific inherent values of technology can only mark its way if they are adapted by its "mature" user. "*Rather than view ICTs as inherently loaded with Western [itmo cultural specific] values, it might be more accurate to say that they are only capable of furthering the agenda of the dominant culture if used to that end"* ([Dyson, 2004], p.69). This is the reason why several authors claim that ICTs should be able to be locally adopted and

---

[3]To underline the appropriateness it should be noted that preliminary results of this research where presented on the Workgroup's Conference in Abuja, Nigeria, in May 2005

enhanced to become useful ([Olivier, 1994], [Mansell & Wehn, 1998], [Audenhove, 2001], [Latu, 2006]). Usefulness of technology is a core element of the Social Construction of Technology approach ([Davis, 1989] cited in [Nocera, 2006]), which focuses on the interpretative flexibility of technology and on how diverse meanings are ascribed to technology. According to this approach usefulness is actively and socially constructed and by the concept of *"technological frame"* (TF) it can be explored how users and technology co-construct each other, pointing towards an understanding of technology culture [Nocera, 2006]. TF states that assumptions, knowledge and expectations, expressed symbolically through language, metaphors and stories influence how people attribute meanings to ICTs ([Nocera, 2006], [Plooy & Roode, 1999]). Orlikowski & Gash introduced the notion of *technological frame* [Orlikowski & Gash, 1994] of reference of individuals, groups and organizations in 1994 and was expanded through Bijker by including the practices and strategies that shape the assumptions, knowledge and expectations of ICT users [Bijker, 1995]. It is assumed that *"people typically use some frame of reference or mental model as a guideline in organising and shaping their interpretation of events"*, when technology is used ([Plooy & Roode, 1999], p.9). Different TFs of individuals (or social groups) lead to distinct and varying degrees of perceived usefulness. As ICTs are human artefacts relying on a technologically developed cultural background, any usage by a person with a differing technological background[4] becomes an intercultural encounter. Nevertheless, *"the non-definitive character of the usefulness of technology is contingent on the intercultural encounters to which it is subject"* ([Nocera, 2006], p.2) and TF differences of technology producers and users have to be resolved. The overall goal is to make differing TFs congruent, which means a optimum utilization of an ICT. Thereby the sine qua non to investigate the way the technology is applied arises. This perspective links directly to the approach by available literature on science teaching which suggests to offer ICTs as an enabling tool [Haidar, 1997] and to make tacit expectations to an issue of discussion and to explain latent conveyed values of ICTs by a *cultural broker* [Aikenhead, 2002]. But let me first enumerate relevant empirical studies of CAtTaC.

Empirical Studies of CAttaC

One case study concentrates on Indigenous Australia and offers interesting perspectives on Indigenous Australian Attitudes to ICTs and on cultural appropriateness of ICTs for Indigenous Australians [Dyson, 2004]. Unfortunately, except Postma, who *"[...] attempts to illustrate how African [itmo some South-African] learners orient themselves towards information in a pre-computerized learning environment"* ([Postma, 2001], p.313), no other case studies dealing with African learners were found in this context. This lack of is underlined by the well-known Charles Ess who presented case studies on cultural conflicts in the domain of CMC technologies. He omits Africa as a continent and groups it together with other indigenous groups [Ess, 2004]. He claims that the articles on black African students [Postma, 2001] *and Filipino virtual com-*

---

[4]Note that this is NOT limited to members of technologically developed societies. Within a culture technology is perceived as a society's subculture [Cobern, 1996] and therefore TF differences are between all users. People with an even more distinct cultural background, bigger TF differences are to be expected.

munities [Sy, 2001] provide detailed insight into cultural groups that have only rarely entered the literature and consideration of IT theorists and researchers. The article on the Kelabit community [Harris et al., 2001] similarly illuminates a cultural group that, to my knowledge "*has yet to be taken up in IT literatures*" ([Ess & Sudweeks, 2001], p.259). The outstanding work by Roger W. Harris was also presented in the "Borneo Research Conference" by the Borneo Research Council in Phillips, U.S.A. which was (though just partly) dealing with ICT in developing countries, its social implications and cultural consequences. As this conference was basically held in the geographical area of Borneo in general, ICT issue received minor attention. Nevertheless it contributes precious studies to the field of interest. In Section 3.2 the important and outstanding contributions of Roger W. Harris are presented and discussed in detail.

3.1.3 Literature on Science Teaching in LDCs with Intercultural Awareness

Implementing a computer project – which means transferring technological knowledge in a LDC - does not stop at the wide ranging discussion about ICT in LDCs and culture. It implies science teaching in a developing country in practice. The **National Association For Research In Science Teaching (NARST)** deals with this issue. A small but global community around G. Aikenhead, O. Jegede, J. Slay, W. Cobern, M. Okawa, M. I.Ogunniyi, A. H. Haidar has been formed to discuss science knowledge transfer between cultures. Research in science education has experienced a remarkable change by the work of Michael Kerney, a cultural anthropologist in which he presented his logico-structural model of world view [Kearney, 1984]. Published in 1984 this model was subsequently adapted for research purposes in science education by [Cobern, 1991] and applied e.g. by [Slay, 2002]. Cobern states that science teaching always takes place in a cultural context as it refers to metaphors and images available to the learner to allow a successful integration into their worldviews [Cobern, 1996]. By Plooy's statement that "*philosophical world views will also influence the technological frame of reference of individuals and groups*" [Plooy & Roode, 1999] it becomes evident that this thread becomes shrunk to above mentioned concept of *technological frame* [Orlikowski & Gash, 1994]. Haidar researched on technology education in Arabian countries. As a result he proposed that science (as a product of a technologically developed and technological orientated culture) should be offered as an enabling technology for those cultures which are not technologically orientated to that extent [Haidar, 1997]. Furthermore, Baker et al. are claiming that there is a need to develop teaching methods that consider students' "cultures" [Baker et al., 1995]. They identified a misfit between science education in technologically developed countries and countries with a different path of development. The Nigerian Jegede commented on the best way to improve science education [Jegede, 1994]. He, similar to Aikenhead and Cushner & Brislin, considers science teachers as facilitators, cultural tour guides, and learners; in short, **culture brokers** ([Aikenhead, 2002], [Cushner & Brislin, 1995]). Though, Jegede's approach generalizes too much, as Africa and its population cannot be seen as a uniform culture. Furthermore, the juxtaposition of an African culture to a unitary "Western"-culture (which does per se not exist) would mean to apply a polarizing concept of encapsulated cultures, which I do not share with. Nevertheless, it gives valuable insights, how this concept can be approached. To Jegede, being a cultural broker means to be in charge of the following tasks:

- *"Generating information about the African environment to explain natural phenomena;*

- *Identifying and using indigenous scientific and technological principles, theories and concepts within the African society; and,*

- *Teaching the values of the typical African humane feelings in relation to, and in the practice of, technology as a human enterprise"* ([Jegede, 1994], p.130).

To quote Aikenhead:

"An effective culture-brokering teacher (Archibald, 1999; Stairs, 1995) clearly identifies the border to be crossed, guides students back and forth across that border, and helps students negotiate cultural conflicts [format changed by the author] *that might arise (Aikenhead, 1997; Jegede & Aikenhead, 1999). Each unit differs slightly in terms of where this border crossing first occurs"* [Aikenhead, 2002].

The "cultural broker" concept confides on the basic assumption of differing world views to which differently developed forms of science knowledge are underlying: the worldview of technologically developed cultures is rather dualistic (cause/effect) and in those where life is "God centred" [Jegede, 1994] it is called monistic-vitalistic [Odhiambo, 1972]. To put it in the polarizing words of the Nigerian Jegede: *"In direct opposition to the Western mechanized view of human in nature, the African is anthropomorphic. The world to the African is full of "life" with every entity having its own type"* [Jegede, 1996].

Accordingly, these anthropomorph cultures hold a metaphysical view of the world ([Odhiambo, 1972], [Ogunniyi, 1988]), whereby reason and faith are dependent and intertwined into one single thought system [Jegede, 1994]. Odhiambo adds that the concept of cause and effect is foreign in these cultures [Odhiambo, 1972]. However, each society has its own indigenous science and technology (e.g. iron smithing [Ssekamwa & Lugumba, 1973], [Nzita, Richard and Mbaga-Niwampa, 1993] p.58), and of some *"its practice is governed by morality[5] and dictated by practicality, acceptability and efficacy"* [Jegede, 1994]. It is also task of the cultural broker to build bridges *"between the indigenous world view and that of modern science with the use of indigenous science knowledge and the comparison of the relative epistemologies of the indigenous culture and modern science"* [Jegede & Aikenhead, 1999]. Jill Slay perceives their positions that *"it is important to make border crossings explicit rather than implicit, with the science teacher taking on the role of the "cultural broker"* ([Slay, 2002], p.6).

By this they try to avoid **assimilation**, which is a result of disrupting the student's view of the world by forcing a new way of (science) knowing. Cobern quotes several epitaphs in available literature on "assimilation" like "educational hegemony", "cultural imperialism", the "arrogance of ethnocentricity", and "racist" and underlines that *"students struggle to negotiate the cultural borders between their indigenous*

---

[5]Scope of morality has decreased in world views of technologically developed societies. The relation between thought and action has been replaced with cause and effect due to socio-economic factors [Ogunniyi, 1988].

subcultures and the subculture of science". Thereby, students reject important aspects of their own natal culture [Cobern & Aikenhead, 1997].

Concluding from above statement it becomes necessary to identify arising cultural differences in the field of computer education in LDCs. TF can help to understand the cultural differences. These can be subsumed to Cushner & Brislin's identified 18 themes of possible intercultural interactions [Cushner & Brislin, 1995] . These themes are divided up into three broad headings, namely *"People's Intense Feelings"*, *"Knowledge Areas"*, and *"Bases of Cultural Differences"*, intercultural phenomena can be assigned to them. Some themes will be recognized as previously presented cultural dimensions of Hall, Hofstede, and Trompenaars - see Page 18.

- **People's Intense Feelings:** anxiety (Hofstede), disconfirmed expectations (Hall), belonging (Hall), ambiguity (Hall), prejudice confrontation (Hall).

- **Knowledge Areas:** work (Hall), time and spatial orientation (Hall), communication & language use (Hall), roles, importance group/individual (Hofstede), rituals versus superstition, hierarchies - class and status, values.

- **Bases of Cultural Differences:** categorization, differentiation, in-group/out-group distinction, learning styles, attribution (Trompenaars).

Memo: In the course of this research project I have already encountered considerable interests on the part of the worldwide dispersed community [Odedra-Straub, 2003]. Actually, all communities' researches have a tremendous influence on the realization of computer projects in LDCs. But although these threads of discussions are inevitably involved with each other, there is hardly any co-operation and mutual discussion among them, e.g. world-view aspects are not considered in available case-studies among the WG 9.4.. So one side-effect of my interdisciplinary research is to demonstrate their separated threads of discussion and to motivate mutual exchange of perspectives and knowledge-sharing between communities. The next section depicts those relevant empirical studies and shows those which present their applied research aims, approach and results in an articulated way.

3.2   Empirical Studies

3.2.1   General Situation

To enable the outcome of relevant answers to the above mentioned research questions I now examine available empirical studies and theoretical approaches they are based on. In literature there are hardly any academic studies at all and there is a lack of case studies

... on social implications at an individual's level

When examining the organization's publications it seems that the argumentation is mainly focused on an organizational level and not on the level of individual experiences. Even well known Chrisanthi Avgerou focuses her "Information Technology in Context" on the macro level: This book elaborates on the organizational, national and sectoral contextual aspects that are implicated in the development and use of information systems. [Avgerou & Walsham, 2001]. Most research is limited to a macro view and very rarely examines the micro level of the individual as to be read in Helana Scheepers' documented experiences about conducting IT training at the SEIDET Centre in Siyabuswa, South Africa [Scheepers & de Villiers, 2000]. Though, I note that if there is a study on an individual's level it considers both micro and macro levels of research. This emphasizes the disregard of case studies on social implications at an individual's level. Since Avgerou's publication in 2001, situation seems not to have change much, when Walsham & Sahay claim: *"more work on individuals, and issues such as identity, would be a valuable direction for future IS research"* ([Walsham & Sahay, 2005], p.14).

... on teaching computers in LDCs

Among found publications the theme "learning & education in DCs" is rarely treated. Scheepers' and Villier's research [Scheepers, 1999] [Scheepers & de Villiers, 2000] and Sandy Turner's "The Use of ICT in Secondary Schools in Ghana" [Turner, 2005] are exceptions. Together with Roger Harris's paper on attitudes towards new technology and computer anxiety among secondary teachers at a school in Borneo [Harris et al., 2001] these case studies are almost predominantly lonesome in the fields of qualitative research. When Scheepers emphasizes in "The Computer-Ndaba experience: Introducing IT in a rural community in South Africa" the importance of the relationship between the "change agent" and participants" ("[...] *the relationship between the change agent and the participants should be open and honest"*) [Scheepers, 1999], she doesn't consider the issue of interculturality during the teaching of computers as Jegede et. al. do. The same situation can be found in Roger Harris's Kelabit study which doesn't discuss the knowledge transfer process itself [Harris et al., 2001]. To my knowledge the initial computer teaching in ICT interventions in LDCs is mainly done by members of technologically developed and orientated cultures. Taking into account the complexity of cultural implications it seems obvious that a solid study can offer new perspectives in teaching computers in LDCs.

... being performed by a scientist with practical experience.

As far as my literature research is concerned, further case studies are based on reports made by administrators to academics. When Odedra-Straub complains that "[...] *there doesn't seem to be much collaboration taking place between the two groups (academics and practitioners)"* ([Odedra-Straub, 2002], p.1) and *"What was more bothering is [...] the little participation from practitioners"* ([ibi.], p.1), it becomes apparent that there is a real lack of case studies being carried out by scientists with practical experience. This allows a skilled and ex-post inspection of (technological) knowledge transfer itself. Starting from this particular point of departure I would like to contribute to the provision of further evidence.

3.2.2 Available Studies

In the following sections I am going to present and briefly discuss (as far as possible) available case studies that are relevant to this research. This means that they deal with technological knowledge transfer in LDCs and cultural issues at least partly:

Challenges and Opportunities in Introducing Information and Communication Technologies to the Kelabit Community of North Central Borneo

In his interesting study Harris investigates attitudes and computer anxiety among teachers at a school in North Central Borneo. On the basis of an action research (see Page 51) approach he applies an interesting combination of quantitative and qualitative methods. Deriving from base-line data from a household survey on "information use" (n=140), he focused on all twelve teachers of a Secondary school. To collect data for his study he combined quantitative and qualitative methods. This includes instruments such as the Computer Attitude Scale (Koh, 1998) and the Computer Anxiety Sub-Scale (Trang, 1999) as well as structured interviews [Harris et al., 2001]. As the study's sample size was below thirty Harris applied the Mann-Whitney U Test and states that there were "[n]o significant differences [...] in the computer attitude or computer anxiety between: (a) male and female teachers, (b) different age groups; (c) those who owned and those who did not own a personal computer; (d) those who never attended any computer courses and those who had attended 1 - 3 computer courses; and, (e) those with different years of teaching experience" ([ibi.], p.13). These results will provide valuable inputs for a comparison to our study. Consequently, Harris substantiates his statistics by individual statements of the research's respondents. Despite the small sampling size of twelve his findings about the respondents level of computer usage are quite interesting: Currently they use computers for record keeping, preparing test papers, preparing their teaching plans, preparing notes and handouts for corresponding purposes. Sadly, lack of time and availability of electricity supply and availability of computers have hindered these uses [ibi.]. Besides, Harris holds inadequate computers for teachers' use accountable for the low level usage of computers for teaching purposes (16.7%). It would be interesting to learn more about the meaning of the term "inadequate computer" ([ibi.], p.15). Further findings among the teased out "pleasures" of the computer were:

1. the computer made their work more presentable, more systematic; and

2. they helped reduce redundancy of repetitive work, thus giving teachers time for other things.

Although Roger Harris's findings are restricted and based on a different sociocultural environments these virtues of the computer are very important to our research, as similar statements by our informants were made (see Chapter 16 for further details)

Teaching of a Computer Literacy Course: A case study Using Traditional and Co-operative Learning

Helen Scheepers' case study compares traditional and co-operative learning methods in a computer literacy course at the Siyabuswa Education Improvement and Development Trust (SEIDET) center in South Africa in the first half of 1998. It focuses on the following research question: "Are there any differences between a traditional learning experience and a cooperative learning experience in a computer literacy course for *teachers at SEIDET (a rural community in South Africa)?"* ([Scheepers & de Villiers, 2000], p.2) Every Saturday afternoon 42 computer-newbies were introduced to certain computer topics. Basically, the co-operative teaching method of Jigsaw (Johnson & Johnson) had been selected for conveying knowledge on word processing and the traditional way of teaching computer literacy had been utilized for MsExcel. Whereas *"the role of the teacher in a co-operative learning environment changes to that of a mediator, facilitator and coach"* ([ibi.], p.5), a teacher relying on the traditional way of teaching computer literacy *"describe[s] the actions that should be taken by the learner in front of the computer"* ([ibi.], p.6]). After the course was finished in July 1998, the participants were handed out a questionnaire with close and open-ended questions. 27 questionnaires were returned and were utilized by the authors for their analysis. While the results of the closed questions are presented quite clearly and thoroughly, the interpretation of the open-ended questions is quite short. Looking at the summary of responses of the open-ended questions in the study's Appendix arouses the reader's curiosity. Most likely the respondents statements could have been investigated much more on a detailed level and could have revealed some further instructive issues; e.g. the cited dislikes about working in groups like "Domination of some tutors in the group" or "Working as equals" might have dampened the credo for co-operative learning methods; by this, conclusions like *"The results about the teachers' experience with the co-operative learning are very positive"* ([ibi.], p.8) are put into perspective.

The main findings for the core research question, "What lessons can be learnt from this learning experience?" are as follows:

- *More time should be spent on co-operative learning to ensure that it becomes "ready-at-hand".*

- *In comparing the results of the test [written, final examination] for MsWord and MsExcel, no significant improvement can be identified. [Score for MsWord: 47%; Score for MsExcel: 46%]*

- *Although the teachers felt uncomfortable with the new method of teaching, they liked it and liked working in groups* ([ibi.], p.8f).

In fact, it is quite interesting that topics taught by co-operative learning methods, which are promoted by donor countries and research institutions (e.g. Foundation for Research and Development, South Africa), did not fare better. Actually, it is difficult and thus problematic to compare MsWord and MsExcel, as they are different programmes which are used for different purposes. Both software packages have underlying similar user/usability concepts, in the meaning of terms: mouse usage, short-cuts, etc. Anyway, Scheeper's study provides on excellent information on different teaching methods within a technological context in Africa. Scheepers completely neglects the role and capability of the teacher in this paper, but tackles it in [Scheepers, 1999] - see Section 3.2.2. In this comparative case study it is not mentioned who the computer teachers had been and if there had been any surprising interaction during the very first "acquaintance" between computers and participants. As 16 out of 42 people did not have any previous computer experience, it would have been quite interesting -

although it might have been too far reaching for this study - if observations similar
to [Nel & Wilkinson, 2001] had been made.

**The Computer-Ndaba Experience: Introducing IT in a Rural Community
to South Africa**

This case study is based on the same teaching experiences as the above mentioned
case study, namely Saturday afternoon from January to July 1998 at the SEIDET, but
this time Scheepers addresses different issues, as she aims to investigate the following
research questions:

1. *"How should IT introduction be done in a rural community center in South Africa
   where there is no previous experience of using IT?*

2. *What are the social and technical risks in such an introduction process and how
   can these be addressed?*

3. *How did the principles for IT development and implementation apply in the
   introduction process and what lessons were learned from them?"* ([Scheepers,
   1999], p.2)

In order to answer these questions she takes from a comparative theoretical perspective, namely the trade union approach, which she describes in [Scheepers & Mathiassen, 1998]. Basically, this is one out of eight Scandinavian approaches to system development. Scheepers applies this Scandinavian approach at the SEIDET, but aligns
the researcher's role to her specific situation. Whereas a researcher within a trade
union project is defined as a resource, *"the researcher in the case of a socioeconomic
development project cannot play this type of role. The knowledge of the participants is
such that the researcher has to play a more leading role, but not a primary role. The
researcher should determine the way in which the project enfolds. This implies that the
researcher should walk a fine line between identifying what should happen and being
lead by the participants of the course"* ([Scheepers, 1999], p.13). Scheepers identifies
the researcher as a change agent, which goes along with her applied methodological
combination of action research [Rapoport, 1970] and Soft Systems Methodology (SSM)
by [Checkland & Scholes, 1990]. Although SSM encourages the researcher to take into
account cultural aspects, Scheepers does not elaborate intercultural aspects at all.
Although emphasizing the importance of relationship between researcher and participants, she skims over hierarchical and possible system (!) influences by further "role
players". Unfortunately, she fails to give hints about those additional "role players".
A further point of interest is that Scheepers describes the Computer-Ndaba as an *"IT*
intervention [changed by the author] *focusing on knowledge building within SEIDET
an example of a socioeconomic development environment"* ([Scheepers, 1999], p.2). To
my understanding it is contradicting to show expectations of participants to computers, (*"A belief that computers are going to solve a lot of problems"* ([Scheepers, 1999],
p.16).) on the one hand and to claim that participants had no prior experiences about
computers on the other hand. Although, "experiences" are supposed to be meant
to be practical ones, expectations are established upon (other forms of) experiences.

It seems as if in order not to dismantle the image of a (closed) system-environment, Scheepers skips discussions on expectations at hand. She names them, but does not put them into a context. This might have supported answering her second research regarding social risks, which she limits to lack of access time to and failure of establishing a collaborative relationship between the participants and the change agent. Besides, technical risks are addressed as "financial resources, unavailability of facilities, bad management of the lab and external factors (theft, vandalism)" [Scheepers, 1999]. The third question is answered throughout the whole text but not very clearly. Even Scheepers does not clearly sum it up by herself, but finally states "*Based on the experiences in this case changes were made to the principles*" ([Scheepers, 1999], p.16). Answers to her first questions are found within the discussion section and are quite instructive to my research and reveal several similarities to the Ugandan project:

- There should be incremental change with difficulty that are levels within reach of the participants. Goals that are too high should not be set, otherwise the participants will not even try to attain these goals.

- The way in which the role players should communicate is very important.

- Opening hours of the lab have to be adjusted to the participants needs.

- It is necessary to place checkpoints in the course to make sure what and how the participants understand.

- "The important aspect of IT diffusion in a development environment is the delinking of the knowledge from the donor and change agent to the participants themselves. The participants should be able to go on with the process and day-to-day activities without the help of other institutions."

Very important to my research are obvious similarities between the SEIDET and Ugandan intervention processes, as e.g. computers were not available at course start, there was a lack of training time identified, and some further similar "problematic situations". Further similarities can be associated to terms like Internet, LAN, homepage and e-mail; or in ways how teachers would use a computer. According to the Computer-Ndaba experience, teachers concentrate more on a personal advantage in terms of job security and support of administrative work than to using a computer for teaching.

To implement the methodology of action research and SSM Helana Scheepers uses well-known methods of diaries, case interviews (3 people every 20 minutes) and questionnaires as mentioned in Section 3.2.2.

All in all, Helana Scheepers's papers represent rare available African case studies, but unfortunately her work has not been continued[6]. Nevertheless, as a former member of the University of Pretoria in South Africa, she contributed to the works of the Department of Informatics under Prof. Dewald Roode and Prof. Carina de Villiers. The numerous, available working papers of this department represent an important knowledge repository in the fields of ICTs in education in LDCs and support this research in different theoretical, methodological and practical issues.

---

[6] According to an e-mail correspondence with Prof. Carina de Villier in July 2005.

"The Use of ICT in Secondary Schools in Ghana" by Sandy Turner

For her study "The Use of ICT in Secondary Schools in Ghana" [Turner, 2005] Sandy Turner has investigated eight secondary schools with computers. All of these schools are not located in major cities of Ghana. Her methodological approach is based on the application of qualitative methods and collection of multiple data sources: she has observed school classrooms, she has had interviews with teachers and headmasters, has performed document analysis and has handed out a questionnaire with three open-ended questions. Her aim of research was focused on *"the teachers - their experiences and the meanings they gave those experiences"* ([ibi.], p.2). By a thematic analysis of data she reveals (among others things) teachers' views of the role of computers in schools, the computer studies' curricula and applied teaching pedagogy. The latter two show that teachers have been using traditional, teacher-centred instructional strategies and have mainly concentrated on Windows, DOS[7] and word processing [ibi.]. Turner proclaims a generally positive view of teachers regarding the role of ICT in schools[8], which leads to a critical methodological aspect: Turner was visiting these schools as an (external) guest. No personal contact (and trust) had been established before, as *"the schools were identified by key informants at the University of Education, Winneba [Ghana]"* ([ibi.], p.3) and therefore, naturally, she was treated like a guest. During her observations she noticed there were age differences among the participating pupils in the computer classroom. When Turner asked for the reason why the pupils were at a different age, they told her that her visit was the reason. So her position as a visitor from the University of Education might have heavily impacted the Schools' self-representation of education and only "adequate" pupils had been chosen[9]. From my point of view, this is understressed, when Turner remarks the demonstration of a variety of teaching strategies besides the located "usual" teaching methods. Especially the emphasis of a teacher not to punish the children (*"If I use IT language you don't understand, just ask. I won't do anything to you!"* ([ibi.], p.8) and some other examples given, lead to the impression of an attempt of "saving ones face" by some teachers. We know that *"corporal punishment is frequent, routine and not administered according to official guidelines"* ([Pryor & Ampiah, 2003], p.25). Discussing this issue on the WG 9.4. conference in Abuja, (Nigeria) in 2005, Turner admitted a similar impression, which underlines the necessity of a discourse on the hierarchical position of evaluators

---

[7]Unfortunately, Turner does not give any information about the computer's systems hardware and software specification, which would have been quite interesting according to the relevance of teaching DOS, particularly because the study was performed in 2003-2004. Turner calls DOS an "outdated topic", which might be generally true from a technological oriented perspective, but from a computer scientist's stance some basic knowledge of DOS might be helpful on equipment up to Windows 98. Due to my personal experience I know that this still occurs several times in LDCs. Nevertheless, I generally agree with Turner, as due to my unsystematic observations in a private learning institution in south-western Uganda, most of the instructional time was spent on DOS, although the installed operation system was Windows 98.

[8]*"Teachers overwhelmingly agreed that it would be appropriate for every school in Ghana to adopt a computer-based curriculum to meet the needs of students"* ([ibi.], p.3).

[9]From the powerful point of view this is no surprise, as it is quite usual to present oneself in the way of being appropriate to the expected wishes of a visitor; e.g. when the U.S. president George W. Bush visited Vienna in June 2006 he discussed political topics only with "carefully selected" students. (see http://www.orf.at/060621-808/809txt_story.html)

in Chapter 16.

Another interesting finding of this study is that none of the examined schools have ICT integrated as a tool for learning across the curriculum. Sadly, Turner fails to mention her assumptions towards this fact and accepts this as a matter of fact. It would have been interesting to go beyond the mere statement. A further issue arises when Turner recognizes that "*without a well defined curriculum, teachers sometimes repeated the same lessons that had been taught to the same students in previous years*" ([Turner, 2005], p.10). Once more, she does not elaborate on this conspicuousness in detail. Nevertheless, this case study poses a highly valuable starting point, which I will frequently refer to in the discussion part (see Chapter 16). Not only the items mentioned above, but rather her impressive findings on teaching methodology itself understress the importance of Sandy Turner's work.

Where is the "Any key", Sir? Experiences of an African Teacher-To-Be

This research illuminates initial problems of computer science students in computer classes in South Africa. It is based on a case study about a student teacher to whom the compulsory computer literacy classes posed a major problem, because he had never touched a computer in his life before. Even after a certain period of time he failed the course and had to repeat it during his second year. Therefore the authors of this study try to answer their basic research question: "*Could other teaching methods and approaches not have given* [them] *the opportunity to adapt to the unfamiliar situation more easily?*" ([Nel & Wilkinson, 2001], p.1). In addition to this main case study, the authors undertook a survey on 150 first-year students in a Computer Literacy class and on 47 second-year Computer Science students. Both were students in a Faculty of Management at a Technikon Free State in South Africa. Regarding their methodology Nel and Wilkinson set up a standardized questionnaire on technological availability and included open-ended questions about initial problems in the compulsory Computer Literacy course. The collected material which allowed them to reveal problems and problem areas which are as follows: "*uncertainty*"; "*keyboard problems*"; "*using the mouse and clicking*"; "*SAVE,OPEN and CLOSE*" and "*impatient assistants and lecturers who do not understand that students had never used a computer before*" ([Nel & Wilkinson, 2001], p.3).

This list gives me a variety of comparable aspects for my analysis. With the use of their results the authors elaborated a seven-point plan which they support to implement and make further investigations in a Computer Literacy class. These seven points are as follows:

Nel and Wilkinson asked for a "refreshed" **awareness** of IT teachers in the nature and extent of the social, economical and cultural problems of students in LDCs. In a period of **orientation**, students should spend enough time to get acquainted with the basical elements of a computer keyboard and the mouse and its usage. Too little time to do the practical work was figured out to pose a problem. Nel and Wilkinson also claimed to have investigated the possibility of alternative **teaching methods** such as co-operative learning in theoretical as well as practical use and to take into account possible **language** problems of students. Due to the fact that few students own computers (have **access**) the research emphasizes sufficient open-lab times regarding arising accommodation, transport and financial problems. The latter problems turned out to be quite relevant as these prevented students from doing extra lessons in the computer lab. The study has also identified that assessment techniques need to be revisited; it claims that **assessment** should be of a continuous nature, with tests in line with what has been done in classes. Finally, **feedback** given by students to their lecturers is identified as being important to improve and adapt lessons and specific teaching methods. Keeping in mind these points will help me to discuss the Ugandan intervention in a better structured way.

Nevertheless, in my opinion, this study could have focused more on "how" social interactions themselves were performed. This might have happened, but it was not possible to reproduce them without detailed information on underlying theoretical and methodological concepts. Furthermore, this study concentrates in an extensive manner on student's problems caused by e.g. "unaware" teachers (or teaching methods), but does not question a computer's appropriateness to local cultures. Another interesting subject area of this research are proposed teaching methods of co-operative learning. As a possibility, how computers can be introduced; the authors suggest collaborative learning ([Walker, 1995], [Salikin and Cummings, 1997]) jigsaw method, and guidelines of the Worcester Polytechnic Institute (WPI 1998). Though, introducing these in a rather short section, it might be useful to interlink these teaching methods with aspects of culture.

Pacific Islanders and ICTs

The Tongan Savae Latu examined Pacific Islander (PI) students at New Zealand's tertiary institutions of Hamilton and Auckland and presented a study on the latter one at the CAtTaC 2006 in Estonia. For the description of his research design he refers to the Hamilton ethnographic case study, which is unfortunately not referenced in the appended bibliography. Nevertheless, I note both studies as quite interesting and I retrieved following cornerstones of the Auckland study, which was a "*redesigned, expanded and adapted*" ([Latu, 2006], p.7) version of the Hamilton study. Though not being precise enough what Latu refers to, the Auckland case study is methodologically based on qualitative, unstructured interviews in the form of informal meetings with ten students. Students were asked to "*express their views and give reasons for studying or not studying ICT*" ([ibi.], p.7) and were audio-taped. Additionally, Latu held informal meetings with scholars parents, as in PI tradition parents are drastically involved in educational aspects [ibi.]. Factors identified by the students were similar to [Nel & Wilkinson, 2001] in terms of language[10], time, teaching methodological[11], and financial aspects ([Latu & Young, 2004], [Latu, 2006]). These were found among some parents as well and ICTs are perceived more as an expense factor and a demoralizing tool than trustworthy information preserving and providing tool. Feedbacks based on unexpected activities "by the computer" as (pornographic) pop-ups and "hidden"

---

[10]This is not limited to English as a foreign language, but pertains also to computer terminologies.

[11]Some students complained that teachers hardly use practical examples to demonstrate and answer questions, which leaves the impression of a teacher centred education. In order to avoid this Latu refers also to the role of a teacher as a cultural facilitator: "*teachers must be aware of their cultural background and expectations*" [Latu & Young, 2004].

Internet connectivity costs arose. Due to misleading information by a teacher, a parent assumed that Internet fee would be NZ\$ 14.50 a month. After a few month the telephone bill increased up to two hundreds dollars/month. This resulted in a mistrust in the Internet, because "the computer makes us pay more money to telecom and we can't afford to pay for me to visit the doctor or to donate money to church and family functions" ([Latu, 2006], p.7). This nourishes two existing assumptions: ICTs as a trap of debt factor [Okee-Obong, 2003] and the fear of being not capable to control the technology. This lead to a disappointment of parents, who had expected a help for all of the family.

As a conclusion Latu argues that "*technologies that are new to any ethnic group should be adapted to suit indigenous communities' prevailing cultures, beliefs, level of skills, and economy instead of using technologies to acculturate[12] weaker cultured societies*" ([Latu, 2006], p.10). Like all the other presented case studies I will incorporate Latu's results in the discussional part of this thesis.

Other Case Studies

I know that there are several further case studies which have not been mentioned in this research, although they could have contributed to this research. I do not neglect those and many of them are very helpful to this research. Nevertheless, I had to limit the list of case study description as the focus of those not-mentioned case studies strays more and more from the underlying research; e.g. Louise Postma deals in an impressive way with information reception and learning in Africa but does not expressively consider ICTs [Postma, 2001]. She has figured out that South African Centres for learning reflect their designer's emphasis on individual and silent learning – in contrast with indigenous preferences for learning in collaborative and often noisy, performative ways. Selinger & Gibson's inputs about technological teaching methods in LDCs at Cisco Systems is valuable as well [Selinger & Gibson, 2004], but papers supported by companies are problematic, as a companies' ultimate goal is to maximize turnover. Cisco Systems holds offices in Lagos (Nigeria) and Johannesburg (South Africa) and is already concentrating on this market[13]. Big firms have investigated this area and they are "*more successful than competitors because of "knowledge, "persuasion" feedback loops from local channels*" and they already make money [Ess, 2004]. Nevertheless, Selinger points out cultural differences, expectations, and traditions surrounding teaching and learning in a variety of contexts, and "*examines how the recognition of cultural preferences can be used to make technology use more relevant. The thesis of this paper recognizes that learners are now faced with a range of appropriate representations of teaching/learning resources and processes that differ from the traditional*" ([Selinger & Gibson, 2004], p.1). This case study still implicitly derives from the stance that people who want to apply technology have to change, but does not take into account the possible misconception of technology to these people. Corresponding to these issues is Peter Sy's "Barangays of IT - Filipinizing mediated communication and digital power" [Sy, 2001] which examines social interactions me-

---

[12]For elaborations on "acculturation" please cp. [Latu, 2006].

[13]Due to my professional experience as a key account manager for Cisco Systems Austria I have attended internal meetings where Africa as a future market was discussed.

diated by information technology and its correlated problems and prospects within the Barangays-subculture in Phillipine society. This paper fails to explore underlying methodological and theoretical concepts, but describes - mainly (though not solely) based on literature research - more or less theoretically (in a expressive manner) how ICTs are perceived as new means of colonization and "proposes to "Filipinize" IT - that is, to creatively and critically appropriate IT into practices congenial to Filipino culture" [Sy, 2001]. Methodological concepts are difficult to derive from, as numerical and personal statements are mixed for argumentation. Further valuable inputs are to be retrieved from IICD's workshop report "Sustaining ICT-enabled Development: Practice makes Perfect?". Because of its non-auditability it cannot be considered as a case study per se, but seems to be pieced together from practitioner reports and contains excellent aspects on ICTs interventions in LDCs (these are considered throughout this research). There are also articles on integrating ICT and multicultural aspects within a classroom like by [Caruana-Dingli, 2005], but at this point I want to emphasize that a research which aims to compare the "how" technological knowledge is conveyed in a technologically developed country to the "how" it is transferred in a LDC would be quite interesting, but is beyond the scope of our research.

In the following I will utilize above mentioned case studies and present obtainable underlying theoretical and methodological approaches. As in social-sciences there are manifold different approaches and derivatives, it is not intended to list all theoretical possibilities, but rather to draw a picture on applied researches to point out possible scarcities for future research. Additionally, these sections illustrate available literature on theoretical and methodological tenets towards ICT-interventions in LDCs.

3.3 Applied Theoretical and Methodological Approaches

3.3.1 General Overview

A wide range of theories are being drawn on by researchers of ICTs in LDCs and any attempt at classifying theories will always be controversial. ICTs are embedded in an interdisciplinary research field and are approached by diverse underlying theories. Available literature is based on several different theories, e.g. development theories by Escobar or Sen or "*theories of society*", which include neocolonialism and post-colonialism. In addition to them, ICTs and LDCs can be discussed from an anthropological (cultural) stance and be placed within the fields of sociology [Walsham & Sahay, 2005]. This includes Giddens' structuration theory, actor-network theory and institutional theory. The latter one is only touched to address issues of development [Walsham & Sahay, 2005]. Figure 3.1 shows the results of [Walsham & Sahay, 2005] who have examined publications on ICTs in LDCs. It shows up numerous different approaches and posts several research areas where they have identified future opportunities.

Whereas Walsham and Sahay have examined research on ICTs in LDCs in general [Walsham & Sahay, 2005], Angelina Totolo has identified four dominant schools of thought in context to Africa [Totolo, 2005]. These are mainly drawn on to explain phenomena of ICT interventions and are as follows: the main school of modernization

Topic	Examples in existing literature	Future opportunities
Development to which ICTs contribute	Use of Internet for sustainable development (Madon, 2000) ICT not instrument for development only within market regime (Avgerou, 2003) Institutional theory to analyze ICTs for development in Chile (Silva & Figueroa, 2002)	• Draw on wider definitions of development, e.g., Sen (1999) • Further work with promising theories, e.g., institutional theory, development economics • Make contributions to related disciplines, e.g., development studies, anthropology
Key issues being studied	Local adaptation and cultivation of ICTs (Bada, 2002; Macome, 2003; D'Mello, 2003) Standardization versus localization of technology (Braa & Hedberg, 2002; Thompson, 2002) In-depth studies of GIS technology (Puri & Sahay, 2003; Barrett et al., 2001)	• Important but neglected topics, e.g., scalability/sustainability • In-depth studies of other technologies: e-government, open-source software • Large-scale infrastructure, e.g., telecommunications • Society-based critical issues, e.g., information related to HIV/AIDS
Theoretical and methodological stance	Wide range of theories drawn on e.g., globalization, post-colonialism, theories of power (Liu & Westrup, 2003; Adam & Myers, 2003; Silva & Backhouse, 2003) Methodology of interconnected levels and in-depth studies more common than decade earlier (Sayed & Westrup, 2003; Walsham, 2002)	• More explicitly critical studies • Need for methodological precision about the nature of interpretive studies • More action research and longitudinal studies • Cross-cultural research teams
Level and focus of analysis	Individual/group/organization (D'Mello, 2003; Okunoye & Karsten, 2003; Bada 2002) Sectoral/national (Mursu et al., 2003; Silva & Figueroa, 2003) Cross-cultural working (Aman & Nicholson, 2003; Adam & Myers, 2003) Public/private sector (Madon, 2003; Rolland & Monteiro, 2002)	• More individual-level studies, e.g., on identity issues • A focus on community in addition to public/private sector • Increased geographical coverage, e.g., China • More locally based research from developing countries

Figure 3.1: Some examples in existing literature on IS in LDCs and future opportunities. [Walsham & Sahay, 2005].

paradigm encompasses the critical and dependency theories as well as structuralism and functionalism [Totolo, 2005]. According to her Tipps claimed that "*One of the most discussed failures of modernization is the transferring of Western [itmo* cultural different] *models of development to* [geographical] *Africa, when in fact* Africa and the West [format changed, *itmo* cultures] *are different in their social, economic and political orientation*" ([Tipps, 1973] cited in ([Totolo, 2005], p.389)). Modernization theories were rejected "*by academia and policy makers because of its positivistic reduction of development to the advancement of Western [itmo* economical orientated] *capitalism*" [Snyman & Hulbert, 2004], and so it became the turn of dependency and critical theory, which is "*evident in works by writers on the information technology phenomena in Africa and other least developed areas*" (Akpan,2000; Oladele,2001; Thapisa, and Birabwa, 1998; Francisco Rodríguez and Wilson, 2000 cited in ([Totolo, 2005], p.389)). Both theories were criticized for their inadequacy to explain phenomena and lead to Giddens's structuration theory, which focuses on both the structure (structuralism) and function (functionalism) of governance in society. It is "*the most dominant paradigm found in the information transfer literature on Africa (in the form of policy, economic and political structures) that enable the diffusion, adoption or transfer of information technology*" ([Heeks, 2002], [Jensen, 2002], [Oladele, 2001], [Obijiofor,1998], [Peterson, 1998], [Roycroft and Anantho, 2003] cited in ([Totolo, 2005], p.390). In the discussion about "agency" and "structure" Giddens takes up a dualistic approach and tries to overcome the division of agency and structure by arguing that "structure" is both the medium and the outcome of the actions which are recursively organized by structures. Since culture can be observed in both, the actions people performed and the way they perceive it, this is another important argument why the intercultural approach applies to this study. Totolo concludes with Tipps' argument that "*modernization theorists have either underestimated or ignored many important external sources of influence upon social change*" ([Tipps, 1973], p.212). Although, recent studies draw upon wider definitions of development ([Scheepers & de Villiers, 2000]) and distance themselves from the modernization paradigm. One can roughly distinguish between positivist, critical, or interpretive tenets though the situation is far more complex and interweaved. **Positivistic** perspectives like diffusion or innovation theories, which are used for the implementation of information systems, have been heavily criticized by [Plooy & Roode, 1999]. They argue the inappropriateness of these perspectives to do research of ICTs in LDCs as they lack of consideration of social interaction and are overly simplistic (even deterministic) in their view of the intervention process, and fail to consider factors such as e.g. power, organizational culture, and ethnic cultures [Plooy & Roode, 1999]. Walsham and Sahay identified only very few studies based on a positivistic approach with stated hypotheses, instruments for data collection, statistical inference etc. [Walsham & Sahay, 2005]. Nowadays, most case studies are based on a socio-constructionistic stance due to their conviction information technology is socially constructed. ""*Social construction" is a term applied to the study of the meanings of technology and how those meanings affect the implementation of technology within the organization*" [Sahay et al., 1994] and environment, respectively. Social constructionism falls within the **interpretive** paradigm. "*Whereas the aim of positivism is to explain, control and predict interpretivists would* *rather understand phenomena through the meaning people assign to them"* ([Plooy & Roode, 1999], p.18). Interpretivists believe more on a individualistic social construction of world and thereby regard cultural factors more than positivistic theorists do. Thereby, interpretivistic studies get closer to anthropology. Though, according to Cobern, I have to stress that *"despite sociologists' appropriation of ideas from cultural anthropology, the two disciplines (sociology and anthropology) differ dramatically, even in their definitions of such fundamental concepts like society, culture, and education"* [Cobern & Aikenhead, 1997]. Nevertheless, anthropology witnessed by the works of Clifford Geertz an *"interpretive turn"* which refocused attention on the concrete varieties of cultural meanings, its complex texture, web of language and symbol [Geertz, 1973]. Thus, it becomes also a good viewpoint to discuss ICT in LDCs with intercultural awareness and despite Coberns criticism gets closer to sociology. In the fields of cultural human-computer interaction (HCI) the most popular source for cultural theories is cross-cultural communication research based on cultural dimensions [Kamppuri & Tukianinen, 2004]. This major theoretical background is accompanied by stances of cross-cultural psychology, genre theories, activity theory (e.g. an application can be found also in the fields of science teaching e.g. Jill Slay who analyzes Chinese and Australian universities), and semiotics [Slay, 2002]. Figure 3.2 gives an overview of the concept of culture in cultural HCI. Examining this figure makes it evident that most draw upon a national scope of culture which is different to my scope. Further the columns depict for each study the original research interest, the applied definition of culture and the study's theoretical background.

Regarding literature on **critical** theory I go along with [Walsham & Sahay, 2005] that as a future opportunity there should be explicitly critical studies.

3.3.2   Specific Perspectives

"It has become something of a commonplace in the social sciences to say that technology is socially constructed" ([Heaton L. and G. Nkunzimana, 2006], p.3). The philosophy of the social construction of technology advises that sampling and data gathering be conducted amongst relevant social groups, rather than aiming at a representative sample of the total population, as would be the case in positivistic research [Sahay et al., 1994]. It is the dominant philosophy in case studies dealing with social implications on an individual's level, too. Anthropological works on ICTs in LDCs are marked by a combination of the frameworks of postmodernism and social constructivism, e.g. Postma's study on some South African learners [Postma, 2001]. Socio-constructivism is also a common stance in literature on scientific-education teaching [Cobern & Aikenhead, 1997], which thereby puts itself forward as a coincidencing theoretical stance. An investigation of a technological knowledge transfer during a computer project is both anthropological, sociological and epistemological.

In fields of IS literature in recent years two approaches became predominant: *"actor-network theory"* and the *"appropriation of uses"* [Heaton L. and G. Nkunzimana, 2006]. *Actor Network Theory* (ANT) [Latour, 1987], which is a semiotic device that builds upon socio-constructionistic principles, became popular not only in LDCs. According to Geoff Walsham and Sundeep Sahay this is due to the fact that ANT *"offers an explicit way of conceptualizing technology as one of the "actors" in the*

Study	Original research interest	Definition of culture	Scope of culture	Theoretical background
Abdelnour Nocera et al., 2003	How people situated in concrete cultural configurations make sense of computer systems	Culture is a web of meanings that man himself has spun; What people say, think or express and also the material dimension	National, workplace, tool related, software developers/customers	Media studies, activity theory, hermeneutics, situated action
Fang and Rau, 2003	Effects of cultural differences on the perceived usability and search performance of web sites	-	National	Cross-cultural psychology
Gobbin, 1998	The effects of cultural fitness on the use of technology tools	-	Organisational	Activity theory
Hall et al., 2003	Validity of guidelines across cultures, design patterns as an aid to the design of culturally localised software	The fabric of meaning in terms of which human beings interpret their experience and guide their action; The way to solve problems	National	Activity theory, cultural dimensions
Mrazek and Baldacchini, 1997	Avoidance of cultural bias that falsely attribute differences in user need to geography or culture	Culture is a multitude of elements and their integration that includes the customary beliefs, social forms, and material traits of a racial, religious, or social group	National	-
Nakakoji, 1996	How to benefit from communication breakdowns that happen while designing for other cultures	The beliefs, value systems, norms, mores, myths and structural elements of a given organization, tribe, or society	National, organisational, software developers / customers	Cross-cultural psychology
Onibere et al., 2001	Attitudes towards localised interface in multi-cultural and multi-lingual country	Organising behaviour and shared beliefs that define a group	National / ethnic, workplace	Semiotics
Sacher and Margolis, 2000	Focus on culture of users and design of conversations	Shared beliefs, values, etiquette, and language of a group	National, subcultures	Semiotics
Smith and Chang, 2003	Significance of Hofstede's cultural dimensions to website acceptability	Selective screen through which we see the world	National, regional, linguistic	Cultural dimensions
Zahedi et al., 2001	How people from diverse cultural backgrounds and with individual characteristics might perceive and use web documents	Learned pattern of thinking, feeling, acting, and values, which are specific to a group or category of people	National	Social constructionism, cultural dimensions

Figure 3.2: Customized overview of the concept of culture as used in HCI studies. [Kamppuri & Tukianinen, 2004].

actor-networks" ([Walsham & Sahay, 2005], p.6). "*Actor-network theory suggests that the destiny of an innovation, its content and its chances of success depend entirely on the choice of representatives that will interact and negotiate to give form to a project and transform it until it finds a home*" ([Heaton L. and G. Nkunzimana, 2006], p.3). Thus, success becomes a matter of the network and not the pertinence of the technical solution. The active use perspective derives from the stance that the "mature" user can adopt or transform planned uses or invent new uses. Despite the users active role and putting the technology transfer in a specific sociopolitical context both are criticized for remaining relatively techno-centric. Heaton suggests to draw upon user-centered theories which attempt to explain how individuals make sense of technologies, by exploring the resources they bring to bear on their activity. Thereby, it gets one closer to examine the context of interactions and how knowledge and understanding can pass from one (cultural) context and person to another [Heaton L. and G. Nkunzimana, 2006]. This corresponds to the basic theoretical stance which is applied in this case study as well - see Chapter 5.

Action Research

Although identified by [Walsham & Sahay, 2005] as a future opportunity, case studies usually draw upon the methodology of Action Research or derivatives[14]. Action research "*aims to contribute to the practical concerns of people in an immediate problematic situation and to the goals of social science by joint collaboration within a mutually acceptable ethical framework*" ([Rapoport, 1970], p.499). Participants of the research are explicitly integrated in its process and thus influence research's and intervention's further progress. There is Snyman and Hulbert's disillusioning[15] longitudinal study on five ICT centers in five rural developing regions of South Africa which criticizes predominant influences by donor-countries [Snyman & Hulbert, 2004]. Scheepers aimed to undertake an action research at IT diffusion in an educational setting in a rural community in South Africa - SEIDET [Scheepers, 1999] and Roger Harris's work is also based on a Rapoports principles and part of an action-research pilot project [Harris et al., 2001].

Tacchi et al. have worked on action research for ICTs in LDCs and have elaborated it to an **Ethnographic Action Research** approach [Tacchi et al., 2003]. They have combined ethnography and action research. According to their understanding "*ethnography is a research approach that has traditionally been used to understand different cultures and* action research *is used to bring about new activities through new understandings of situations*" ([Tacchi et al., 2003], p.1). Despite Okunoye - by referring on Creswell - considers ethnography and action research as different in ontological, epistemological, axiological, rhetorical, and methodological characteristics [Okunoye, 2003], [Tacchi et al., 2003] use ethnography to guide the research process and action research to establish a circular relationship between project plans and activities. The

---

[14] This might be subject to the fact that Geoff Walsham and Sundeep Sahay investigated papers in journals in which empirical works are rare.

[15] I apply the term "*disillusioning*" as I am deriving from a optimistic believe that ICT centres in LDCs can be integrated into everyday's life, but which is resolved by his study which reported a more or less disastrous outcome.

methodology of *Participatory Appraisal* (PAR) is also close to Action Research and was developed by Paolo Freire in the late 60's [Krummacher, 2004]. Freire enforced an active participation of all involved actors for developmental interventions. Recently, an enhanced variant of PAR is becoming increasingly popular: Participatory Rural Appraisal (PRA) combines ethnology, "Rapid Rural Appraisal" and PAR [Krummacher, 2004], [Burton, 2001]. A further Action Research related methodological stance was applied by Glen Aikenhead's cross-cultural science teaching study: "Rekindling Traditions for Aboriginal Students" and is called *Research & Development* (R&D). "*In an R&D study, research is undertaken and data are collected to be fed directly into improving the product of the study or into initiating practice related to the product [...] R&D is an emerging methodology that can yield useful curriculum materials and instruction practices for classroom use*" ([Aikenhead, Glen, 2001], p.2).

Institutional Theory

Within organization theory I have identified institutional theory, which is descriptive and demonstrates associations between structural and contextual variables rather than explaining causal connections. An application of it would be the methodology of SSM, which Scheepers & Mathiassen utilize in their depiction of the "Ndaba Experience" [Scheepers & Mathiassen, 1998]. Methodologically, SSM consists of seven stages and was developed by [Checkland & Scholes, 1990]. Based on the principles of the mnemonic CATWOE a system of a problematic area is made out and operationalized. CATWOE stands for:

> "* *Customer: everyone who stands to gain benefits from a system is considered as a customer of the system. If the system involves sacrifices such as lay offs, then those victims must also be counted as customers.*
> * *Actor: The actors perform the activities defined in the system.*
> * *Transformation process: This is shown as the conversion of input to output.*
> * *Weltanschauung: The German expression for world view. This world view makes the transformation process meaningful in context.*
> * *Owner: Every system has some proprietor, who has the power to start up and shut down the system.*
> * *Environmental constraints: External elements exist outside the system which it takes as given. These constraints include organizational policies as well as legal and ethical matters*" [Couprie et al., 2006].

SSM is mainly being expressed in first-world countries, but according to [Plooy & Roode, 1999] it has the "ability to enable developing countries to avoid the errors of more developed countries as far as the introduction of information technology is concerned" [Plooy & Roode, 1999].

3.4 Methodological Aspects

Quantitative and qualitative are the two dominant groups of research methods in information systems (IS). Actually, quantitative methods were originally developed

in the natural sciences to study natural phenomena, but they have been applied in
social sciences through survey methods, laboratory experiments, formal methods and
numerical methods. Qualitative methods were developed in the social sciences to
enable researchers to study social and cultural phenomena. These methods are useful
in understanding people and the social and cultural contexts within which they live.
Qualitative methods include action research, case study research and ethnography.
Each of these methods has different ontological, epistemological, axiological, rhetorical,
and methodological characteristics ((Creswell 1994, 1998) cited in [Okunoye, 2003]).

Cross-checking available literature on research methodologies of social implications
of ICTs in LDCs with a view to intercultural awareness results in a methodological
pluralism. Actually, most publications are theoretically focused and (academic) case
studies are rarely published (see Chapter 3.2). Theoretical researches tend to promote
action research [Walsham & Sahay, 2005], ethnographic research and co-operation with
local researchers [Snyman & Hulbert, 2004], [Walsham & Sahay, 2005], [Tacchi et al.,
2003] to gain deeper insights into local sociocultural structures. Among empirical
works, qualitative interviews are often involved, but no major methodological stance
is predominant. What is most noticeable are frequent combinations of quantitative
and qualitative methods ([Harris et al., 2001], [Nel & Wilkinson, 2001]). Table 3.1
gives an overview on prevalent methods in relation to the above described case studies:

Author	Qual. Instruments	Quan. Instruments	Methodology
Harris et al., 2001	Interviews	Mann-Whitney-U Test	E,AR,CS
Latu, 2006	Unstructured interviews		E,CS
Nel et al., 2001	Open-ended questions	Statistics on closed, Questionnaire	E,CS
Scheepers, 1999	Case interviews	Statistics, Questionnaire	SSM,CS,AR
Scheepers et al., 2000	Field notes, Open-ended questions	Statistics on closed, Questionna	E,CS
Snyman, 2004	Informal interviews, Focus group discussions	Questionnaire	CS,E
Turner, 2005	Participant observation, Document analysis	Questionnaire	CS,E

AR=Action Research CS=Case Study SSM=Soft Systems Methodology E=Ethnography

Table 3.1: Case studies and their methods.

3.5 Summary

After a general introduction to literature on ICTs in LDCs (Section 3.1) I focused
on the most relevant research communities and literature within the fields of "Science

Teaching in LDCs with Intercultural Awareness" and "Culture and ICTs (in LDCs)". Their perspectives cover "mostly[16]" relevant issues to technological knowledge transfer in LDCs at an individual's level, e.g. teaching methods, cultural relevance of technology, different technology adoption due to differing worldviews. Afterwards I turned to the proclaimed dearth of case studies and described the shortcomings in current literature: studies on an individual's level and those being performed by a practitioner. Scientific studies by practitioners were identified by Odedra-Straub as a shortcoming in scientific literature [Odedra-Straub, 2003]. I presented the communities' theoretical perspectives which was followed by the depiction of available case studies. Results on teaching ICTs in LDCs on an individual's level were discussed more in detail, whereas those touching teaching ICTs in LDCs in a limited way, just briefly mentioned. I adumbrated also those case studies which I could not figure out underlying methodological tenets but seemed to be important to my research to a certain extent - see Section 3.2. By this I aimed to give the reader insight on available case studies' stylings. Unique case studies (but similar in terms of dealing on an individual's level) on teaching ICTs in LDCs pose a good opportunity to compare results and methodological approaches. A comparison of methodological approaches (Table 3.1) at the end of this chapter shows that ethnographical instruments are widely applied, though mainly in a combination with quantitative methods. Beside others, the works of [Walsham & Sahay, 2005], [Totolo, 2005], and [Kamppuri & Tukianinen, 2004] were utilized to show contemporary theoretical and methodological tenets in literature. As I revealed in Section 3.3 - in addition to the prior two sections - the interpretative paradigm is predominant and most studies focus on both micro and macro levels of research, the way to the theoretical perspective I am building upon, an actor-oriented socio-constructionistic stance, is paved. As it will be depicted in Chapter 5 and 6, I also draw upon the interpretative paradigm, but my investigation is conducted by a different and new methodological approach. Due to my conviction that culture cannot be restricted to a national scope I focus on the impact and integration of ICTs into a participant's everyday life and thus concentrate on an individual's level. Walsham & Sahay explicitly claim more studies on an individual's level in [Walsham & Sahay, 2005]. Thereby, my research is different from most cultural related HCI studies - see Figure 3.2 - which applied a rather fixed scope of culture.

According to my believe an intervention implies social interaction which means cultural change. Leaving the assumption that an evaluation of an intervention (i.e. cultural change) should investigate the technology's integration into everyday life, I aim to *understand* the way "how" the technology is integrated into everyday life by considering computers as a previously identified problem area. Furthermore, in opposition to most other case studies, to me it is not important "how many" participants had a certain experience with the computer. Even if a participant had a problem "only" once, becomes an important fact and is not less important than experiences which were named several times. Thereby, this Ugandan case study abandons quantitative methods as methodological instruments and focuses on the understanding of people's social interaction. This is a unique methodological combination of Witzel's problem cen-

---

[16] According to my literature research most scientists dealing with case studies on technology transfer in LDCs on a micro level (with a view to culture) can be assigned either to IFIP, CatTAC, or NARST.

tred interview [Witzel, 1985] and objective hermeneutics by Oevermann [Oevermann, 2002].

Thereby, this study gains a foothold in its scientific environment and is able to contribute further insights on ICTs and their integration into everyday life to the pre-compiled results of other previous results.

tred interview [Witzel, 1985] and objective hermeneutics by Oevermann [Oevermann, 2002].

Thereby, this study gains a foothold in its scientific environment and is able to contribute further insights on ICTs and their integration into everyday life to the precompiled results of other previous results.

Chapter 4

Aim of Research

4.1 Premises

The overall goal of this research is to generate a *substantive theory* (sensu [Glaser & Strauss, 1967]) which contributes to the dearth of academic empirical case studies of ICT projects in LDCs especially in the geographical area of Africa. As I have shown above case studies on an individual level are as rare as those which are explicitly intercultural aware - see Section 3.2. Due to the lack of financial and personal resources this study would ideally be continued in a second volume to develop from a substantive level to a more formal one. Thus, it goes along with Glaser who suggests "[...] *first (to) publish a monograph about a substantive field and then go on to a second volume dealing with related formal theory*" ([Glaser & Strauss, 1967], p.257).

The successful generation of a substantive theory is subdued to four interrelated properties, which are "*Fitness*", "*Understanding*", "*Generality*" and "*Control*" ([Glaser & Strauss, 1967], p. 238). Doing justice to these properties means that sociologists of any view-point to students and significant laymen have to understand the theory and its duct. Such a theory has to fit the situation being researched, and work when put into use. Accordingly, generality implies that a vast number of diverse qualitative "facts" on many (different) situations has to be general enough to be applicable to the total picture of the theory. To be in charge of the theory during its application is what they mean by the term "control". The theory is supposed to be transparent in a predictive way, to "*control consequences both for the object of change and for other parts of the total situation that will be affected*" ([Glaser & Strauss, 1967], p.238) and to give the practitioner a controllable theoretical foothold in diverse situations. This underlines the importance and indispensability of "general", "fit" and "understand-able" as preconditions for "control". For researches in ICTs in LDCs some additional factors have to be considered to "nail down" upcoming research results in order to clarify them. I agree with Geoff Walsham and Sundeep Sahay that all research studies on ICTs in developing countries should address the following four questions and aim to give detailed answers in relevant sections:

1. *What is the "development" to which ICTs aim is to contribute?*

2. *What are the key issues being studied related to ICTs?*

3. *What is the theoretical and methodological stance?*

4. *What level and focus of analysis is being adopted?* [Walsham & Sahay, 2005]

In this case study these questions are understood as follows:

ad 1) In Chapter 2 I have depicted how I understand the term "development". If ICTs contribute to this meaning of development it is to be investigated within this research.

ad 2) The relevant key issue of this study is to illuminate the "problematic object" "computer" in a Ugandan sociocultural environment. This shall be analyzed due to the fact that the intervention was undertaken by a person with a different cultural background from the participants' one.

ad 3) This research is based on an actor-oriented socio-constructionistic view and draws upon hermeneutical reasoning [Oevermann, 2002] and Witzel's programme of problem centred analysis [Witzel, 1985]. Within this methodological framework I aim to focus on social interactions and intercultural phenomena. Thereby, I apply the methods of hermeneutics.

ad 4) The level and focus of analysis is based on social interaction between individuals and can be basically seen as a micro one. However, as to my theoretical stance an actor performs within certain social constraints, I investigate and depict possible constraints to understand the actor's activities in a wider context - see Chapter 5.

4.2   Research Questions of Case Study

In Chapter 3 I have sketched the literature on ICTs and LDCs and shown future opportunities for research. Evaluating a successful transfer of computer knowledge in LDCs by a teacher with a different cultural background claims to find out the social function of computers within the social environment. As " [...] *every social group is characterized by processes of conflict and struggle and by mechanisms for resolution and mediation; between different social categories based on class, race, gender, age and other aspects of group formation*" ([Hodge & Kress, 1988], p.266), the impact of computer teaching within a social environment should be examined:

- What are the social functions of computers within the geographical area of Kabale (Uganda) and (how) is a computer integrated into a participants' everyday life? Does the fact of "having computer access" influence power relationships among pupils, staff, non-teaching staff and other schools as well as between them? How is computer knowledge represented and posed?

Teaching leads to social interactions (see Chapter 5) between social actors. An important aspect within these interactions is power, which is negotiated during social interactions - for further details please see Point 8 in Chapter 5.

- In which way does power manifest itself in computer teaching? Which signs of social function can be spotted among various social actors? I.e. a computer teacher or a teacher at a primary school.

In order to identify possible cultural conflicts as mentioned by [Aikenhead, 2002] and to support (science/computer) teachers to become "cultural brokers" the author's (my) field notes shall be analyzed. This claims to shift to a meta-level as the researcher's influence on the intervention progress is taken into account as well. At first they have to be perceived and then, secondly, assigned to Cushner and Brislin's general phenomena of intercultural interactions [Cushner & Brislin, 1995]. Thereby, other computer teachers are made **aware** of possible arising misunderstandings during the intervention. Thus, the resulting research questions can be summarized as follows:

- What kind of intercultural phenomena can be observed? When and how is referred to different (cultural) systems?

A presented lack of available case study researches is the one being performed by a practitioner. A critical point of this case study is definitely the multifunctional role I, as e.g. a colleague, interviewer, teacher, donor, researcher have been taking up. Pros like pre-established trustfulness are to be balanced against cons like role confusion. Summarized my point of interest is:

- In what way does the multifunctional role as an interviewer, teacher, colleague and donor influence the progression of interviews? Should and can donors evaluate such interventions on their own?

There are answers to these questions which I will present after depicting the context of the intervention's background and its ongoings. This is followed by a discussion on possible consequences in the Chapters 16 and 17.

4.3 Summary

Aims of research and its underlying theoretical contexts were the focus of this Chapter. They were evolved from former "rough" research statements and lead to well-formed questions to clarify my research aims. The answers to these questions tend to create awareness among decision makers who deal with ICT related interventions in LDCs. The achievement of the research goals accomplishes a case study on technological knowledge transfer at an individual's level in LDCs by a scientific practitioner with a technological oriented cultural background. Consequently the importance of trustfulness as a researcher shall be considered, which puts forth to what extent it is important and possible to establish a close relationship with the intervention's participants to explore the integration of computers in the participants' everyday lives. Thus, this research contributes threefold to existing shortcomings in current literature: an explicit study by a practitioner within a scientific framework, a study on an individual's level and a study which reaches its targets by a unique nesting of two methodological approaches: objective hermeneutics by Oevermann [Oevermann, 2002] and problem centred analysis by Witzel [Witzel, 1985].

In the next two Chapters I am going to derive my underlying theoretical and methodological stances. This is necessary to allocate my research in the diverse fields of theoretical paradigms and so I will present the theoretical framework of the Ugandan case study. Methodological instruments will be derived from its methodological stances and are depicted in detail to allow a transparent and meaningful auditability of how I am going to approach the points of interests named above.

Chapter 5

Theoretical Approach of Actor-Oriented Social Constructionism

5.1 Introduction

Concluding from the above section one sees that this thesis is actually philosophically grounded in an actor-oriented social constructionist view of change and continuity. Talking about social constructionism one notes that constructionism goes beyond constructivism as it is more general and includes both the cognitive and the social dimensions of behavior and social practice [Long, 2001]. Norman Long draws upon Burr and explains the social constructionist view thus: *"The goings-on between people in the course of their everyday lives are seen as the practices during which our shared versions of knowledge are constructed. Therefore what we regard as "truth' (which of course varies historically and culturally), i.e. our currently accepted ways of understanding the world, is a product not of objective observation of the world, but of the social processes and interactions in which people are constantly engaged with each other"* ([Burr, 1995] cited in [Long, 2001], p.244). A "successful" transfer of technological knowledge conveys some further knowledge and, *"is that which manages to insert itself into a series of practices"* [Heaton L. and G. Nkunzimana, 2006]. It implies social processes and interactions. The following sections substantiate the appropriateness of this approach for my research:

5.2 Background of Social Constructionism

Taking a look at the available literature on development and social change one recognizes two theoretical mainstreams: on the one hand studies applying a "macro' view, as they are dealing with aggregated or large scale structures and trends; on the other hand there is extensive work which takes up a "micro" view, which examines profoundly the nature of changes at the level of operating or acting units. Until recently the focus of the sociology of development was put into two structural ('macro') models, namely those based on modernization theory and on neo-Marxistic theory of political economy. Although they seem to be quite different at a first glance (a liberal stand-point vs. a standpoint considering "development" as an unequal process involving continuous exploitation of "peripheral"societies and "marginalized" populations) they are similarly subjected to the same basic assumptions: "*Both see development and social change emanating primarily from external centres of power via interventions by the state or international bodies, and following some broadly determined developmental path, signposted by "stages of development' or the succession of different regimes of capitalism*" ([Long, 2001], p.11).

This means that both hold a deterministic, externalist and linear view of social change. During their first researches the supporters of these theories realized its limitations, as they did not do justice to the social heterogeneity appearing in environments with "similar circumstances" and "social structures". Consequently, they attempted to reformulate structural analysis as Preston, Buttel et al. did, but still kept the focus on "big players" such as the *International Monetary Fund* (*IMF*) or the World Trade Organization are. The theorists shifted from a structural view to a more or less "protagonistic" view, but were still neglecting less "powerful" actors participating in developmental processes. So this string of literature and theories was still concentrated on the concept of external determination. [Long, 2001].

As to my persuasion "*all forms of external intervention necessarily enter the existing lifeworlds of the individuals and social groups affected, and in this way they are mediated and transformed by these same actors and structures*" ([Long, 2001], p.13)., any external intervention is internalized and thereby loses its external status. By this, external factors "*come to mean different things to different interest groups or to the different individual actors involved, whether they were implementers, clients or by-standers*" ([Long, 2001], p.31). In other words: depending on the inputs' recipient, the input will be transformed differently depending on the actor and his/her social context. One has to keep that in mind. Drawing upon the assumption that "*the incorporation of new information and new discursive or cultural frames can only take place on the basis of already-existing knowledge frames and evaluative modes, which are themselves re-shaped through the communicative process*" it means that to get hold of this emerging knowledge, "*as a product of interaction, dialogue, reflexivity and contests of meaning*" and something which "*involves aspects of control, authority and power*", it is mandatory to apply a "bottom-up" perspective concentrating on the "actors" involved and their executed "activities" ([Long, 2001], p.70). At the heart of an actor-oriented sociology of development is the characterization of social action as implying both social meaning and social practice [Long, 2002].

5.3   Agency

The existence of social actors - which is obvious - is inevitably related to agency (*itmo* human agency) which "*attributes to the individual actor the capacity to process social experience and to devise ways of coping with life, even under the most extreme forms of coercion. Within the limits of information, uncertainty and other constraints (e.g. physical, normative or politico-economic) that exist, social actors possess "knowledge-*

ability[1] *and "capability*[2]. *They [the actors] attempt to solve problems, learn how to intervene in the flow of social events around them and to a degree they monitor their own actions, observing how others react to their behavior and taking note of the various contingent circumstances"*(([Giddens, 1984], p.1-16) cited in ([Long, 2002], p.16)). Social processes and, thus, cultures are constantly changing and never fixed. By observing and reflecting their own actions the actors do not remain constant across all social contexts. Long's key tenet is more or less the one of a "mature" actor who enrols others in his/her "project". An actor runs several "projects" at the same time and by "human agency" he/she enrols them. This enrolment stands for social interaction and negotiation, because any *"face-to-face situations [...] is the prototypical case of social interaction"* ([Berger & Luckmann, 1966], p.28)[3]. This does not mean thatIt is important to keep in mind that human agency is not restricted to single individuals and implies also such entities having the possibility of formulating and carrying out decisions, e.g. state agencies, church organizations, political parties (([Hindess, 1986], p.115) cited in [Long, 2002]). Actors are highly heterogenous and to understand their differing ways of enrolment, *"it becomes necessary, therefore, to identify the conditions under which particular definitions of reality and visions of the future are upheld, to analyse the interplay of cultural and ideological oppositions"* ([Long, 2001], p.70).

5.4   Social Actors

Social actors are related to social structures of the environment, and both the actors create social structures by their actions, and social constraints influence their social actions. Therefore, a solid study demands considering both, actors and structures. This goes along with the dualistic view of Berger and Luckmann, who emphasized and derived social interactionism [Berger & Luckmann, 1966]. Only bearing in mind certain social constraints allows for a holistic investigation of appearing phenomena.

To investigate how ICTs are perceived by participants of interventions and "how" they become (perhaps not) integrated into their lifeworlds relies on these theoretical foundations. Following cornerstones of an actor-oriented approach were derived by Norman Long and represent the basis of my approach:

1. *Social life is heterogeneous. It comprises a wide diversity of social forms and cultural repertoires, even under seemingly homogeneous circumstances.*

2. *It is necessary to study how such differences are produced, reproduced, consolidated and transformed, and to identify the social processes involved, not merely the structural outcomes.*

3. *Such a perspective requires a theory of agency based upon the capacity of actors to process their and others' experiences and desires are reflexibly interpreted and*

---

[1]Knowledgeability means that experiences and desires are reflexibly accorded to meanings and purposes [Long, 2002].

[2]Capability of an actor means that the actor is in charge of relevant skills, can access resources of various kinds, and engages in particular organizing practices [Long, 2002].

[3]Please remember that the actor can only act within environmental social constraints - see Section 5.4.

internalised (consciously or otherwise), and the capability to command relevant skills, access to material and non-material resources and engage in particular organising practices.

4. *Social action is never an individual ego-centred pursuit. It takes place within networks of relations (involving human and non-human components), is shaped by both routine and explorative organising practices, and is bounded by certain social conventions, values and power relations.*

5. *But it would be misleading to assume that such social and institutional constraints can be reduced to general sociological categories and hierarchies based on class, gender, status, ethnicity, etc. Social action and interpretation are context-specific and contextually generated. Boundary markers are specific to particular domains, arenas and fields of social action and should not be prejudged analytically.*

6. *Meanings, values and interpretations are culturally constructed but they are differentially applied and reinterpreted in accordance with existing behavioural possibilities or changed circumstances, sometimes generating "new" cultural "standards".*

7. *Related to these processes is the question of scale, by which I refer to the ways in which "micro-scale" interactional settings and localised arenas are connected to wider "macro scale" phenomena and vice versa. Rather than seeing the "local" as shaped by the "global" or the "global" as an aggregation of the "local", an actor perspective aims to elucidate the precise sets of interlocking relationships, actor "projects" and social practices that interpenetrate various social, symbolic and geographical spaces.*

8. *In order to examine these interrelations it is useful to work with the concept of "social interface"[4] which explores how discrepancies of social interest, cultural interpretation, knowledge and power are mediated and perpetuated or transformed at critical points of linkage or confrontation. These interfaces need to be identified ethnographically, not presumed on the basis of predetermined categories.*

9. *Thus the major challenge is to delineate the contours and contents of diverse social forms, explain their genesis and trace out their implications for strategic action and modes of consciousness. That is, we need to understand how these forms take shape under specific conditions and in relation to past configurations, with a view to examining their viability, self-generating capacities and wider ramifications.* ([Long, 2001], p.49f).

As this kind of analysis must not base on predefined social categories (see Point 8), it has to be performed ethnographically. Key methods of ethnographic research like participant observation or interviews enable to illuminate the complete range of social relationships and processes within which an intervention takes place. These methods

---

[4] A *social interface* is a critical point of intersection between different social systems, fields, domains or levels of social order where social discontinuities, based upon discrepancies in values, interests, knowledge and power, are most likely to be located [Long, 2001 - Appendix].

are social interactions per se and so a proper methodological approach has to meet the requirements of a profound analysis. These requirements appear to consist of three major claims:

- **Problem perspective** - interested in social implications of introducing technology urges for an approach which allows to focus on social interactions within the "battlefield" ICT.

- **Process perspective** - in order to be able to react to new emerging inputs during the research process no hypothesis' are built in advance. No outcome is expected and a methodological stance which is similar to [Glaser & Strauss, 1967] theoretical sampling is necessary. This enables a researcher to incorporate reflected results both during the interviews and during its interpretative process.

- **"Substantive perspective"** - depending on situational contexts the methodological grid may be adapted during the research's progress.

5.5   Summary

This chapter outlined the theoretical stance of the Ugandan case study. It depicted the underlying actor-oriented social constructionistic point of view and showed that this case study mainly draws upon the assumption that both actors and social structures mutually reconstruct each other in everyday life. Putting action, which is social in any case, into a context means that it is cultural as well. Thereby, I take up the previously proclaimed intercultural perspective of this study. From a methodological point of view this theoretical stance asks to perform ethnographical research with the regard to three perspectives (problem perspective, process perspective and substantive perspective) to investigate the everyday life of the involved actors. This investigation goes also along with the goal to understand people's everyday life, which I will explore with regard to hermeneutics. Thus, practises and its symbols, heroes, rituals, and values can be identified and scrutinized. This is exactly by which culture manifests itself. The following chapter deals with the methodological aspects in more detail.

Chapter 6

Methodology

6.1 Methodological Basis

The attitude of analysis is mainly based on the ideas of the hermeneutic school [Gadamer, 1960], [Oevermann, 2002]. More specifically spoken I build on recent "*objective hermeneutics*" by Oevermann [Oevermann, 2002] and not on "*classical hermeneutics*" based on Gadamer [Gadamer, 1960].

It is my aim to understand which meanings actors associate with - by their actions among themselves and/or by their interactions with artifacts, which e.g. in my case are computers. It has to be assumed that these meanings and structures of meanings are hyper-individual and hyper-subjective. These meanings are independent from a subjective meaning an actor associates with his/her own actions. It is not sufficient to simply ask actors for latent meanings, because in most cases they are not even aware of them.

Actions and comments/expressions of acting subjects are embedded in general meanings of context and structure. These do not determine social action by themselves, but coin them in a crucial manner and are both constructed and reproduced during the course of actions. As mentioned above actions are social and within a context culture manifests itself. Hermeneutics aims to understand

"[...] *processes* [orig. "Verfahren"], *rules* [orig. "Regeln"], *patterns* [orig. "Muster"], *implicit premises, socializationarily* [orig. "sozialisatorisch"] *conveyed interpretational and understanding ways of acquirement, instruction and transmission*" ([Soeffner, 2003], p.165, [my translation]). Meaning cannot be observed directly, as it is often latent and has to be tapped in a time consuming process. An approach which devotes itself to understand such latently emerging, but objectively effective structures of meaning are hermeneutics. Hermeneutics represents one of the most applied and adopted approaches of qualitative research in German speaking area.

Whereas Scheepers' [Scheepers, 1999] and Roger Harris's [Harris et al., 2001], studies are based on the method of action research [Rapoport, 1970] this research adopts a different approach, as the whole transfer is mainly analyzed after its intervention process. From a methodological stance this suits the material even better as hermeneutical analysis has to be applied after an intervention, as a thorough and multilayered interpretation of social meaning in the hermeneutic tradition requires data in a fixed form, which transient observations cannot provide. Performing a scientific analysis afterwards does not mean, "[...] *that a social scientist should avoid a direct experience of the case events* ["Fallgeschehen"]. *Just the opposite. He/she should approach as closely as possible: in order to collect good data for the subsequent interpretation*" ([Kurt, 2004], p.237, [my translation]). Putting this into practice for this research by teaching and living on the school's compound lead to embedment of the researcher in the social environment. Such kind of progress goes also along with [Glaser & Strauss, 1967] who support a previous empirical period to get close to relevant issues and topics. By this the researcher gains "*insights*" ([Glaser & Strauss, 1967], p.251f) which are needed to be transformed into relevant categories, properties, and hypotheses. In this study I do not concentrate on building hypotheses, but rather on problematic situations - which were experienced - in fields of ICTs, on which actor-oriented research also puts its focus. "*A fundamental principle of actor-oriented research is that it must be based on actor-defined issues or problematic situations; whether defined by policy-makers, researchers, intervening private or public agents or local actors, and whatever the spatial, cultural, institutional and power arenas involved. Such issues or situations are often, of course, perceived, and their implications interpreted, very differently by the various parties or actors involved*" ([Long, 2002], p.58).

Tracing the social constructed lifeworlds of actors is, according to [Berger & Luckmann, 1966], closely connected to the understanding of "language[1]". This means that both verbal and non-verbal communication are taken into consideration as structures of social meaning and social practices are objectified in textual artifacts. I do not investigate acts of non-verbal communication in depth though, as no visual records have been made. Additionally, emerging cultural/social differences within the social interactions of non-verbal forms of communication (like gestures or random moves) are more difficult to gather than recorded "language-based" interactions.

Another issue to be mentioned is the one of reflexivity: The problem and actor-centred approach of this study encourages to reflect and influence the progress of interviews (which per se is a social interaction). Procedures may be changed/adapted advisedly to enable a deeper understanding of the "problem"[2]. As an isolated single on the spot, for the researcher it is virtually impossible to reflect both verbal and non-verbal communication and influence the process of research.

6.2   Qualitative vs. Quantitative Research

As I have deduced in preceding sections this research focuses on the subjective relevance of computer and arising social actions and therefore the selected methods are qualitative ones. Furthermore, the Bakiga (the ethnic group this case study focuses on) tradition in transferring knowledge is an oral one. Jan Vansina has explored the Rwandan oral tradition of transferring knowledge [Vansina, 1973] and as the Bakiga are Rwandan descendants [Ngologoza, 1967], [Nzita, Richard and Mbaga-Niwampa,

---

[1]It would be interesting to concentrate on the role of language during the teaching of computers, but is beyond my scope of work. The particular language in texts (e.g. transcribed interviews) can appear as genre [Kress, 1989], register [Zwicky & Zwicky, 1982], or jargon [Polenz, 1981].

[2]Actually, when using "problem" we think of external influences - in form of computers - entering the local actors lifeworld.

1993] quantitative methods would be unusual to the local communication structures[3]. Talking is part of daily life and by interviews a daily life situation is established. Thus, barriers can be dismantled. Besides **qualitative interviews** I have enriched the gathered interview material with expert discussions and the method of **participant observation** to gain a deeper understanding of the sociocultural environment. These relevant papers are put into practise dually: on the one hand they are used as additional sources to underline or "de"-emphasize derived theories [Glaser & Strauss, 1967] and on the other hand to subsume intercultural interactions to Cushner and Brislin's themes [Cushner & Brislin, 1995].

In the following sections I want to sketch the methodological programme which has been ubiquitous during the research progress.

6.3 Framework of Methods

The methodological steps of my research - data collection, analysis and interpretation - are closely connected to Norman Long's theoretical assumptions: "*In an attempt to avoid teleological trap of predefining "system goals", Long suggests that the focus of systems research should be on "problem situations" in which there are felt to be unstructured problems, ones in which the designation of objectives is itself problematic*" ([Long, 2001], p.173) As Witzel's method of the "*problem centred interview*" [orig. "Das problemzentrierte Interview"] [Witzel, 1985] enables the exploration of "problem areas" (as "computers" seemed to me to be in 2002) I decided to apply Witzel's method for the research period of 2003. The *problem centred interview* deliberately abstains from pre-built hypotheses to avoid a preformed categorical framework. It aims to look for explanations of social actions and of intentions instead of distinct factors. Furthermore, it attempts to reveal explanatory reasons by a systematical interpretation of texts ([Witzel, 1985], p.228, [my translation]). The whole methodological programme is subjected to the "*principle of openness*" [orig. "Prinzip der Offenheit"] and belongs to the category of open, half-standardized investigation methods. It is based on three basic principles, which are a "*problem perspective*", a "*subject perspective*", and a "*process perspective*".

6.3.1 Problem Perspective

Witzel asks the researcher to turn his/her attention to problematic situations and to present and systemize his/her knowledge background, in order to get closer to an emerging social relevant problem. This requires that the researcher is to gain both structural everyday life characteristics of involved social actors and its social environments. By focusing on a problem a researcher is able to scan a thematic area in full and to reveal and to explore by further questions "*curtly, stereotype or contradicting explications*" ([Witzel, 1985], p.235, [my translation]). In order to reach relevant results the dedication to both relevant theories and social constraints are necessary and by explorative fieldwork and integrated expertise it is aimed to regard

---

[3]Nevertheless, surprisingly most case studies used written questionnaires to collect their research data - see Tab 3.2.

as much social context as possible. The researcher is expected to gather sufficient information to take up two extreme positions: the researcher should (theoretically speaking) be able to answer like an interviewee who is participating in the fieldwork, but the researcher is also supposed to act as an expert during the analytical period [Witzel, 1985]. This contradicting situation of "*unknowing interviewee*" and "*informed theorist*" is resolved by Blumer's "*sensityzing concepts*" ([Blumer, 1954], p.7) which are understood by Witzel as follows:

1. *Temporary phrasing of the problem area to sensitize the researcher's perception.*

2. *But it remains important to keep previous knowledge open for empirical experiences, in order to be controlled by its theoretical and conceptual definitions respectively. This, in turn, constitutes re-newed, but profound knowledge, which has to be used as previous knowledge in the course of the following empirical process until one assumes that the object of research has been grasped adequately.* ([Witzel, 1985], p.231, [my translation]).

"Practical" and theoretical knowledge are exchanged in an elastic and dualistic way. Thereby, knowledge is both refined and created and enables the researcher to immerse him/herself in the area of interest ("problem") until he/she considers to have gained sufficient knowledge.

6.3.2   Subject Perspective

This second criteria allows the (de-)selection of appropriate methods during the research process depending on "*examined relevant objects*" [*orig.* "Untersuchungsgegenstände"]. This is opposite to a normative-deductive "instrument perspective" which aims to break up theoretical assumptions into measurable and separated variables, unfortunately this contributes little to clarify "how" social reality is constructed. Of course, this perspective does not claim for a careless application of research methods, but quite the reverse. Instead of being fixed to a certain predefined utilization of certain research methods, the researcher who has thoroughly preconsidered applicable methods, can - depending on the research subject - decide on their positioning and their importance and degree of modification during the analytical period; e.g. in the interpretational process of this study I concentrate more on hermeneutics when analyzing transcribed interviews to reveal latent structures of meaning. For analyzing available meta-documents I assign intercultural interactions to Cushner and Brislins's framework [Cushner & Brislin, 1995]. This is supported by Witzel, who encourages the use of different methods to enhance our research malleability to the research context.

6.3.3   Process Perspective

Witzel's third perspective is based upon the principles of Glasers "*grounded theory approach*". Theories are not built ex ante, but are developed during the process of data collection and interpretation. First theoretical concepts are induced during the period of data collection and its interpretation. These concepts help to refine and deduce further examinations, which again enhance (induce!) the building of adapted (new) theoretical concepts. Being very close to theoretical sampling [Glaser & Strauss, 1967] I refer to Glaser to sum up this mutual interaction with the words: "*Deduction serves further induction*" ([Glaser & Strauss, 1967], p.38). Witzel shifts his methodological point of view on a meta-level and tries to prolong the induction-deduction interplay on a more detailed level when he claims: "*Thus, the criteria of process orientation is concerned with the over-all design of the research process and the development of the communicative exchange during the interview. Besides it regards the aspect of developing the process of understanding during the interview and is applied up to a controlled coverage and extension of interpretation in a scientific context*" ([Witzel, 1985], p.234, [my translation]).

Drawing upon the above statement I want to build upon the scalable character of this criterion. Even though it implies adapting the whole research progress to its context, it encourages me as a researcher to alter interview proceedings in order to reach a higher level of understanding. At first glance this might seem haphazardly, but keeping in mind my research goals such methodological shifts allow to gain deeper insights of the researchers social interactions[4].

The following paragraphs will briefly depict the applied processes of data collection (see Section 6.4.1), analysis and interpretation (see Section 6.4.2).

6.4   Methodology

The listing below outlines the research design which was applied in this study at hand. A qualitative research process is usually marked by the steps of data collection, analysis and interpretation. These steps are not to been considered as distinct and isolated milestones, but rather as circular. Thus, it follows Glaser's approach of *grounded theory* (sensu [Glaser & Strauss, 1967]) as the relevant subject is defined and enhanced by time and intense devotion to problem areas.

- Data has been gathered in an unsystematic way; written texts, similar to a *thick description* [Geertz, 1983] and field notes of participant observations have been compiled.

- During the collection process the problem area has been identified and relevant issues and topics sampled theoretically; by using these, interview guidelines can be developed.

- Interviews have been transcribed and hermeneutically interpreted in discussions with heterogenous[5] groups.

6.4.1   Data Collection

Different methods were applied to substantiate the claim of prevalence in this study: During my stay in July 2003 systematic data collection and gathering of relevant ma-

---

[4]I have already stated above that an interview represents a social interaction itself by which culture manifests.

[5]To assure "neutral" results a heterogenous discussion group has interpreted relevant texts - see Part II.

terial was carried out. I have interviewed different types of informant groups (teachers, pupils, non-teaching staff) and applied the method of participant observation during the computer teaching. All in all I have interviewed 28 participants of the intervention. All together, 14 adults (the whole teaching and non-teaching staff (matrons, secretaries, purser) and 14 teenagers (pupils) have participated and agreed to share their knowledge. On the basis of Witzel, the problems, difficulties and particularities of the school staff and pupils in their interaction with computer teaching and computers are analyzed, whereby computers represented the centred problem. As mentioned above this method explicitly asks to gather as much context information as possible to enable the interviewer to subtly introduce his profound knowledge about e.g. the people, institution, project into the interviews. So while I was staying at the school as a researcher, I was exploring the regional sociocultural context as much as possible, to conduct a problem centred interview appropriate to the region. The exploration of this context serves as an additional reference for the systematic analysis of the gathered data. According to Witzel, it is possible to stray from the interview guidelines if it seems necessary to the interviewer to illuminate special aspects in order to focus on the "problem". 27 interviews have been audio recorded and transcribed. Most of them (22) have been conducted in the computer class room intending to lever up computer oriented associations. The rest was conducted in different places within the compound. The interview length varies from 14 minutes to 82 minutes. Whereas an adults interview lasted 41 minutes on average, a pupil's interview took only 21 minutes. This might be due to the fact that children felt more uncomfortable than adults, which is deduced in Part .

Meta-Documents & Socio-regional Cultural Context

As the careful researcher is looking for as many comparisons as possible [Glaser & Strauss, 1967] other sources of categories were used: reading available newspapers, watching political developments and visits to other schools aimed to enrich my understanding of the field of interest. Additionally, talks to local cultural experts facilitated a deeper understanding of the sociocultural environment in which the intervention took place. Historical data and informal interviews on local developments affirmed or rebutted insights.

Participant Observation

When I started with the computer intervention in 2002 I realized that obviously there were problems in applying and conveying the technology. Throughout my stay, misunderstandings led to an unsystematic documentation of relevant data, in terms of cultural and ethnological. This data was partly completed in December 2002 and compiled to a text about the first computer intervention's experiences – similar to Geertzs' suggested *thick description* [Geertz, 1983]. A thick description is an ethnographic text which depicts the layered, rich and contextual description of events or social scenes. Geertz claims that ethnographers need not to be worried about subjectivism of their ethnographic observations and produced texts as subjectivism is not avoidable anyway. This results from a conceptual translation of observed situations, which depend on the author's cultural context. Geertz concludes that any written text contains within itself interpretational aspects, which thereby offers to reveal new insights as it puts the ethnographers' culture in contrast to the described culture and makes consequently BOTH cultures more visible. Thus it appears that the document of this research is also supposed to show the teacher's (ethnographer's) feelings, attitudes and experienced intercultural phenomena.

In 2003 computer classroom activities were observed and field notes taken. Generally speaking, participatory observation offers to gain new "insights" which are to be substantiated in research. Whereas Glaser and Strauss ask to transform those "insights" into relevant categories, properties, and hypotheses, I try to avoid premature propositions and take them as further valuable information to develop a deeper understanding of the intervention's process ([Glaser & Strauss, 1967], p.251). Putting these "insights" in correlation to available interview material does not only underline or weaken prior assumptions, but rather allows to adopt and refine those.

Problem Centred Interviews/Qualitative Interview

Witzel's approach draws upon four elements which can both be modified and applied in an individual order and are as follows: qualitative interview, biographical method, "case-study" [*orig.* "Einzelfallapproach"] and group discussions. When these elements are packed together they represented one possible concept of a solid problem centred interview approach. This unsystematic interconnection of the elements and their loose coupling are criticized and considered as limits of this approach [Ellinger, 2006]. Although, in contradiction to Ellinger, applying Witzel's subject perspective to this research means that some elements take a back seat and the attention is directed mainly on the element of qualitative interview.

To depict Witzel's explanations on qualitative interviews I have split it up under temporal aspects: before, during and after the interviews.

1. (a) **Preparation of Interviews**

    The way an interview is shaped depends a lot on its preparation and the way interviewees are selected and enrolled to participate. As mentioned above, for a problem centred interview the interviewer's task is to gather as much knowledge on the "problematic" context as possible. Informal talks, newspaper, etc. are supposed to enlighten the interviewer on certain problem areas and to get a deeper understanding of the (local) social environment.

   (b) **During Interviews**

    The interview process is based on four instruments: *a postscript, an interview road map, a short questionnaire* and *audio-recordings*. At the beginning the interviewer tries to establish a comfortable situation in order to enable a free talk. A short questionnaire containing questions on e.g. biographical data, can ease the development of comfortable situation and animate one's mind towards the interrogated problem area. While interviewing, keeping in mind the "problem area" helps the interviewer. Of course he/she can talk around the subject in order to elucidate certain issues of interest but he/she also has to come back to the basic ideas of research. This is assured by a prepared *interview road map*, which contains relevant questions and thereby supports the interviewer to keep in mind predefined problematic aspects.

(c) **After Interviews**

All interviews are *recorded audily* and completely transcribed afterwards to obtain not only already perceived answers of interviewees during the interview but further to reflect on the interviewers influence on the interview process. Transcribed interviews allow to reveal suggestive question and depict the interviews process as a whole. Witzel considers video-recording as an unnecessary effort, because it is the researcher's task to verbalize non-verbal forms of communication and to make it a subject of discussion e.g. "Why are you smiling, how can I understand this?". Additionally, so as to avoid isolated interviews and to put them in a wider context, the interviewer notes down a *postscript* after each interview [Witzel, 1985]. He/She takes notes on the situation immediately, before and right after the interview. This allows personal impressions on the interview process and other relevant[6] circumstances e.g. information how the informant was contacted, what expectations to an interview were perceived and so on. By contextualizing the interview, the researcher intends to gain a deeper understanding of problematic situations as a whole and even of individual text snippets. The interviews where done in English and my transcriptions were partly supported by a Ugandan sociologist to assure the quality of transcriptions. This assisted in cross-checking how local verbal sounds are annotated and in giving explanations on occurring local words.

According to Witzel's demand for a solid research as much information as possible should be collected. I have decided to interview all fourteen adults and older pupils of P4 to P7 (10 to 13 years old). During the first period of intervention younger ones (aged 7 to 9) did not attend computer lessons to a sufficient[7] extent to give an interview focusing on a "problematic situation". Investigating the social meaning a computer makes to them to them would have required different methodological concepts, e.g. drawings could have been. Deriving from experiences of the first period, younger pupils tended to be shyer than elder ones. Therefore an interview did not seem to be feasible. Possible reasons like e.g. language problems and further underlying assumptions are discussed in detail in Chapter 16. Witzel assigns qualitative interviews to the *Biographical Method* as they are frequently applied for biographical ethnographic research.

Basically, the biographical method neglects interactional interview situations and so Witzel postulates to let the interviewee talk and to minimize the influence on his/her statements. This aims to provoke long narrative statements and to establish a spanning interview context and avoids isolated answers to iso-

---

[6]It is up to the researcher to decide if something is considered to be "relevant" for the research progress. Thereby, it gives us additional insights about the researchers attitude.

[7]Younger pupils have been taught, but introductions to computer has been limited - due to lack of time. They have been taught the to main components of the computer (mouse, tower, screen, keyboard), how to switch a PC on and off, how to move the cursor and to type in their name.

lated questions. I have put the heading in inverted commas, because generally I have tried to base interviews on this approach. But: if an interviewee does not turn to relevant problems, it is up to the interviewer to ask further questions and to strive in his/her knowledge on the problem area. Flows of narration [*orig.* "Erzählfluss"] are scanned regarding core conflicts and *"communication strategies, which orient themselves on the flow of narration, are complemented by a function which* creates understanding. *This aims to leave core conflicts and problem areas not a superficial level with contradicting, curtly and stereotype presented events and self-perceptions"* ([Witzel, 1985], p.239, [my translation]). Thus, the biographical approach is adapted and becomes less important to the interview progress.

Through the *Case-Study approach* (*orig.* "Einzelfallapproach") the qualitative interview retrieves further impulses. Actually, the biographical method can already be regarded as a case-study per se, but is not within the focus of this research. Witzel considers a case-study as a stand-alone approach as well and reasons/originates its major advantage in reaching more distinguished and complex results by concentrating on just one subject or a small group of people. Abels criticizes that more data does not imply better results and claims that *"instead to concentrate on more and more conceptional generalizations (which is aimed to find by bigger and bigger data collections), we should attempt to gather data by intensive case studies. This data permits statements on palpable reality and the perception of this reality by concrete persons."*([Abels, 1975], cited in ([Witzel, 1985], p.239, [my translation])). Thus, focusing on a small group of people and understanding their constructed reality in a holistic way is preferred to a shallow interpretation of a bunch of interviews. Nevertheless, generalization can still be achieved by comparative systematization, which contains as many relevant conditions and aspects of the problem area as possible. *"Systemization aims at the more general level of collective action- and interpretation-patterns in its single versions. These results, which can be generalized, are gained by the inductive form of a systematic and comparing looking through of thoroughly interpreted single interviews"* ([Witzel, 1985], p.244, [my translation]). This indicates the last step of "systemizing", when analysis and interpreting interviews has been performed. Due to our methodological understanding such kind of systemizing is not necessary in a quantitative manner, as a meaning which has allocated once is not less valuable than one which has been met 25 times. Though it might show more acute aspects, by focusing on them, it neglects beginning trends, which could have been set by the introduction of e.g. the new technology [Oevermann, 2002]. Nevertheless, *"methods, which operate frequency analytically and on samples with standardized investigations, can complement and complete research approaches in a very effective way. Namely in those areas where they are needed"* ([Oevermann, 2002], p.31, [my translation]).

6.4.2   Interpretation & Analysis

Verbal data can be interpreted and analyzed in a variety of ways depending on the research goals and methodological stance. Witzel's approach is actually based on a variant of the "biographical method" (see above) and encourages a "sentence-by-sentence" analysis, which is supported by e.g. Mayring. Due to the fact that my research goals are also to reveal latent structures of social meaning and the social function of a computer within its geographical area under aspects of power and hierarchical positions, the approach of social hermeneutics by Oevermann seems to be even more appropriate.

Hermeneutics

As mentioned in the introduction I build on recent "*objective hermeneutics*" by [Oevermann, 2002] and not on "*classical hermeneutics*" based on [Gadamer, 1960]. Hermeneutics is characterized by its possibility to be flexibly adapted to the subject. Furthermore, in order not to restrict quality and manifoldness of interpretations, this approach does not "deliver" technical action guidelines. Fundamentally, it aims to reveal and reconstruct the context of creation and perception of social-texts[8]. Oevermann renounces the ex-ante building of categories as it restricts the researcher [Oevermann, 2002] and can therefore be ideally combined with the theoretical perspective mentioned above (see Chapter 5).

Objective hermeneutics is a method to reconstruct latent structures of meaning, and does not, in the first instance, interpret texts through a perspective of motifs and intentions of actors (cp. [Wernet, 2006], p.18). This does not mean that latent and manifest meanings are indifferent to each other and to reveal latent structures of meaning does not imply that objective hermeneutics is not interested in the world of mental representations [*orig.* "mentale Repräsentanz"]. "*Just the opposite: the self-perception of the acting subject* [and his/her intentions of acting], *in the meaning of identified differences* [between the conveyed subjective-intentional manifest representations and objective latent structures of meanings], *essentially go into the reconstruction of latent structures of meanings*" ([Wernet, 2006], p.18, [my translation]). Though, this becomes meaningful only after the level of latent structure of meaning is revealed.

To achieve this, Oevermann draws upon the method of "*sequence analysis*" [*orig.* "Sequenzanalyse"] instead of classification. Based on the assumption that a text is sequential like human action, text statements are not studied in parallel, but rely on the fact that any latter text snippet is in context to the prior one. Thus, options, assumptions or explanations (which were established in prior passages in the text) are closed through latter text snippets and/or further options are opened up. "*The sequence analysis calls attention to open and close each practice* [*orig.* "Praxis"] *of human life, in order to be able to negotiate bindingly and in a structured way*" ([Oevermann, 2002], p.6, [my translation]). Oevermann encourages to investigate the greeting of texts, as it constitutes both opening and closing procedures [*orig.* "Eröffnungs-" und "Schließungsprozeduren"]. During the greeting a "practice-space-time frame" [*orig.* "Praxis-Raum-Zeitlichkeit"] is negotiated; the frame within the rest of the text consequently is placed within. Afterwards, every further "sequence passage" [*orig.* "Sequenzstelle"] closes still existing options and opens up new ones. Thereby, this approach enables me as a researcher to focus on latent hierarchical relationships and

---

[8] Oevermann2002 prefers "protocol" to "text", as any interview represents also a space-time dimension. In order to avoid confusion and to maintain readability I leave the term "text" unchanged.

to reveal the (latent) social functions of computers within social environments. Please note that this is not contradicting to Witzel's framework as he encourages a methodological adoption in suggesting a process perspective, as it is up to the researcher to decide on appropriate interpretational tenets.

For this study the following three steps were conducted: data collection, analysis, interpretation. These steps are not to been considered as distinct and isolated milestones, but rather as circular. By the ideas of the hermeneutic school, "*it would be rather pointless to pretend that the qualitative data analysis stage be separated from the gathering process, since the hermeneutical attitude guides not only data interpretation but also data collection. The community's interaction is taken as a text to be analyzed and, above all, understood*" [Nocera, 2002]. Hermeneutics emphasizes on reflection throughout the research process, which increased its appeal to me as I am as well as interviewer, researcher, and analyst unified. The multiple role disables isolated threads during an interview and by objective hermeneutics latent social functions of the researcher are carved out. To avoid contextual influences during the revealing process of the latent structures of meaning I withdrew myself from the analysis as much as possible and the method of sequence analysis was performed by groups of four to five people who drew upon the five principles of objective hermeneutics. These principles and further details on how the method was applied in this context, is described in the Chapter 11. In Chapter 16, though, these results will be combined with all the gathered context knowledge.

My text on the first intervention's experiences in 2002 and its intercultural phenomena is close to Geertz's *thick description [Geertz, 1983]*. Culture is seen as text – an inscribed discourse for the translation of cultures into written text that is understood and interpreted by the reader. Geertz claims that the only way to describe cultural phenomena is to interpret them because cultural phenomena are "*vehicles of meaning, they are signs, messages, texts – the culture of a people ... an ensemble of texts*" ([Geertz, 1973], p.9) - a *context*. I discuss those phenomena one by one and by doing so, I unravel complex situations, put them in context to literature and customize it for my readers.

6.5 Summary

The attitude to analyze the available research data is going along with recent "objective hermeneutics" by Oevermann [Oevermann, 2002]. Hermeneutics are used in a qualitative research, which is evidently useful in an epistemological oral culture. This suits my theoretical stance of an actor-oriented socio-constructionistic view, which claims for ethnographic instruments and enables the researcher to reveal latent meanings and thereby the social function of computers within a certain sociocultural environment. From a methodological point of view I showed that the interpretation of (con)texts (collected by such ethnographic instruments!) is a way how culture becomes uncovered. I have described that social actors and hierarchical structures can be identified and understood. Continuing with a differentiation to other available literature I presented Witzel's methodological programme of "problem centred interviews" and explained that whereas other qualitative methodologies aim to avoid influencing the research process, this method explicitly asks to collect as much context (i.e. cultural) information as possible to enable the interviewer to bring in his knowledge on the problem area about e.g. the people, institution, and the intervention's progress in order to elucidate certain critical issues. Both approaches, hermeneutics and problem centred analysis, are applied in a nested way and are intertwined as they mutually influence their progress. This section is followed by methods applied in this research with a view to its underlying interpretational tenets. The pillars of this qualitative research, namely data collection, interpretation and analysis are presented, whereas it is crucial to keep in mind that these steps are not applied one by one but in a circular manner. The revealing of latent structures of meaning was achieved through discussions of discussion groups, which had no context knowledge on the intervention. These groups carved out (objective latent) structures of meanings and in selected text passages. As it would have been impossible to neglect my own teaching experiences and the process of interviewing the people, I withdrew myself as much as possible from these discussions and interfered only whenever utopian sociocultural constraints were assumed. In Chapter 16 I will assemble available context knowledge and revealed structures of meaning, explicate and discuss those.

Chapter 7

ICTs in Developing Countries

7.1 Millennium Development Goals

The *Millennium Development Goals* (MDGs) are the result of manifold UN resolutions and conferences that took place mainly in the 1990s. At the end of the UN Millennium Summit, 189 countries signed the final declaration and thereby committed themselves to a specific agenda for reducing global poverty by half by 2015. Years have passed, but at the current rate of progress, many countries and regions will not reach the MDGs by 2015 [Gerster & Zimmermann, 2003]. Yet, these goals have become a frame of reference for development activity and ICTs are seen as tools for achieving them. In Target 18 of MDG Nr. 8 the signing countries have explicitly mentioned ICTs and its use: "*In cooperation with the private sector, make available the benefits of new technologies, especially information and communications*" [World Bank, 2000]. Despite the fact that this target is criticized for being without a more specific thematic focus it undoubtfully represents the predominant opinion that ICTs are supposed to support the achievement of poverty reduction. Commonly, literature addresses ICTs as a cross-sectoral theme with the potential to promote development as an "enabler" in the context of the social themes/priorities where it can have the most impact [Haqqani, 2003]. Education, health, supporting public services, fostering employment, economical growth and entrepreneurship have been identified as sectors where ICTs can enhance MDGs ([Haqqani, 2003], p.83). Kenny, Navas-Sabter, et. al. have additionally identified some further themes which are depicted in Figure 7.1.

One might say that ICTs are considered as the panacea for LDCs and therefore uncountable[1] donor agencies invest in ICT projects and "fire" plenty of interventions and spend huge sums of money[2] thereby; e.g. [Haqqani, 2003] estimates that there are hundreds of supporting initiatives only within the **entrepreneurship** sector. So ICTs are

---

[1] Though "uncountable" might sound too exaggerated, it is not possible to estimate numbers on involved organizations, be it governmental or non-governmental ones, as no global records on development activities were taken.

[2] No figures of expenditure on ICT interventions in LDCs are available, but it would be most likely beyond the hundred million US-Dollar barrier, e.g. the "UNDP THEMATIC Trust Fund Information and Communication Technology for Development" announced an " Indicative Contribution Target 2001-2003" with a total volume of $ 30 mio. [UNDP, Thematic Trust Fund, 2001].

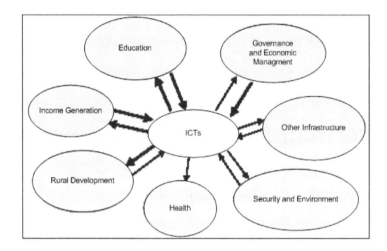

Figure 7.1: ICTs and Poverty. [Kenny & Qiang, 2000].

considered as an important resource in socioeconomic development in LDCs [Avgerou & Madon, 1993], and "*the main reason for introducing IT in these countries [LDCs] is to help them with the development process*" ([Samli, 1991], p.10). This statement clearly depicts the prevalent opinion on ICTs and its assumed *value* for LDCs. But: This concept implies that LDCs look up to superior, "developed" countries. These call themselves information or knowledge societies. In fact "[...], *information has begun* [in the 60s] *to be projected as the "engine" of development commensurate with traditional stages of growth definition of economic development*" ([Madon, 2000], p.4). Thus, information is the engine which levers development, and in order to get it in motion, technological products have to be transferred to enhance communication. This tenet is based on a consequent technology variant of modernization [Audenhove, 2001] and carries three latent assumptions: technology is neutral, technology is transferable and technology shows the same effects in any society. From a sociocultural view, technology is produced by interconnecting activities of certain actors (inventors, producers, and users), coincidence and socio-technological constraints. Therefore, technology is definitely not neutral and depends drastically on its "original" sociocultural context [Audenhove, 2001]. Technology is inseparable from the social, cultural, historical and political context which produced it. It is "*part of a social environment, an agent of social change, the physical medium through which symbolic values are expressed, the trace of a civilization*" (Martinand 1995, p.59 cited in [Dyson, 2004]).

However, especially within the domain of computers, the technological general physical appropriateness to different cultural contexts is rare in discussions on technology transfer in LDCs. Physical characteristics like the principal division of a "computer" into input devices (mouse, keyboard), output devices (e.g. screen) and a processing unit ("tower") are tacitly presumed and ubiquitous[3]. It seems as if this technology with its design is commonly accepted or discussions on it (consciously) neglected, which makes a rather hegemonic impression on me. Interestingly, Michelle Selinger from Cisco Systems Europe touches upon the topic of culturally specific appropriateness of technology use. She invokes on oral cultures like the United Arab Emirates (UAE) and Australian aboriginal cultural with a predominantly oral learning tradition. Selinger refers on Alvin Toffler and says: "*Here, the predominantly text based focus of commonly available technologies precludes cultural relevance in the transmission of the history and learnings of the various cultures and traditions of the hugely variable panoply of aboriginal family and tribal groupings. Further, given the longevity of this culture and the recognition of the focus on oral stories and artistic renditions of history, it is clear that ignoring other variables or certain technologies would not be immediately appropriate or acceptable in these cultures without significant unlearning and relearning*" ([Toffler, 1995] cited in [Selinger & Gibson, 2004]). Some cultures e.g. like the Bakiga descendents of Rwanda refer also to a mainly oral tradition of handing over knowledge [Vansina, 1973]. By that, information carriers which are relying on visuals (e.g. icons, graphs) and written text to transfer knowledge appear to be inappropriate (at least not without changing the prevalent learning culture). This shows that the composition of computers should not be considered as the best solution worldwide. Many studies have shown that a technology transfer can only happen if it is adopted and enhanced by the implementing country [Audenhove, 2001] and the way in which the technology is implemented could be by delinking the technology from the source and allowing the local context to determine the implementation [Jegede, 1994], [Scheepers & Mathiassen, 1998].

Another squalid (design) issue is that literacy is generally be taken as a granted capability. Though, the "Simputer" project has developed "*a low cost portable alternative to PCs, by which the benefits of IT can reach the common man*" [Trust, 2006]. The underlying concept has been to develop a device for illiterate people in India, but due to the Website's inactivity[4] I conclude that this engaged initiative has expired. As a consequence to the recent withdrawal of governments in R&D (neo-liberalism), idealistic initiatives are becoming rare. This bears the risk that social innovations become stunted. Private funds have to be raised to develop the idea, but investors expect a return on investment and inevitably demand the idea's readiness for the market.

Computers in LDCs are not only subjected to its logical application and physical appropriateness, but also to its physical maintenance. Power unsteadiness, dust, and brightness are just some critical factors to be mentioned. There are different attempts to design "computers" which are supposed to fit better to environmental conditions. One is the "One Laptop per Child"-initiative (olpc) by the Media Lab at the Massachusetts Institute of Technology (MIT) which builds upon black-white displays for enhanced usability in the sun and man-made power supply. The first laptops are planned for 2007 at a price of $ 135 [PMZ, 2006]. Interestingly, the notebooks will not be available for (public) sale and only children in LDCs will get it. "*The laptops*

---

[3]Of course product designers of ICTs develop tools which correspond best to their favors and cultural values and communicative preferences.

[4]As the official website of "Simputer" (www.simputer.org) has apparently not been updated since 2004 it appears to me as an indication of less activities.

will only be distributed to schools directly through large government initiatives" [olpc, 2006]. It is not my aim to discuss olpc's reasonability due to lack of detailed available information, but one arising concern involves ethnocentrism of this initiative. If olpc really restricts the device's availability to children I sincerely doubt its success in those cultures where the role of children is different to that where olpc is initiated. The olpc-responsible people latently take it for granted that for educational purposes it is aimed to protect a child as long as possible from unpleasantness to allow its maturing within a secure environment. In other traditions, e.g. in some Nigerian places [Jegede, 1994] or among some Ugandan cultures [Ngologoza, 1967], children are prepared for life and its peculiarities from the very beginning. By doing small tasks, like fetching water, or gathering wood daily necessities are conveyed and children are maturing. Whereas elderly in differing cultures, like in Austria, are spoon-fed by the younger generation, there are cultural characteristics in different cultures, where the *"words of our elders are greater than an amulet or are words of wisdom"* [Jegede, 1994] dominates. My fear concerning this issue is that families, who have e.g. in Uganda an average income of $ 2 a day [ÖFSE, 2006] are going to sell the laptop to assure an additional income.

Apparently, above cogitations seem to be ignored widely in literature, and we receive messages about the failing of ICT projects quite often [Avgerou & Walsham, 2000]; e.g.: Proenza reports from an Argentinian initiative where after the first year of intervention only 72% of 1281 telecenters were still up and running [Proenza, 2003]. Richard Heeks estimates the failure rate of ICT interventions in LDCs to be at least 80%(!). Of course, no public rates are available as investors try to avoid admittance of failures in order to raise new funds. Partly successful projects, whatever the circumstances have been, have to be justified and tend (naturally) to be presented in a more positive manner [Heeks, 2000]. Development collaboration has become a vast business[5] and donor agencies are criticized for not exhaustively aiming to "really" develop LDCs in order to keep their business running e.g. rumor has it that the Austrian development organization Horizont3000 has selected its name, because they want to exist still in the year 3000 AD.

Another striking issue is that in current publications computers and its functionality are equated with the Internet. The device itself is almost neglected and melts together with its main application "the Internet". This is a logical consequence to the fact that information is seen as the engine of development. When Press pointed out a positive correlation between the number of Internet hosts in a country and the UNDP Human Development Index, the fire was set and ICT initiatives were setup [Press, 1997]. Soon it became apparent that human development does not depend on the existence of Internet connections, but rather on the technology's acquisition. This puts our focus again on issues like learning and knowledge [Madon, 2000].

---

[5] In the year 2002 Austria spent 21 Mio. Euro on development co-operation [BMAA, 2002]. In average each EU country has agreed to spend about 0,8% of its annual GNP on development cooperation.

7.2   Stakeholders of ICTs in LDCs

In December 2001 the UN decided to hold a *World Summit on the Information Society*
(WSIS) to show up chances and risks of upcoming developments, which was held in
Geneva in 2003 and in Tunisia in 2005. Interestingly, the ITU as a technical organization was selected to organize it. According to Ludwig, the core competence would be
at the UNESCO [Ludwig, 2003]. "*Though, apparently the UN did not want to catch
again caveats, oppositions, or even public conflicts with leading members like USA,
Canada or Great Britain. This was feared if the UN had defined the importance of
the information society's core questions in a more political and consequent manner*"
([Ludwig, 2003], p. 95, [my translation]). Basically, the WSIS has drawn upon a
"multi-stakeholder-dialog": Governments, Economy, Civil Society and International
Organisations were allowed to participate. I do not want to discuss relevant outcomes
of the WSIS as they are far from my interests, but I will depict the interests of the
participating stakeholders to give an overview of their interests. The content basis on
a German Website on the WSIS [Heinrich-Böll-Stiftung, 2003] and was synopsized and
translated by myself:

7.2.1   Government

The Southern Governments underlines their interest of being better connected to global
communication channels. Access to information and knowledge about technical infrastructure and their appliance for development is highly prioritized. Nevertheless,
they aim to support local contents to maintain their cultural diversity. Northern Governments additionally underline topics like e-government, citizen-participation, education and creation of an investment-friendly environment.

7.2.2   Economy

For the first time not only big international economic institutions like the International
Chamber of Commerce or the World Economic Forum were represented, but also single companies, mostly big *multinational companies* (MNCs) such as Cisco Systems,
Microsoft are. This meant a double representation of economy and has been criticized. This is nourished by the fact that NGOs and other civil society actors where
left partly uninformed by the WSIS-secretary and not adequately supported. The
economic institutions grouped themselves together to become the Coordinating Committee of Business Interlocutors (CCBI) and emphasize the positive role of information
technology for development and economic growth and demand easier investments, an
establishment of market economy and legal protection of investments.

7.2.3   Civil Society

The Civil Society cares less about technical requirements, but more on social questions
and necessities, which are supposed to be a basis of the information society. They claim
to accredit communication as a human right, the "Global Information Commons" as a
counterpart to commercial utilization of information, freedom of opinion, the central
role of local communities and support of collective initiatives.

7.2.4   International Organizations

Besides UNESCO which stands for similar topics as the civil society does, and the Europarat who (in alliance with some others) advocates for civil freedom in information society like data protection. There are technical, economy oriented organizations who ask for global technical standards and investment reliefs. [Heinrich-Böll-Stiftung, 2003].

Memo: An interested reader might look up Wolf Ludwig's critical essay on the WSIS in Widerspruch 45/03 ([Ludwig, 2003] in German).

As the MDG asks for a co-operation between the private sector, NGOs and governmental policies this view can only represent a simplified representation and reality proves to be much more complex and interweaved. Therefore differing, partly overlapping and/or disagreeing interests within each stakeholder group have hampered discussions and finally limited the decent outcomes of preconferences and of the WSIS itself. The following example aims to clarify how initiatives can look like: The Imfundo Project is a unique partnership between the British Governmental Development Department (DFID), Internet giants Cisco Systems, Marconi and Virgin One and aims to bridge the digital divide between the world's rich and poor countries [IMFUNDO PROJECT TEAM, 2000].

Note: Interestingly, one year later the following statement was published: *"The primary goal of the Imfundo Project will be to support education in developing countries in a cost-effective way. It will not be tasked with the wider agenda of "bridging the digital divide"*, which is being taken on within the G8 by the DOT task force [Dot force initiative, http://www.dotforce.org], and within DFID by the Bridging the Digital Divide programme (which the Imfundo Project will support) [...] The goal is *""Education for All" - because education is the surest way to enable people worldwide to improve their own lives and the lives of those around them"* [DFID - Department for International Development, 2001]. Perhaps the "true" goal is none of above mentioned idealistically presented bold teasers, but the minor emphasized statement to "[...] *increase business involvement in Africa, link the local private sector with international firms and enhance the commercial viability [...]"* [DFID - Department for International Development, 2001]. This is emphasized by Charles Ess who claims that the big firms like Cisco Systems are already making money in LDCs [Ess, 2004] - see Footnote on Page 45. This is nothing wrong per se, but some MNCs utilize idealistic teasers as a disguise for economic motives. Thus, such initiatives taste a bit mouldy.

Anyway, together with developing countries governments (which will also have to participate in the project's realization) one realizes the heterogenous composition of actors. On the one hand this offers new opportunities to merge experiences and knowledge from all actors, but on the other hand the practical implementation of differing interests asks to be subjected to a longlasting negotiation process to allow a collaborative action. Some further and similar constellations are listed below in Chapter 7.4.

7.3   ICTs in Africa

In 1996 the *ITU*, the *UNESCO*, the *IDRC* and *The World Bank* set up the *African Information Society Initiative* (AISI), a programme for the application of ICTs. Its main document called "An Action Framework to Build Africa's Information and Communication Infrastructure" "*to build Africa's information and communications infrastructure*" [Miller, 2006] and is regarded as "*a guiding framework on which to base other information and communication activities in Africa*" ([Audenhove, 2001], p.280). The AISI programme calls for: "*The elaboration and implementation of national information and communication infrastructure plans involving development of institutional frameworks, human, information and technological resources in all African countries and the pursuit of priority strategies, programmes and projects which can assist in the sustainable build up of an information society in African countries*" ([Miller, 2006], p.17).

Together with a ITU Green Paper on Africa's ICT infrastructure these documents pose the framework of Africa's ICT policy. Though this framework is not compulsory many major donor agencies consider these documents as a guiding framework for the accomplishment of their initiatives (e.g. World Bank's infoDEV does, but not IDRC). African country leaders have set up the *New Partnership for African Development* (NEPAD) which strongly builds on ICT and formed the *eAfrica Commission* (ECA) to promote ICT activities. At the African Development Forum "*Globalisation and the Information Economy: Challenges and Opportunities for Africa*" in Addis Ababa in 1999 the ECA AISI group issued a widely-accepted "Common Position for Africa's Digital Inclusion" which contains four pillars for African ICT policy:

- Creating the Enabling Policy Environment

- ICT, Youth and Education

- ICT and Health

- Electronic Commerce.

Therefore the ECA is also concerned with the area of ICT in African schools (namely the pillar "ICT, Youth and Education"). Together with representatives from UNESCO, UNDP, the UN ICT Task Force and major developmental initiatives a pan-African and international network of schoolnet practitioners and policymakers, donor organizations, development agencies and private sector representatives has been consolidated by a Workshop in Botswana in 2003 [SchoolNet Africa et al., 2003]. Major international players from the donor and development sector are as follows:

- *Development Agency (CIDA)*

- *the Commonwealth of Learning (COL)*

- *the Department for International Development (DFID) Imfundo Project*

- *the Open Society Institute (OSI)*

- *the International Development Research Centre (IDRC)*

- *the International Institute for Communication and Development (IICD)*

- *SchoolNet Africa*

- *the World Bank Institute (WBI)* [SchoolNet Africa et al., 2003].

It would take too long to describe all various policies which are currently put in place or in process, but in general most activities in ICT policy making throughout Africa are restricted to the formation of national ICT commissions and central implementation agencies. It can further be stated that African countries promote telecommunication liberalization and privatization, but are still tasked with the realization of its national policies [Miller, 2006].

Practical ICT initiatives started at least as early as 1996 when the IDRC, ITU and UNESCO proposed a Multipurpose Community Telecentre Pilot Project in five African countries to the previously formed AISI with a budget ranging from about \$ 900,000 to about \$ 3.5 million [IDRC et al., 1997].

7.4 ICT in Uganda

7.4.1 Overview

The way Uganda's territory covers today was established in 1894 as a British protectorate. Uganda proclaimed its independence from Great Britain on the 9th of October 1962. Since 1967 Uganda has become a presidial republic. After its independence in 1962 living conditions started to improve, but under the dictatorship in the 70's and 80's the country's situation worsened and left it in devastation. After the revolution in 1986 Yoweri Museveni started to govern the National Resistance Council (with members of the *National Resistance Army* (NRA) and *National Resistance Movements* (NRM). Like in a "grassroot democracy" local representatives were elected into the NRM, which was the leading corporation, but actually the centre of power was the military and its NRA. This situation has not changed yet. Table 7.1 shows some facts[6] on Uganda compiled of [Niavarani, 2005], [Nzita, Richard and Mbaga-Niwampa, 1993], [UBOS, 2002].

Since 1994 Museveni was supposed to enable a multi-party system. Due to more or less opaque reasons Museveni postponed the first free elections up to Spring 2006 (!). Uganda can be considered as the "star pupil" of developmental aid and more than 50% [Ayodele et al., 2005] of the nations' Annual Federal Budget is raised by

---

[6]Please note that figures and especially the naming of ethnic groups is problematic, because no ethnic distinction line can be drawn. Nevertheless, I aim to give a (generalizing) overview to give the interested reader a feeling for Uganda as a "unified" country. Therefore, I use present four ethnic communities accordingly to Nzita and Mbaga-Niwampa, who claim that *"these ethnic communities could conveniently be divided into four broad linguistic categories"* ([Nzita, Richard and Mbaga-Niwampa, 1993], p.2). As mentioned earlier statistics tend to vary and so differing sources for the figures of this table might be found elsewhere. Though, I have decided on those sources which seem most reliable to me.

Fact	Figure
Country size	241,548 kmš (about 3x Austria)
Population (2004) [U]	26,699 mio. (13,419 mio. female/ 13,280 mio. male)
Population Growth [U]	3.3 % (annual)
Religion	40% Cath., 39.5% Anglican, 10.5% Muslim; Indigenious religions
Officicial Languages [Nz]	English, Swahili, Luganda
Main ethnic/language groups [U], [Nz]	Bantu (50%), Atekerin (13%), Luo (5%), Sudanic (5%)
Urbanization [U]	12.3%

[Ni]=[Niavarani,2005]; [Nz]=[Nzita, Richard and Mbaga-Niwampa, 1993]; [U]=[UBOS, 2002]

Table 7.1: Some facts and figures on Uganda.[Niavarani, 2005], [Nzita, Richard and Mbaga-Niwampa, 1993], [UBOS, 2002].

donor countries. As "The Economist"[7] pointed out, "*Cynics might say that Uganda can hold the world to ransom because The World Bank, the IMF and the other foreign donors cannot afford to let their star pupil go under*". Museveni squirmed and writhed, but finally international pressure was big enough and he was coerced to exclaim the first multi-party elections (after twelve(!) years). In this election Museveni was acknowledged as Uganda's President, but doubtful actions like imprisoning his biggest counterpart Mr. Besigye during the election campaign in November 2005 left a bad taste[8]. Nevertheless, free elections finally took place and the international donor community seem to be satisfied. Donor and development agencies have a long experience in working together with Museveni and know what to (not) expect from his politics. The Ugandan Okoth recalls on two geopolitical reasons why the donor countries had decided to keep quiet about Museveni's procedure for such a long period:

1. Uganda's position is geopolitically crucial to the U.S.[9], Britain and France. As such Uganda has been used as a barrier against the influence of Islam.

2. France and the USA have used Uganda to try to stop the spread of revolutions from Uganda itself to neig[h]bouring countries [Okoth, 1995].

For more than 15 years the (Christian) Ugandan army has been fighting against the Islamic *Lord's Resistance Army* (LRA) in Northern Uganda. The goal of the LRA is to establish an independent Islamic nation. Though this seems as a substitutional war between Christianity and the Islamic world, it cannot be taken for granted if this is so. This conflict could also be a utilization of religious aspects[10] to hide e.g. economical or raw material based motives. Actually, Museveni gains profit from this

---

[7] The Economist, 12. 02. 2000; p.61

[8] URL: http://news.bbc.co.uk/1/hi/world/africa/country_profiles/1069181.stm

[9] Uganda's position as a U.S. "darling" is indicated by the fact that former President Bill Clinton visited Uganda twice in four years (1998 and 2002) It took 20 years for Austria!

[10] Nowadays, it seems as if any conflict roots on religious motivations. Most likely situation is far more complex and cannot be reduced to such polarizing concepts.

war and it is a well-know political principle he acts upon: Museveni has identified the rebel leader Kony (LRA) as the Nr. 1 public enemy. Thereby, his fellow countrymen are unified and Museveni covers his political back. But also countries, which have been selling military hardware through out Uganda's existence [Okoth, 1995] are on the winning side. By selling arms to Uganda some countries retrieve some of their developmental aid. This poses latent subvention for their own arms industry[11]! Of course, uncountable idealistic NGOs and their staff are without any burden and abhor such political intentions and intrigues. Nevertheless, this has been a popular practice and one should be aware that there are always winners, even in times of war.

Despite its claimed preferred status, Uganda IS still one of the poorest countries worldwide and 55% of its population live below the national poverty line. Which means that the average Ugandan has less than $ 2 per day[12], 36.7% have to live on less than $ 1 a day [ÖFSE, 2006]. Recently, the situation has improved and there has been a net growth of net income of 3.9% from 1990 to 2003. Between 1997-2000 the top 10% earners could increase their income by 20%, the lowest tenth just by 8% [ÖFSE, 2006]. This divide within Uganda is underlined by the change of the Gini coefficient from 34 to 42 since 1997. This indicates that Uganda is about to become a high-inequality country [UNDP, 2005]. "*Over the past decade Uganda has experienced sustained economic growth and made important advances in human development*" ([UNDP, 2005], p.47). Since 1990 Uganda increased its *Human Development Index* (HDI)[13] score by about 20% which is mainly due to the progress in school enrolment (see Page 95). Although poverty reduction has been a national priority the success remains fragile. Since 2000 income growth has slowed and poverty has risen, and the MDG achievement of halving poverty in 2015 is endangered. Also in the area of health Uganda has become off track. Being one of the strongest MDG performers in Africa in recent years [UNDP, 2005], life expectancy at birth has been 47.3 years in 2005. This means that it has hardly increased (though there is economic growth) and is close to the level of 1970-1975 when it amounted to 46.1 years [UNDP, Thematic Trust Fund, 2001]. Another striking issue is that under-five child and maternal mortality has not been significantly reduced (see Figure 7.2)

7.4.2   Uganda's ICT Politics

In 1995 Yoweri Museveni identified five factors needed for Africa to assert itself globally. These factors included, "*an educated population that will have the capacity to utilize technology in order to transform our natural resources into wealth*" [Museveni, 1995]. These words can be seen as a starting signal for an ICT revolution in Uganda, because since the mid 90's two major developments have taken place.

---

[11]Development aid has always been implying economic components and e.g. in Austria politicians explicitly demanded development aid to develop new markets and to support artificial demand for national unwanted products by stat-aid. Development interventions are bound to an obligation to buy certain products. Thus national companies get latent subventions ([Hödl, 2003], p.34f).

[12]This was defined as the level for poverty, defined by the UN.

[13]"*A composite index measuring average achievement in three basic dimensions of human development: a long and healthy life, knowledge and a decent standard of living. It is used as an index by the UN as a key figure to compare countries among each others*" [UNDP, 2005].

Indicator	1992	2002
Income poverty (%)	56	38
Gini coefficient	36	42
Children under age 5 under weight for age (%)	62	86
Under-five mortality rate (per 1,000 live births)	187 a	152
Maternal mortality ratio (per 100,000 live births)	523	505

a Data are for 1990.

Figure 7.2: Uganda's mixed performance on human development. [UNDP, 2005].

Firstly, on governmental side, after the preceding liberalization of the telecommu-
nication sector in 1996[14] and the forming of the Ugandan Communications Commision
(UCC) in 1997 [Mugisha, 2002], Uganda has developed a "NATIONAL INFORMA-
TION AND COMMUNICATION TECHNOLOGY POLICY" which was published in
July 2002[15] [Ministry of Works, Housing and Communications et al., 2002]. Afterwards more and more institutions were formed and bureaucratic situations worsened.
Tusubira listed inside are critics of the policy; eight different institutions in the fields
of Uganda's ICT development [Tusubira, 2004]. To take charge of this confusing situation Uganda has put in place an extra Ministry of Information and Communication
by Spring 2006. Thus, a single and consistent voice that speaks for the sector at a
political level (as demanded in [Tusubira, 2004]) has been established. Time will prove
if Uganda succeeds in a better coordination of all relevant activities.

Secondly, "*a proliferation of ICT initiatives, programs and projects as well as
matching increase in the ICT infrastructure spearheaded by the NGOs and the private
sector respectively*" ([Mugisha, 2002], p.2). Uganda's policy is based on the vision of
"a Uganda where national development, especially human development and good governance, are sustainable enhanced, promoted and accelerated by efficient application
and use of ICT, including timely access to information" and the goal "*to promote the
development and effective utilization of ICT such that quantifiable impact is achieved
throughout the country within the next 10 years*" [Ministry of Works, Housing and
Communications et al., 2002].

Nowadays, most ICT initiatives are currently undertaken through donor funding
by government, quasi government, and private institutions. In Uganda's ICT policy
most actual projects are included, which are as follows:

- The ACACIA project for pilot Telecenters at Nakaseke, Buwama and Nabweeru

- The DANIDA Local Government Project in Rakai District

- The Infodev Information Infrastructure Agenda at the Institute of Computer
  Science at Makerere University.

- The Inter-Ministerial Mapping and Geographic Information System (GIS)

- The Academic Research Network Project

---

[14] Thereby it followed the policy of the ITU.
[15] The first country to engage in ICT policy formation was Rwanda in 1998.

- Initiative to Create an ICT Resource Centre and Internet Café

- The Local Area Network and Internet Connectivity Project for the Parliament of Uganda

- The Campus Network Project for Makerere University

[Ministry of Works, Housing and Communications et al., 2002]

This list is not exhaustive, because there "*are a number of donor supported ICT projects in a number of public service organizations, which are not publicly documented*" ([Ministry of Works, Housing and Communications et al., 2002], p.30). How many ICT related interventions have been and are being undertaken in Uganda is not possible to estimate as ICTs are put in place cross-sectorally and a request at the Austrian Foreign Ministry for some estimated data only on Austria's ICT expenditures came to nothing, although I know that there are supported projects in Uganda.

What have all those initiatives achieved so far? As discussed it is difficult to find out (especially as we do not know about all activities), but by numerical statistics some interesting facts for comparison can at least be retrieved. According to [UNDP, 2005] combining telephone mainlines and cellular subscribers form **an** indicator for MDG Nr. 8. In 2003 five per mille of Ugandans were Internet users[16], 30 per mille were cellular subscribers[17], and two out of a thousand had telephone mainlines [UNDP, 2005][18]. At a first glance the rate of cellular subscribers sounds quite fair, but looking at Figure 7.3 makes it evident that Uganda, compared to its HDI ranking neighbors, is currently performing quite low.

7.4.3   ICT & Education Policy in Uganda

As previously indicated the extraordinary status of Uganda (see Section 7.4) in developmental activities is underlined by the fact that big transnational ICT initiatives were setup in Uganda from the very first moment: The World Program (now SchoolNet) began as a pilot project in Uganda [Kozma, 2002], the Grameen Telecom's Village Phone Project [Stuart Mathison, 2003] was implemented in Uganda as the second country after Bangladesh, and Uganda was one of the original countries of Connect-ED (EDDI) as early as in 1998, when U.S. President Clinton pledged to strengthen African education systems. Thus "The *Education for Development and Democracy*

---

[16]Internet users are people with access to the worldwide network [UNDP, 2005].

[17]Cellular subscribers (also referred to as cellular mobile subscribers) are subscribers to an automatic public mobile telephone service that provides access to the public switched telephone network using cellular technology. Systems can be analogue or digital [UNDP, 2005].

[18]Interestingly, all three figures together form **the** indicator of MDG Nr. 8 - Target 18 (see Page 79) and are measured by the ITU as an agency for the UN [UNESCO, 2003]. If this combination really can indicate whether cooperations with private sectors are established has to be discussed, but is not within my scope; also the fact that the ITU on the one hand has developed AISI and measures thereby its own policy programme leaves a bad taste in my mouth. Too much focus is put on sheer numbers, which underlines the predominant and recently revived and enforced position of quantitative methods. The interested user might look up [Denzin & Lincoln, 2005] on this trend.

Initiative (EDDI)" was created as a partnership among the State Department, US-AID, the Peace Corps, and other U.S. agencies. Connect-ED's Mission Statement is: *"To enhance the quality of teaching and learning in primary schools through the use of Information Communications Technology (ICT) and empower teacher professional development while promoting a sustainable environment"* ([Cisler & Yocam, 2003], p.14). "Schoolnet Africa" and "Imfundo" have catalogued ICT in education initiatives all over Africa, but most data appears to be a few years old [infoDev, 2005b]. Thus, I am also coerced to refer to IICD's list about ongoing ICT and Education programmes in Uganda:

1. *Educational Management Information System (EMIS)*

2. *SchoolNet*

3. *School Broadcasting*

4. *Education For Development And Democratic Initiative (EDDI)*

5. *Enhancing Government and Management of Private schools in Uganda - British Council*

6. *World Links (World Bank) (now SchoolNet (Uganda)*

7. *Developing an Information Infrastructure Agenda for Uganda*

     [IICD, 2000]

Besides these programmes there are further[19] (mainly private) ICT interventions as the one of Projekt=Uganda has been. These interventions are mainly directly established on the basis of a personal relationship of e.g. school parents interlink themselves with NGOs in donor countries. These relationships are motivated by different kind of reasons, be it religious, political or personal. This means that several initiatives are not embedded in a wider, more co-ordinated context and so it happens that the intervention's process becomes confused and even unguided. This does not necessarily imply that the success (*itmo* integration into everyday life) would be better, if they were placed within a nationwide ("official") programme; but let me set an elucidatory example: During my first stay in Uganda in 2002 the Headmaster of a neighboring school asked us (my colleague and me) to set up their computer room. The situation was as follows: An U.S. based mission sent 15 PCs to a neighboring school, but no local staff assembled those. When we arrived eight months after the PCs arrival, we were asked to do so.

Orivel has figured out that computer supported education can help to save money in countries with a GDP (per person) higher than $ 7,300 ([Orivel, 2000] cited in [Afemann, 2003]) (in 2003 Uganda's GDP (per person) was $ 1,457 [UNDP, 2005]. Well, such figures need to be discussed as other sources claim that there is no guidance on how to conduct cost assessments at all [infoDev, 2005b]. As a matter of fact ICTs

---

[19]Due to my personal experience I know that there are such further isolated initiatives.

are expensive in general and computers in LDCs are even more costly[20]. Nevertheless, most problematic is not to obtain computers or conduct initial preparations as the provision of a lockable and adequately equipped computer room is. Maintenance costs pose to be highly expensive and most troublesome. Ruth & Shi declare maintenance costs of a computer supported education with $ 100 per year and pupil ([Ruth and Shi, 2001] cited in [Afemann, 2003]). According to their calculation printing is about 6 US-Cents per page. This sounds reasonable as in Uganda already a ream of paper is $ 7.5, a blank CD is $ 2.7, a black printer cartridge is $ 65 or the Bushnet Internet charge is $ 350 per month [Cisler & Yocam, 2003]. Data varies, but according to a report of UNESCO and ITU Uganda's annual dial-up Internet tariffs accounts for about 3-12 times of the annual GDP (([UNESCO, 2002], p. 69), ([ITU, 2001], p.21)). Regional deviances in the "at hand" case study are discussed in Chapter 9.

7.5   Summary

Chapter 7 outlines the general discussion on ICTs in LDCs. As ICT *for* LDCs has become a huge business many different stakeholders (with different and partly over-lapping interests) participate in a discussion on "who shall profit to what extent" of this developmental aid. Stakeholders, which are powerful and influential to a different extent, negotiate possible cash flows and politicians struggle to get a piece of the cake - see Section 7.2. This composition of stakeholders and their discussion leaves the impression as if the "receiver" of this cash flows do not have the chances to participate in an equal way. The one who pays the piper calls the tune. Keeping this code of practice in mind allows to discuss aspects of power and the hierarchical position of us Austrian computer teachers, who were perceived as donors as well. This becomes evident especially in practices of conflict avoidance by the "receivers".

Afterwards, I briefly present framing (national) ICT policies (7.3, 7.4) and depict its educational policy context within Uganda (7.4.3). This is important, because it helps to understand what kind of expectations towards ICTs are conveyed by politicians and educational policies and downright loaded on an intervention's participant. Though, in the following discussion I will also carve out that Uganda's current education policy prevents integrating computer teaching into the standard curriculum and limits the school in the utilization of the computers to a certain extent. Whereas the list on Uganda's nationwide ICT and Education programmes makes it evident that the context of the intervention to be examined is different, as it is not embedded in nationwide activities. If a one-time intervention necessarily implies a worse integration into the participants' everyday lives because of an alleged limited professional knowledge is not within the scope of this work; but it is a point of interest to explore the underlying relationships to answer the question if donors can and should evaluate their interventions on their own. Discussing the success of an intervention, does also imply matters of possible maintenance. Therefore, I presented some rough numbers on ICT supplies to show that expenditures to maintain a computer lab have to be quite similar to those in Austria, though incomes are much lower.

---

[20]In 2003 the price of a computer in Uganda was double the price of a computer in Austria. This is due to high import taxes.

		MDG		MDG		MDG	
		Telephone mainlines [a] (per 1,000 people)		Cellular subscribers [a] (per 1,000 people)		Internet users (per 1,000 people)	
HDI rank		1990	2003	1990	2003	1990	2003
139	Bangladesh	2	5	0	10	0	2
140	Timor-Leste	..	..	0	..	0	..
141	Sudan	3	27	0	20	0	9
142	Congo	7	2	0	94	0	4
143	Togo	3	12	0	44	0	42
144	Uganda	2	2	0	30	0	5
145	Zimbabwe	13	..	0	..	0	..
LOW HUMAN DEVELOPMENT							
146	Madagascar	3	4	0	17	0	4
147	Swaziland	17	44	0	84	0	26
148	Cameroon	3	..	0	66	0	..
149	Lesotho	7	16	0	47	0	14
150	Djibouti	11	15	0	34	0	10
151	Yemen	11	..	0	35	0	..
152	Mauritania	3	14	0	127	0	4
153	Haiti	7	17	0	38	0	18
154	Kenya	8	10	0	50	0	..
155	Gambia	7	..	0	..	0	..
156	Guinea	2	3	0	14	0	5
157	Senegal	6	22	0	56	0	22
158	Nigeria	3	7	0	26	0	6
159	Rwanda	2	..	0	16	0	..
160	Angola	8	7	0	..	0	..
161	Eritrea	..	9	0	0	0	7
162	Benin	3	9	0	34	0	10
163	Côte d'Ivoire	6	14	0	77	0	14

Figure 7.3: Uganda's MDG key figures in HDI context. [UNDP, 2005].

Chapter 8

Education

8.1 ICT & Education in LDCs

The major stance in literature is that introducing ICTs within an educational system demands a reform in education e.g. [DFID - Department for International Development, 2001], [infoDev, 2005a], [infoDev, 2005b], [Kozma, 2002], [Hawkins, 2002]. Hawkins states that "*linking ICT and education efforts to broaden education reforms*" [Hawkins, 2002]. These reforms are difficult to bring through, because "*many ministries of education consider computers as a stand-alone subject*" [ibi., p.41]. Hawkins concludes that "*ministries must make a commitment to helping teachers effectively integrate computers and Internet technologies into their schools by aligning curriculums, exams, and incentives with the educational outcomes that they hope to gain*" [ibi., p.42]. According to the way of transferring computers he underlines to consider computers as a tool like many others, which is also supported by [Jegede & Aikenhead, 1999]. It is important to assure that the ways the tool can be applied are conveyed properly; e.g. the tool "literacy" was considered to be transferred successfully, but there are Ugandans who blame the Wazungu[1] that "[they] *have taught us to read and write, but they didn't teach us how to use this knowledge*" [Karwemera, 2003]. If knowledge transfer in literacy evidently lacks meaningfulness and knowledge, how can it be applied? Obviously, even more complex efforts have to be put in to show up the possibilities and usefulness of ICTs.

From an educational point of view student centred teaching is encouraged to make an opposite to traditional teacher centred teaching [Pryor & Ampiah, 2003]. A reason for teacher centred learning is that "*teacher centred learning seems to be the easy way out in teachers' perspectives teaching children, most of whom cannot be engaged in any meaningful verbal interactions in English [...] on topics in the curriculum*" ([Pryor & Ampiah, 2003], p.38f). So student centred teaching is strongly dependent on a common language base. For introducing ICTs in education[2], student centred pedagogy is the most advised and encouraged approach e.g. [Kozma, 2002], [infoDev, 2005b], [Mansell & Wehn, 1998]. Bolander perceives student centred learning as "*an approach on teaching which recognizes the student as an individual and his/her per-*

---

[1] Swahili plural of Mzungu for "European".
[2] Please note that this is not limited to LDCs.

sonal development as an important factor" [Bolander, 2003]. She emphasizes such teacher-student relationships, which are *"characterized by collaboration, consultation and negotiation where students are seen as a learning resource and participants in a clear and transparent process"* [Bolander, 2003]. [Menon & Naidoo, 2003], who investigated experiences of ICT education in Africa, claims that *"collaborative reflective practice"* has turned out to be the *"core of the training process"* [Menon & Naidoo, 2003]. Teacher centred pedagogy is also seen as a coping strategy for those who do not grasp the content of the lesson [Pryor & Ampiah, 2003] and apparently the more pupils a teacher has to teach, the more difficult it is to remove subject matter related ambiguities. This attaches conditions to the achievement of an adequate teacher pupil ratio (see below in Section 8.2) and on qualified and capable teachers. The UNESCO has stressed latter argument and clearly states:

"With the emerging new technologies, the teaching profession is evolving from an emphasis on teacher-centred, lecture-based instruction to student-centred, interactive learning environments. Designing and implementing successful ICT-enabled teacher education programmes is the key to fundamental, wideranging educational reforms."

The UNESCO[3] has had thoughts on these reforms and suggests those as follows:

"Teacher education institutions may either assume a leadership role in the transformation of education or be left behind in the swirl of rapid technological change. For education to reap the full benefits of ICTs in learning, it is essential that pre- and in-service teachers are able to effectively use these new tools for learning. Teacher education institutions and programmes must provide the leadership for pre- and in-service teachers and model the new pedagogies and tools for learning. They must also provide leadership in determining how the new technologies can best be used in the context of the culture, needs and economic conditions within their country. To accomplish these goals, teacher education institutions must work closely and effectively with teachers and administrators, national or state educational agencies, teacher unions, business and community organisations, politicians Towards a Strategy on Developing African Teacher Capabilities in ICT and other important stakeholders in the educational system. Teacher education institutions also need to develop strategies and plans to enhance the teaching learning process within teacher education programmes and to assure that all future teachers are well prepared to use the new tools for learning" [UNESCO, 2002].

Thereby student centred education becomes less a single teacher's issue than once more a political one. Table 8.1 is not only for the sake of completeness, but presents identified student centred teaching methods and the students corresponding roles as participants [Bolander, 2003]. It makes the teacher-pupil relationship pillars *"collaboration, consultation and negotiation"* evident and underlines that, indeed, sufficient facilities have to be supplied and structural constraints unravelled.

---

[3]For more details, please see the UNESCO-report by [Haddad & Draxler, 2000]: Technologies for Education: Potentials, Parameters and Prospects. Available at: http://www.aed.org/ToolsandPublications/upload/TechEdBook.pdf. It aims to help educational decision makers survey the technological landscape and its relevance to educational reform and attempts to understand how technological tools can better contribute to educational goals. It looks at how technology can promote improvements in distribution and delivery, content, learning outcomes, management of systems, teaching, and pertinence ([Haddad & Draxler, 2000], p.7).

Teaching methods	Students as participants
Active teaching methods	Students identify learning needs
Active learning, student activity	Identify own strengths and weaknesses
Learning by doing	Students' needs in focus
Problem based - not didactic	Feedback
Varied teaching methods flexibility	Students responsibility for shaping and completing task
Computer Assisted Learning (CAL)	Reflection on learning process
Group work	Development of skills

Table 8.1: Student centred education. [Bolander, 2003].

In many countries of Africa the prevalent school system is an "achievement" of colonial days, when colonial education was introduced as something superior compared to the community related indigenous tradition of transferring knowledge (see below Section 8.2). Thus, *returning* to "collaborative" related learning fits quite well, as learners with a traditional differing learning culture feel isolated when they learn individually. This relies on the fact that individual excellence is still regarded as (former) colonialists preference [Postma, 2001]. She concludes that ICT "*can serve more communitarian preferences and learning styles if students are first taught to use them in collaborative, rather than primarily individual ways*" [Ess & Sudweeks, 2001]. Otherwise ICTs will "*serve as technologies of cultural imperialism, rather than as technologies of liberation and cultural diversity*" ([ibi.], p.266).

8.2 Education in Uganda

In the 70's education's role for development was discovered ([Watkins, 2000, p.15] cited in ([Gotschi, 2003], p.27)) and lead to the belief - not only in modernization theory - that education and "*increasing economic growth go hand in hand*" (Webster, 1984, p.114 cited in ([Gotschi, 2003], p.27)) or that "*education has a critical role to play in economic development*" (Chung, 1996, p.207 cited in ([Gotschi, 2003], p.27)). Social change takes place through education, which also represents a recreation of society and its reigning hierarchical structures (cp. [Bourdieu, 1970], p.192f). Consequently, education can be seen as an instrument of politics. Education does not necessarily always have the same meaning, since there can be identified three different forms of education:

- Formal education
  This type is nowadays typical for Uganda and is now what is considered as education. Formal education builds on institutions like schools, where teachers prepare lessons thoroughly and students expect to be tested in order to receive certificates. These help governments and employers to distribute the young in employment to where they fit best.

- Non-formal education
  This form of education refers to what happens outside the formal structures an

is especially targeted to skill training. It is less structured and by application the non-compulsory education form is more flexible.

- Informal education
  Actually, this was the former indigenous common Ugandan form of education. It existed in everyday life. No certificates were needed and knowledge was transferred unconsciously by showing and explaining things in daily situations. Knowledge was for example transferred by dramatic performance of indigenous stories, plays, or legends ([Gotschi, 2003], p.27f).

8.2.1   Indigenous Education

Indigenous traditional education systems have existed for hundreds of years, and were seen in terms of its ability to solve problems. Be it present or future problems, one of the aims of education was the development of the learner's ability to deal with daily situations. At least in Uganda the set-up was that the problems which had to be solved were those which occurred in daily life. There were no isolated teaching classes and learning did not take place in "*isolation from the child's environment*" ([Mwanahewa, 1995], p.100), it was the child's environment! Education was placed within the responsibility of parents and the whole family aimed to achieve educational goals ([Ssekamwa & Lugumba, 1971], p.22f). Even society took responsibility of others in cases of social attitudes and behaviors and elder people rebuked younger people. By that elder ones served as models and so he/she had to follow those values to demonstrate them to the young. "*This meant also order and stability within that single group of people*" ([ibi.], p.22). Within the community certain rare skills (to become a skilled iron smith, skilled potter) were transferred not only within the family, but also to other community members in order to have enough skilled workers to satisfy the community's needs. These women and men one could call "*professional teachers*" (([ibi.], p.22) or "*experts*" ([Kwitonda, 1995], p.226f)). When colonialists arrived in the 19th Century they completely failed to perceive the indigenous form of education and even denied it. In "Kenya Hanging in the Middle Way" it is stated: "*Education was first brought to the Africans [itmo Ugandans] by the Christian Missionary agencies [...]*" [Cole, 1959]. Soon after the colonialists arrived, first schools were established, and from then on school and daily life got separated, not only temporal, but also spatially.

By this Ugandans got disoriented and failed to combine both educational approaches. Of course, as mentioned above, their education was not considered as a form of education at all. The colonialists presented themselves as superior and educated. They called the indigenous population "*ignorant*", "*uncultured*", and "*uncivilised*" ([Mwanahewa, 1995], p.100) or even a sinner, a heathen, a pagan or lazy, indolent, a thief, and inferior being ([Kwitonda, 1995], p.228). Colonialists took over superior positions and Ugandans "*were given little opportunity to reconstruct and reorganise their own experiences*". Some Ugandans felt that colonial cultures were superior to others and abandoned theirs in favor of colonialistic ones ([Nzita, Richard and Mbaga-Niwampa, 1993], p.v). Thus, education fell victim to a new hegemonic culture ([Mwanahewa, 1995], p.100f). "*It is then that the young African [itmo Ugandan] who went to the missionary school started to develop a sense of guilt and an inferior-*

ity complex" and "[...] *gradually* [the majority] *became clowns who imitated the white man just as a monkey will imitate a man"* ([Kwitonda, 1995], p.228). Mwanahewa attributes this to the fact that "*Uganda's cultural practices to a large extent functioned on the logic of* either/or *and not the logic of* fuzz" ([Mwanahewa, 1995], p.98). As an example he states that a Ugandan belonged to a group of one particular totem and one clan. Within this social structure "*there was little or no freedom of choice as regards cultural maxims and edicts. This apparent rigidity should not be interpreted to mean or imply stagnation; it was rather meant to ensure sustained identity of the cultural practices"* ([Mwanahewa, 1995], p.100). As "*it is an indisputable fact that we East Africans [itmo* Ugandans] *have accepted it* [the foreign school system] *[...]"* ([Ssekamwa & Lugumba, 1973], p.vii) and that the indigenous education fell behind[4], in the next paragraphs I will sketch the development of this foreign school system (which became equal to (formal) education strategies) in Uganda to dismantle its historical context to gain a better understanding of actual problems.

8.2.2   Colonial Education

As with many other LDCs, in the educational system of Uganda there exists a leftover of colonialism. When Uganda was annexed by Great Britain in 1894, missionaries did no fail to appear and established the first school in 1902. Protestant and Catholic missionaries strived for "*uneducated"* ([Mwanahewa, 1995], p.100) Ugandans with the aim of turning them into good believers. The biblical way of teaching by the missionaries was a quite authoritarian technique[5], where the indigenous people must keep quiet, listen carefully, and then later try to do what has been taught at the church, on their own. Due to the fact that the missionaries were also the teachers, the same approach was used for schooling. Moreover, presumable because of above mentioned logical confusion of "*either/or"* vs. "*fuzz"* ([Mwanahewa, 1995], p.98) the missionaries words were taken for granted, and therefore the indigenous people naively believed that every word spoken by a missionaries was true and final. Religious groupings rivaled[6], which lead to separate construction of schools, without any central coordination. Until 1927 all schools were private and by "The Education Ordinance 1927" the whole education system was brought under government direction and control. With this ordinance the government provided detached supervision and minimum grants-in-aid. Government could direct and determine what the owners could do in their schools and employ teachers, but did not own and manage schools. The ordinance can be seen as the beginning of the partnership between missions and government, but nevertheless "*religious bias as stressed by the missions, remained an essential factor in education"* ([Ssekamwa & Lugumba, 1973], p.50). Each racial and religious grouping, be it Asian, Muslim or Christian had its own schools. In May 1931 Uganda decided on English[7] as the lingua franca for the educated East Africans [Ssekamwa &

---

[4]Similar effects were perceived among the indigenous population of Pacific Islanders - cp. [Latu & Young, 2004].

[5]Latu, who describes similar circumstances in Pacific Islands, calls this the military's 'command-and-do' manner [Latu & Young, 2004].

[6]Karwemera even talks about armed conflicts in some regions [Karwemera, 2003].

[7]Competing Kiswahili was taught as a second language, but was declared to be the national language for Uganda in 1973, soon after Idi Amin had taken over in 1971. Thus the country was brought

Lugumba, 1973]. Only one language for formal education was used, so as to unify the country politically and culturally[8]. Not until 1942, when "The 1942 Education Ordinance" was published did government encroach on school management. The "new" technical education[9] started with the education of Ugandans as mechanics in World War II and it intensified from 1950 to 1960, but never took root. Superior technical positions were held by Asians and Europeans who held lower qualifications. As no career with a technical degree was possible, technical education became unpopular ([Ssekamwa & Lugumba, 1973], p.68). Ugandans were mainly placed in the ranks of auxiliaries or sub-ordinates, while the colonialists played the roles of manager, director or supervisor [Mwanahewa, 1995]. Consequently, the courses though desirable became unattractive, which was "supported" by the shockingly mediocre "*calibre of instructors*" ([Ssekamwa & Lugumba, 1973], p.70) who seemed to be unable to teach the students effectively.

According to Mwanahewa, especially by independence in 1962, "*education was expected to provide answers to the social, economic and political problems. [...] It was expected to liberate the population from the hostile environment by eradicating negative outlooks on life, enabling the population not only to survive, but also to live comfortably*" ([Mwanahewa, 1995], p.223). In 1964 a further reform was necessary, because supervising mechanisms also did not work properly and racially and ethnically separated schools made establishing an overall education policy difficult. As a consequence to the "Castle Report" of 1964 all separate racial and religious systems were integrated and all matters concerning primary and secondary schools in the country came under the direct control of Central Government in 1964 [Ssekamwa & Lugumba, 1971]. Consequently, more and more pupils wanted to attend secondary schools[10] and pupil-teacher ratios increased from 25 to 40%. When government could not satisfy the high demand for secondary education manifold private institutions were setup, mainly by Asians. This was because of two reasons: Running schools turned out to be quite lucrative (1) and many Asian children did not get a place in secondary schools now, as low scoring children would not have been admitted to governmental schools (2) [Ssekamwa & Lugumba, 1973]. One may or may not consider this private initiative as a good and smart solution, but due to school fees the poorest population was even more marginalized and quality of education worsened. As a matter of fact not enough qualified teachers were available. Uganda's government suggested that at least 25% of the teaching staff should be registered teachers - which would have guaranteed a minimum level in teacher education - but private schools still employed underqualified teachers. Tricks were played because "*in order to gain above percentages they* [Asians]

---

in line with the rest of East Africa in terms of language, but Kiswahili is considered to be Idi Amins language and lost its official and national status in the 1995 Constitution. In September 2005, the Ugandan Parliament voted to make Swahili the second official national language once again.

[8] This approves the low attitude colonialists hold on Ugandan cultures and contributed to its cultural disorientation and neglect [[Mwanahewa, 1995], p. 100].

[9] There had always been indigenious technical education like iron melting and vocational training to become a blacksmith - see ([Mwanahewa, 1995], p.226f).

[10] Secondary education is based on at least four years of previous instruction at the first level and provides general or specialized instruction, or both, at such institutions as middle schools, secondary schools, high schools, teacher training schools at this level and vocational or technical schools [UNDP, 2005]. In Uganda primary level starts at the age of six and lasts for seven years.

*employed teachers with poor academic backgrounds although on the official register
they had names of teachers duly qualified but were never in school"* ([Ssekamwa &
Lugumba, 1971], p.154). This means that the quality of teaching posed a problem
and nevertheless granting-aids by the government remained low. During the times of
oppression and dictatorship in the 70's Uganda's economic situation worsened and it
became difficult to get school materials, scientific equipment and even the construction
of schools was stopped.

In the 1980's the education sector was not promising: enrolment ratios in primary
schools were 66%. School books, which were explicitly required by the curriculum
were not yet printed, instruction material was hardly available. A lack of bad teachers
and a steadily growing population made school places limited. It became apparent
that a further reform had to be done. In 1989, after a period of three years the
Education Policy Renew Commission presented a report called "Education for National Integration and Development" to the Ministry of Education and Sports. The
reports wide-ranging recommendations were to be implemented partly immediately
(1990-1992), some over a medium-term period (1993-1996) or even later. The break-
through came with presidential elections in 1996 when President Museveni promised
the introduction of free[11] and compulsory education and announced *universal primary
education* (UPE) [Gotschi, 2003]. By UPE it is believed that it is one key requirement
for achieving high economic growth rates (cp. Watkins, 2000, p.299 cited in ([Gotschi,
2003], p.45), ([Ministry of Finance, 2001], p.42)).

8.2.3   Recent Education

*"One of the Millennium Development Goals (MDGs) is achievement of universal pri-
mary education by 2015. We must ensure that information and communication technologies (ICT) are used to help unlock the door to education"* [Annan, 2005].

In order to achieve this MDG, Uganda setup the current UPE initiative in 1997,
which grants free school attendance for up to four children (two boys, two girls) per
household. With additional U.S. loans new schools were built and as a consequence
the number of attending pupils increased from $ 5.3 mio. in 1997 to $ 6.6 mio. in
1999. However, it decreased to $ 6.1 mio. in 2000. Nevertheless, this is considered the
major reason why Uganda has increased its HDI in recent years and has improved its
ranking from 141st out of 162 nations in 2001, to 144th out of 177 nations [UNDP,
2005]. Apparently, UPE is costly and education took 19.8% of the GDP in 1994
and 26.5% in 2002 (without Official Development Assistance) [UNDP, 2005]. When
enrolment increased Uganda's government were faced with other (well-known) problem
as still not enough secondary educational schools were available and class sizes in
primary schools increased to an average of 75 pupils per teacher [Cisler & Yocam,
2003]. This means that *"the PEAP/PRSP targets for pupil-teacher, pupil-classroom
and pupil-textbook ratios for 2000 were missed"* [Ministry of Finance, Planning and
Economic Development, 2001]. From a financial point of view it was estimated that
implementing UPE in Uganda would mean that a total of 135,430 classrooms would

---

[11]Nevertheless, there are some additional costs for parents, as they have to provide for pencils,
meals, clothing and "teachers' funerals, yearend celebrations, classroom construction and telephone
connections" [UNESCO, 2004].

have to be constructed. Another problem was that new established schools were not usable (as cheap building material had been used) or due to corruption the school was not built at all[12]. The Education Management Information System Census 2000 figured out that only 66 percent of the schools were meeting their minimum quality standard requirements. According to the Community Survey of 1999/2000 only a quarter of schools were considered in good condition [Ministry of Finance, Planning and Economic Development, 2001]. This concluded in a deterioration of school quality and NGOs were worried about the working conditions of new teachers as they were insufficiently paid or not remunerated for a longer period of time at all. In general, the salaries of teachers which were paid by government were still very low or even below livelihood. Therefore the Parents Teacher Associations supplemented in many schools the salaries of the teacher, which lead to different quality levels and competition between schools.

A further problem was and still is also that teachers are hardly unionized [ÖFSE, 2006]. Another issue is still the lack of governmental support for secondary education [ÖFSE, 2006]. The majority of parents (80%) sell their property to allow their children to be educated ([Bosworth, 1995], p.194).Gross enrolment ratio for secondary schools was 3.8% lowest quantile (20%) of population in the year 2000 [ÖFSE, 2006].

So UPE aims to create adequate educational infrastructure and though initial key figures have been good, it struggles with other (unexpected) social factors which have to be bettered as well.

8.2.4 What Kind of Ugandans Should Schools Produce?

Nowadays Uganda's school system is based on the colonial system and aims to "produce" as many (formally) educated pupils as possible, which is heavily criticized and attacked by [Kwitonda, 1995] and [Ssekamwa & Lugumba, 1971]. Some critics are concerned with the individualistic nature of schools in general. Thereby, "*it [the colonial school system] undermined not only the indigenous systems of education, but also idealized Western [itmo colonialistic] civilization*" ([Kwitonda, 1995], p.220), which deprived the peoples self-confidence, as discussed above. It "*promotes individualistic attitudes, i.e., the practice of hiding books and journals so that one is the first and only one to achieve the highest goal in school*" ([ibi.], p.222) and runs counter the long evolved traditional practice of collaborative societal living, teaching and learning. One might consider this development as nothing wrong, because culture (and education is part of culture ([ibi.], p.221)) is constantly changed and revised, but the problem is that due to the system's "artificial caused manner" other correlated structural preconditions are not yet set: There are hardly adequate job opportunities for graduates available and most of them have not obtained enough skills to initiate jobs by themselves. It is also criticized that during their education they have internalized the "*same colonial mentality of getting a white collar job*" ([Kwitonda, 1995], p.225) and refuse to pick up agricultural related jobs [Kwitonda, 1995], [Ssekamwa & Lugumba, 1971], where they could gainfully be employed (Many[13] people are employed in the agricul-

---

[12] According to an information of Konstantin Huber, former Austrian representative in Uganda.

[13] Here I had inteded to show figures on sectoral employment and GDP percentages. Numbers within the Ugandan Bureau of Statistics, the CIA-Factbook, and other relevant resources differ so drastically

tural sector). "*People have come to take book knowledge to be so important and to think that an educated person who has gone through school is so valuable and brittle that he cannot do rough/menial work*" ([Kwitonda, 1995], p.225). They consider themselves as educated and "*despise other members of the society whose skills are not based on academic abilities*" ([ibi.], p.225). This has lead to an examination-oriented[14] educational system, by which the development of moral and ethical values, sound physical health, practical skill, participation in social and cultural activities, etc. has been neglected, since they have not been influencing students' achievements [Kwitonda, 1995]. Furthermore, rare job opportunities induce "brain drain"[15] which means that highly qualified people leave the country to get better jobs in a more promising high–income countries [Mansell & Wehn, 1998]. This is not necessarily restricted to countries, but can also occur within a country, which is called a "local brain drain".

So Ssekamwas's question from 1971 remains yet unanswered: "*What kind of Ugandans should schools and University help to produce?*" ([Ssekamwa & Lugumba, 1971], p.61). In the context of this study it is assigned with the questions, "*Can ICT related education bring back Uganda's collaborative education or will it widen the gulf between so called "educated" and "uneducated"?*" The following chapters seek to provide an answer to these questions.

8.3 Summary

The general introduction to ICT & Education in LDCs formed the first part of this forerun chapter. Thereby, I presented the major stance which is that the introduction of ICTs demands educational reform. Most literature also supports the shift from "classical colonial" teacher centred methods to student centred ones, which I identified as actually more appropriate to former indigenous education. This chapter dealt also with the historical development and change of education in Uganda. Starting from the possible forms of education I described former indigenous education which

---

that I omitted them; e.g. the quoted percentage of population which is employed in agricultural sector varies from 40% to 95%! The situation is similiar according to the relevance of agriculture to Uganda's GDP - cp. [CIA, 2006]; [Uganda Bureau of Statistics, 2004], [Kwitonda, 1995].

[14] In this context I note that Uganda relies on a nationwide unit exam. Good marks are needed for entry to the next level of education. On the one hand this enables enrolement transparency, but on the other hand limits the focus of teaching. This represents a political instrument which controls content and orientation of subject matters. I know from own experiences at the school in Uganda that unit exams of the last years where analysed and assumptions about the upcoming exams were undertaken. Other relevant - perhaps more up to date - topics were neglected.

[15] The World Bank study, "International Migration, Remittances & the Brain Drain," found that from a quarter to almost half of the college-educated nationals of countries like Ghana, Mozambique, Kenya, Uganda and El Salvador live abroad in an OECD country [Schiff & Ozden, 2005]. The "brain drain"-phenomenon is widely discussed and pros and cons are juxtaposed in opposition. Whereas one group sees above mentioned negative effects of losing high qualified manpower, others claim that these expatriates send back huge sums of money ($ 650 mio. [Okee-Obong & Langthaler, 2006]). The Ugandan Parliament enacted a law which enables the export of high qualified labour forces in order to achieve this cash flow. The dark side of this stream of cash is that expatriates mainly support their families and a family related inner-national divide grows. Most activities are uncoordinated and not arranged and sunk costs interventions are the order of the day (cp. [Okee-Obong & Langthaler, 2006]).

was discarded by the Ugandan population by the arrival of the foreign colonial school system. Afterwards, I presented Museveni's current plan of UPE to eradicate poverty and its problems. Finally, I closed this chapter with critical aspects from Ugandan literature in which the colonial system per se was discussed as an inappropriate and misleading achievement to the country's development. All in all, this chapter outlined the sociocultural (education) schooling context of this case study. It described how colonial education has become that important and that parents sacrifice much to make a "proper" education to their children possible. The elaborations of this chapter will influence the empirical part with its discussion in several ways: one issue will be to show that though student centred and collaborative learning is promoted in literature and would be appropriate to (former) indigenous education, the students' expectation towards the way of ICT teaching is a teacher centred and authoritarian one. So it becomes an issue of discussion if ICTs will indeed "*serve as technologies of cultural imperialism, rather than as technologies of liberation and cultural diversity*" ([Ess & Sudweeks, 2001], p.266), when computers are introduced in primarily individual ways. Another issue of this chapter is related to education and contributes to the discussion about the extent the participants aim to present themselves as "educated" and if the fact to have computer lessons influences their hierarchical status. Furthermore, educational measurements of sanctions at the examined primary school involve corporal punishment and even fines which have to be expected if something is damaged. I will elaborate on the fact that instead of exploring the technology on their own by "trial and error", the pupils avoided to explore the technology on their own in fear of such consequences. This seems to be counterproductive for the acquisition of technological knowledge. If this is really true, will be clarified in Chapter 11.

Part II

Empirical Study - Preconditions

Chapter 9

The South-Western Region of Uganda

In a previous Chapter (7), I have discussed a variety of different factors related to the technological knowledge transfer in LDCs. I have depicted policies and some rough numbers on Uganda, the country of interest. However, as derived in Chapter 5 it is due to my theoretical stance that interactions of actors are shaped by certain socio-cultural constraints which have to be explored. This chapter aims to do justice to this claim. Nonetheless there is no single Ugandan culture and the name Uganda is just derived from a geographically central located former kingdom and largest single ethnic group [Nzita, Richard and Mbaga-Niwampa, 1993] named Buganda. When Uganda became a "country" (actually it became a protectorate) in 1894, more than 63 ethnic groups ([Okoth, 1995], p.xiii) with at least more than 33 different languages ([Nzita, Richard and Mbaga-Niwampa, 1993] p. iii) got united and restructured in 56 political districts - see Figure 9.1. Uganda's geographical and ethnical diversity is overwhelming and as geographical preconditions shape people's development (cp. [Diamond, 1999]) people from dry land developed other norms and values than those e.g. living in wet land. So it would be frivolous to describe "Uganda's culture" as a whole in one single Chapter. In fact, I limit my presentation on Kigezi, the mountainous geographical area around Kabale in south-western Uganda - the town where the case study took place.

In the following paragraphs I briefly depict and illuminate Kigezi both by its eth-nological (Section 9.1), historical (Section 9.2), and economical (Section 9.3) context as far as accessible and graspable to me as a Mzungu. Of course, such a description can never be considered as adequate and sufficient, but I hope to get the interested reader closer to the people's lifeworld and everyday lives.

9.1 The Bakiga

Generally, this case study focuses on the area's main ethnical group, the Bakiga, who are "descendants" of the Hamites[1]. In 2002 they represented about 8.3% of the

---

[1] Please note that this presentation is ovesimplified as no closed group like "the Hamites" exists. This depiction draws upon the literature of [African, Demographics, 2004].

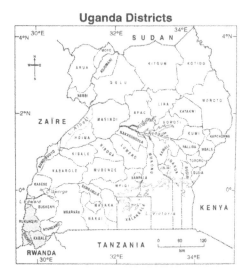

Figure 9.1: Uganda and its districts. [Nzita, Richard and Mbaga-Niwampa, 1993].

Ugandan population [African, Demographics, 2004]. Their language Rukiga, a Bantu-Language, was studied in detail by B. K. Taylor [Taylor, 1969] and in the late fifties Charles Taylor started implementing the Kiga Dictionary which contains more than 12000 entries [Taylor, 1985], [Hyman & Lowe, 1959]. The local cultural expert Festo Karwemera has written down an introduction to Kigezi's local languages Runyankore and Rukiga [Karwemera, 2000]. Furthermore, Bosworth's "Land and Society in the South Kigezi" [Bosworth, 1995] and publications by Freedman and Baxter provide solid information about the Kiga, its cultural manifestations and its context, [Freedman, 1976b], [Freedman, 1976a], [Baxter, 1960]. Beside these researches, there are several publications about the agricultural and economical situation of the Kiga [Chemonics International Inc. Washington, 2001], [Heidenreich, 1994], [NEMA, 1997], [NEMA, 2001], [NEMA, 2004], [Ngologoza, 1967], [Nzita, Richard and Mbaga-Niwampa, 1993]. Further socio anthropologistic studies like e.g. Espelands research on "The relation-ship between Bakigas and Batoros in Kyenjojo District, Uganda" [Espeland, 2003] or Luig's "Preliminary Observations on Kinship, Friendship and Voluntary Associations among the Kiga in Mulago" [Luig, 1968] are also available. All in all there is a variety of literature dealing with the Kiga in the extreme south west of Uganda.

It is not for sure when first settlers arrived in Kigezi, but certainly the Bakiga settled in Kigezi before 1500 AD. These Bakiga, Rwandan and Congolesan families, were structured in clans and probably looking for some fertile land. Kigezi's country-side is cleft and mobility is arduous, but at that time the country was basically quite fertile though subjected to some limitations on natural ressources, e.g. important raw materials like salt had to be collected from distant Katwe's salt lake which lasted a whole change of the moon[2] to get it. This journey coerced men to travel through dangerous wild life areas and thus clans bundled their forces by blood brotherhood. Usually clans lived separately, and as the countryside is quite hilly *"one can say that each clan was on his own hill"* [Karwemera, 2003]. A clan, it is estimated that by the turn of the century there were about 180 clans in Kigezi highland ([Bosworth, 1995], p.31), consisted of several lineages and political authority rested in the hands of the lineage leaders, commonly the elders [Nzita, Richard and Mbaga-Niwampa, 1993]. A man was not allowed to marry from his clan and traditionally, no marriage could be honoured without the payment of bridewealth, which was paid by the man's father. Actually, *"the Bakiga are a very polygamous society; the number of wives was only limited by the availability of land and bridewealth obligations"* ([Nzita, Richard and Mbaga-Niwampa, 1993], p.57). An interesting fact is that on the opposite divorce was also quite common among the Bakiga. A wife or husband could bring in a petition for divorce with reasons like barrenness, laziness or other matters of misunderstanding. This would-be instances of divorce were settled by the elders. Consequently, bridewealth decreased as the woman was no virgin any longer, which was highly esteemed. If a girl got pregnant before marriage, *"she would either be taken to a forest and be tied to a tree and left for wild animals or, she would be tied feet and arms and thrown over a cliff"* ([Nzita, Richard and Mbaga-Niwampa, 1993], p.60)[3]. Other anti-social activities were also heavily punished. Among these were stealing, murder, sorcery, blocking paths, and night dancing. According to Nzita, a murderer was buried alive in the same grave as his victim [Nzita, Richard and Mbaga-Niwampa, 1993]. This changed when the external religions like Christianity and Islame were introduced. *"The exponents of these religions created no room for religous compromise"* ([Mwanahewa, 1995], p.101) and for the moment, as mentioned above (see Section 8.2) the logic of *either/or* worked (instead of *fuzz*), which *"deprived a large section of Ugandans of their natural freedom to embrace, develop and consolidate the traditional religions which were engraved in their hearts"* ([Mwanahewa, 1995], p.101). As discussed elsewhere (Section 8.2) religion and morality were intertwined. Consequently the external cultures enabled (through its tripartite in nature) that anti-social activities like steal-

---

[2] Up to the arrival of colonialists the indigenous population orientated by the lunar cycle. Daily life, which was not divided into distinct timeframes, was interrupted by happenings and events as weddings were. A day began at dawn (about 6 a.m.) and ended at sunset, e.g. the 6th hour of the day was noon [Karwemera, 2003]. Sometimes this sense of time is still prevalent when one arranges a meeting. Time is counted differently also in Kiswahili and the start of the daily time system is at dawn.

[3] This asks to discuss the social role of women among the Bakiga. Obviously, power is traditionally distributed in an unequal way, although women have the possibility to get divorced, their *"value"* is estimated and settled by others. Factors like virginity and bridewealth obligations are utilizied to decide on their future, but not the women themselves. This indicates a lower social status than men. However, the colonial education system introduced women in the role of teachers which indicated a change in the role (and power) of women. Traditionally, the investigated school was managed by colonial Headmistresses and even nowadays the school is also managed by a local woman. Generally, the social status of a teacher is high (see Page 102) and at the investigated schools the numbers of male and female teachers are the same, which shows that the educational system has obviously also changed the relationship between man and women. As I will present on Page 112 power relations did not only change by the educational shift but also by the introduction of money.

ing could be regarded as *"exclusively legal or seclusively moral or exclusively a religious felony"* ([Mwanahewa, 1995], p.102). Mwanahewa considers this as the begin when central authority of a traditional elder began to be undermined. This coincided with the construction of schools, which claimed to bring education to the "uneducated" indigenous population. Nevertheless, many Ugandans are uncomfortably divided and cultural disorientated. Someone may be a constant churchgoer while secretly at night he goes to worship *Nyabingi*. In those times, *"the Bakiga believed in a supreme being,* Ruhanga, *the Creator of all things earthly and heavenly. At a lower level, they believed in the cult of* Nyabingi*" [Nzita, Richard and Mbaga-Niwampa, 1993]*, which was considered to be a satan like cult by the colonial emporers. This is contradictive to Festo Karwemera, who emphasizes the "harmlessness" by the citation of a typical prayer, which he had learnt from his mother: *"As I go and now, keep me safely, protect me from deseases, until I come back safely"* [Karwemera, 2003].

9.2   Historical Development

Kigezi consists of the present three districts of Kabale, Kisoro and Rukungiri and constitute the very South-West of Uganda - see Figure 9.1. It covers around 5000 km², is circumvented by high mountains (above 4000m) and is placed on an average altitutude of 2000m above sea level. This lead to the nickname "Switzerland of Africa" and implied limited mobility and hampered trade with far-off markets since ever. The first colonialists entered this territory in 1891 and the Muganda[4] Joash Sssebalijja lead surveyors to Kigezi not before 1908. Due to the prevalent local social structure of clans and lineages each clan was reigning over its own territory and no cities like Kabale existed yet. Kabale was just an important trading point between Rwanda, Congo and Uganda's lowland, and means "smallstone" [Ngologoza, 1967], [NEMA, 2004]. Not before the British administration decided to establish its administration office at the location Kabale, a small town developed [Ngologoza, 1967].

In 1911 the roman-catholic missionary Yowana Kitagana founded the first Missionary's department and church. Already in 1913 Anglican missionaries followed with the construction of their own church. Both communities touted for new believers and struggled against each other. A thing they had in common was that they denied former religious aspects and named former "small gods" a satan like religion[5]. Of course, these religious struggles were not limited to Kigezi and happened all over Uganda. Nevertheless, the spatial diversion is still evident, because until today in Kabale town there is a Catholic hill which is strictly separated from the Anglican hill, both operating their own educational and clerical institutions [Karwemera, 2003]. Some Muslim schools were also built, but *"it seems that the Islamic religion had not attracted as many people, as did other religions, and even now there are few Muslims in Kigezi"* ([Ngologoza, 1967], p.74). In 2002 44% named themselves as Roman Catholics, 53.5%

---

[4]In Bantu languages the singular form is formed by the prefix "Mu" and the plural by "Ba". Therefore one person of Uganda becomes a Muganda and more Baganda. The same rule has to be applied on Kiga: Mukiga, Bakiga.

[5]In 1912, the Witchcraft Ordinance was enacted. Anyone who was accused of practising "witchcraft", which could be interpreted to mean Nyabingi religion, coluld be arrested and detained ([Bosworth, 1995], p.40).

as believers of the Ugandan Church (Protestants) and only 0.7% as Muslims [UBOS, 2002].

In those times the land of Kigezi was still fertile and in 1931 there were 226,214 Bakiga, 33 Europeans, 86 Indians und 6 Arabs in Kigezi [NEMA, 2004]. Despite the fact that almost the whole population constisted of Bakiga, Kigezi was not reigned by its indigenous population up to 1929. Bakiga people kept to prefer to settle in clan structures and it was untypical that one Mukiga should jugde upon another ([Ngologoza, 1967], p.82).

Population kept growing and population pressure was increasing, because fertility was high and after a famine in neighboring Rwanda in 1927 starving people migrated to Kigezi's fertile area [Ngologoza, 1967]. As a consequence swamps (peat soil) were drained to achieve more fertile land. Though, plants like sorghum grew less and nutritious vegetables vanished [Bosworth, 1995]. Nevertheless, the population increased and 1949 population pressure was insomuch high (a population of 419,588 meant a density 85 heads/km^2, which was highest in Uganda) that the first migrations from Kigezi to less populated areas in Uganda[6] impended and were arranged by the local administration. [Ngologoza, 1967], [Bosworth, 1995]. This aimed to prevent from upcoming famines. Another reason for the dense population was that besides a high natural fertility and migrants from neighbouring Rwanda, additionally local chiefs (leader of clans or lineages) attracted people from other areas. This was due to the fact that a chief was paid by the people, and the payment varied according to the number of people a chief was administering ([Ngologoza, 1967], p.76f). All together from 1946 until the end of 1960 the total number of official registered migrants was 42,108 ([Ngologoza, 1967], p.99). Yet, British and Baganda held superior administrative positions and not until 1964 "Rutakirwa Engabo ya Kigezi" (the shield of Kigezi) was elected as the constitutional head of Kigezi district. Two years after Uganda's independence from Great Britain and about 50 years after the take over by the Europeans. Under the dictatorship of 70's and 80's Uganda's Asians were expelled[7], although they run four out of five businesses and contributed much to the GDP of Uganda ([Allen, 2000], p.327) - see Section 9.3.3. According to Baker, the Madhivani family had established an industrial empire (all over Uganda) that became central to Uganda's economy, which contributed over 18% of GDP in the 1970s [Baker, 2001].

Taking a look on the diary of Sir Peter Allen, Chief Magistrate and Chief of Justice of Uganda from 1955-1986 makes evident that in this time there were also outstanding high numbers of rapes and defilement reported ([Allen, 2000], p.359). The reasons for these reported[8] crimes remain unclear, but becomes an interesting fact as it is one of Allen's rare comments on Kabale. When Museveni took over, economic situation was still difficult and the country was lying down. When neighbouring Rwanda was haunted by its terrifying genocide, Kabale reported some profits by illegal cross-border trade, as the district borders to Rwanda (25 km). Since 1991 Kabale's (without Rukungiri and Kigezi) population increased from 417,218 to 458,107 in 2002, which is a

---

[6]In 1969 Uganda's population density was 48 [Uganda Bureau of Statistics, 2004].

[7]"*Amin's regime fanned negative ethnic and xenophobic sentiments in 1972 when he expelled over 60,000 Asians by giving them 90 days' notice*" ([Baker, 2001], p.6).

[8]It is important to notice that Sir Peter Allen refers to reported crimes, as Uganda's household surveys have shown that only about every second crime is reported [Ministry of Finance, 2001].

population density of about 281 heads/km^2 [UBOS, 2002]^9 ranks "*fifth highest district in terms of in terms of population density and high incidence of fertility of women at 35 years and above*" ([NEMA, 2005], p.10). As a town Kabale municipality is the most densely populated area within the district with about 620 persons/km^2 [NEMA, 2004] and has a population of about 41,500 [UBOS, 2002]. Actually, 55,000 inhabitants belong to the growing urban area [Uganda Bureau of Statistics, 2004]. Generally, Uganda is getting urbanized (3.8% in 2002 [UBOS, 2002]) and taking a look at the economical situation will show that increasing income is to be expected in urban areas. By a look at Kabale's fertility rate [Myuganda, 2004] shows that this situation might not change soon as even Kabale's fertility rate of 7.2 is above the Ugandan average (6.9).

9.3   Economical Situation

9.3.1   Money and Labour Migration

By the British manner of indirect rule Baganda were put in to take over minor administrative tasks: by the introduction of unit taxes in 1915 it became task of the Baganda to collect taxes. At that time money had been recently introduced and rare. The Baganda utilized this fact and pretended awry quotas. They claimed that one goat is equivalent to one rupee instead of three. In northern Buganda even better results could be obtained and one goat was sold for six rupees. Many Bakiga could not cope with such high taxes and therefore male family members left their clans and offered themselves as labour forces in Buganda. Thus, their families were freed from the tax burden ([Ngologoza, 1967], p.64). According to Joanne Bosworth's study 50% of male villagers were engaged in labour migration during the 1950s and 1960s. This lead to the division between men as migrants or traders and women as subsistence cultivators. When public enterprises and private concerns broke down, women were drawn further into cash economy. They had to sell their products on the market and thus "*the distinction between subsistence farming and the sale of crops for cash was eroded*" ([Bosworth, 1995], p.225f). It was a dilemma: on the one hand crops were too low to nourish all family members sufficiently and on the other hand incomes from sold products were too little. Bosworth sees the consequence thus: "*The sale of labour has now become unavoidable for many households. This introduces an arena for struggle between the control of those resources acquired through waged labour force participation and through the family labour pool*" ([Bosworth, 1995], p.226). Thereby, cash became more important and "*has become the object of new divisions of resource control*" ([Bosworth, 1995], p.226). During the Rwandan warfare from 1990-1994, Kabale district^{10} profited through its nearby border by the illegal cross-border trade ([Bosworth, 1995], p.122). Despite the fact that during that time more households got engaged in trading activities and labour migration got temporarily less, young and educated men failed to find an opportunity to put their education to use in employment and investment and left Kabale. This situation has not changed much as

---

^{9}In 2002 Uganda's overall population density was 124 [UBOS, 2002].

^{10}As Kigezi grew, Rukingiri which was part of Kigezi was made a separate district in 1974. Kisoro became an own district in 1991, too.

Kabale's sex ratio[11] of 88 is second lowest[12] in Uganda and indicates that for 88 men there are 100 women in this area [Uganda Bureau of Statistics, 2004]. Rural people are more affected by this phenomenon where the sex ratio comes to 86, whereas in urban area it is 99. Thus, the emerged readiness to leave Kabale influences the town's development even today. Economic possibilities in Kabale are rare and therefore they leave the district and move to bigger cities to support their families financially. This is also due to the fact as educational facilities are better in bigger cities [Okee-Obong, 2003]. Nowadays, one can find many leading politicians and business men in Kampala from Kabale district, which can be called a "*local brain drain*" [Okee-Obong, 2003].

9.3.2 Agriculture

The Bakiga were basically agriculturalists growing mainly sorghum, peas, millet, and beans, which were mainly used on a subsistence level [Nzita, Richard and Mbaga-Niwampa, 1993]. Bananas were introduced by the Baganda administration and most important crop nowadays are potatoes[13]. Until today a high percentage of the population of Kabale is working within the agricultural sector (see Footnote on Page 102). Cattle ranching is due to narrow fields limited to subsistence level, nevertheless cows, goats, pigs, sheep, and poultry are common. In average one farmer can plow 1.2 ha, but due to the country's fragmentation, fields are dispersed (5 km distance is common) and one field covers in average only 0.4 ha [Bosworth, 1995]. The soils of Kabale District (~1.827 km^2) are mainly volcanic, ferralitic and peat soils. The volcanic soils are mainly found in a sub-county in Kabale District. The ferralitic soils are the most widespread in the district and are in the advanced stage of weathering and have little or no mineral reserves to draw on. Peat soils are found in the leftover of swamps. Soil around Kabale is quite sour (SOIL PH 4.7-6.4)[14], and has been worsened by the leaching of the soils in the area [NEMA, 1997]. The country's clefts have also hampered mobility since ever and therefore it has meant a great effort to sell products on bigger markets. In the 70's and in the beginning 80's the national Kigezi Cooperation Unit organized collective transports of goods to the 400 km distant capital. This non-profit organization was victimized to liberalization efforts. Therefore the ambition to sell potatoes via a bad infrastructure on markets on which hardly higher prices can be achieved is naturally low. The expensive transport to bigger markets leads to increasing local consumption [Okee-Obong, 2003]. Nowadays, bad stocking possibilities, inpredictable weather and soil erosion worsens the regional economical situation. 90% of Kabale's land area is affected by soil erosion, which is mainly caused by slopes, population pressure, deforestation, poor farming, and vulnerable soil [NEMA, 2001]. Estimations proceed on the assumption that the area of cultivable land decreases from 0.27 ha/head in 2000 to 0.19 ha/head in 2010 meaning a loss of 30% [Chemonics In-

---

[11] The sex ratio defined as number of males per 100 females is an index for comparing the numerical balance betweeen the two sexes.

[12] Only Kisoro, which was part of Kabale until 1991 has an even lower one (83) [Uganda Bureau of Statistics, 2004].

[13] In 1993 more than 50% of yielded crops in Kabale were potatoes (sweet and irish) [NEMA, 1997].

[14] Most of the nutrients have their maximum availability in the ph-range of 6.0 - 7.5. At ph-values below 5.5 primary and secondary nutrients become less available especially phosphorous [NEMA, 1997].

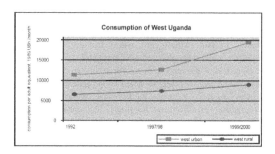

Figure 9.2: Consumption of West Uganda. Adapted from [Chemonics International Inc. Washington, 2001].

ternational Inc. Washington, 2001]! Considering the fact that one person needs to cultivate 0.3 ha/head to achieve food self sufficiency underlines this worrying situation ([Bosworth, 1995], p.123). Some might suggest to clear woodland to generate fields, but only 470 ha woodland in Kabale district are left, which is the smallest area in whole Uganda [Chemonics International Inc. Washington, 2001].

Figure 9.2 shows the development of consumption of Western[15] Uganda from 1992 to 1999/2000 and compares rural and urban areas. This graphic depicts nominal, inflation-adjusted data on the price basis of 1989. It clearly shows that whereas rural area expenditures increase slowly, available household expenditures of urban population grow faster. However, there has been a reversal of trend in the period from 2000 to 2003 when Western urban area, in opposite to the rest to Uganda registered a significant decline of about 19% [UBOS, 2003]. At the price basis of 1999/2000 Western rural area's monthly real mean per capita expenditure increased from USh 21,500 in 19990/2000 to USh 22,500 in 2002/2003. Urban area decreased from USh 58,700 to USh 47,500. Nevertheless, there is huge income disparity between rural and urban areas. In Kabale, generally one can say that agricultural situation is at best stagnating - household incomes within agricultural households has remained the same from 1991 to 2001 [Appleton, 1998], [Chemonics International Inc. Washington, 2001], whereas Kabale's urban area household expenditures are four times of those on the countryside [Uganda Bureau of Statistics, 2004]. Evidently, it is more attractive to look for employment in town than to remain a farmer. Day-laborer can be hired for $ 0.8 a day, whereas a doctor charges $ 6.4 per visit, which underlines income disparities.

9.3.3 Trade

At the end of the 19th century Ugandan Asians (Gujarati or Hindu) arrived to work for the railway line and by 1969 about 70.000 Ugandan Asians were counted. Historically,

---

[15]The western region consists of the districts of Bundibugyo, Bushenyi, Hoima, Kabale, Kabarole, Kamwenge, Kanungu, Kases, Kibaale, Kisoro, Keynjojo, Masindi, Mbarara, Ntungamo, and Rukungiri.

trade was conducted by Uganda's Asians, which had been the only chance for them to do their own business. They were regarded as "foreigners" and therefore had no rights to acquire land, which was entitled to formal Ugandans. Thereby, they were forced to live in urban areas - "*a factor that caused them to be thought unwillingly to integrate with other Ugandans*" ([Baker, 2001], p.26). They started to concentrate on trade and commerce, gained control of retail and wholesale trades, including coffee and sugar processing (ibidem). During Amin the Ugandan Asians were expulsed and expropriated. More than 20 years later, under the new government of Museveni, some Ugandan Asians returned and claimed compensation for their dispossessed properties, e.g. buildings, estates, and factories. Museveni relied on Ugandan's Asian to play a crucial role in Uganda's economic revival. Nevertheless, Baker reports a "*unwillingness on the part of the general public to end all forms of discrimination and exclusion against these groups* [Ugandan Asians and Banyarwanda]" ([Baker, 2001], p.20), which impedes on this hope. Due to the fact that no Figures exist on their percentage of local population[16]. To my personal perception, when one walks through Kabale town many businesses are run by Ugandan Asians. Especially imported goods from MNCs, like glue, paper, scissors, tapes, batteries, rice, or sirup, are almost exclusively found in stores of Uganda's Asians. The combination of running businesses in urban areas and living there induces the notion that this group has bigger household expenditures than other ethnical groups. To my notion prominent hotels, restaurants, groceries, and telecommunication shops are controlled by a few big families by which tourists[17] and other well situated Ugandans are obviously attracted. Thus, only a limited group profits from this currency import and cash flow. Whereas it is easier to those who have already some property to increase it by taking out a loan, chances for day-laborer in town are limited. Most try their luck by buying a bicycle and offer their services as "Boda-Boda drivers[18]". For a ten minute ride one has to pay about $ 0.07, which is still more than working on a construction site, but does not enable property building leapfrogs, which would mean more social security.

9.4 ICT Related Facts

Until now, computers work only with power, to be more specific, electricity. Currently, 99.63% of the population of Kabale district depend on wood fuel for domestic puprposes and mere 0.05% on electricity or gas. 0.11% use paraffin and 0.21% use other sources of energy ([NEMA, 2005], p.10). Kabale retrieves its electricity from the Owen Falls Dam via Kampala to Kasese then to Kabale through the 33KV line from Kasese. Another transmission is from the Maziba power station in Kabale district.

---

[16] In Uganda's household surveys no figures on Uganda's Asians are available as this ethnic minority is categorized among "Other Ugandan" [Uganda Bureau of Statistics, 2004], which is critizised by [Baker, 2001]. The interest reader is encouraged to read Baker's essay on ethnic minorities in Uganda "Uganda:The Marginalization of Minorities" [Baker, 2001].

[17] Kabale aims to improve its economical situation by the promotion of its touristic attractions to national and international tourists, which are the famous mountain gorillas in the National Parks of Bwindi Impenetrable Forest and Mgahinga or nearby Lake Bunyoni.

[18] Actually, "Boda-Boda" stands for the transport from "border-to-border". It means the utiliziation of a bicycle as a taxi service.

Sometimes the district experiences power failures, which is mainly due to the long old lines between Kampala and Kabale. As Kabale is at the end of the line, any disturbance along the line affects electricity supply to Kabale [NEMA, 2004]. There exist power generators, but to my experience power supply remains insecure as sometimes no fuel for power generators is available. Interestingly, those with access to electricity in 1992 were predicted to have income growth of between 2 and 4.5 percentage points higher than those with no access. Electricity is supposed to interact with (formal) education in enhancing income growth. [Ministry of Finance, 2001]. Figures on available (private) computers in Kabale seem not to exist. Regarding public Internet station Kabale was among the first towns where the IDRC initiated the establishment of Multipurpose Community Telecentres by African highlands Initiatives [Mugisha, 2002]. Thus Kabale got, together with the local post office, connected to the Internet. A further World Bank project "Kigezi High School Based Telecenter" in 2000 [Mayanja, Meddie, 2003], with seven computers followed. Some business men started teaching computers as well and offered Internet access. Though, rumour had it that those got in financial trouble, as online fee escalated. This indicates that ICTs might be seen as a new trap of debts [Okee-Obong, 2003]. In general the basis Internet account in Kabale in 2002/2003 was $ 20 per month for a basic e-mail access and $ 30 per month for web-access. Nonetheless additinal fees arised as one has to dialup via land line in Kampala and additional $ 11.4 per hour had to be paid (on working days)[19]. Other maintainance costs are equivalent to those mentioned in Section 7.4.3. In 2003 the Radio Voice of Kigezi opened a new center and a user costs decreased to a fee of $ 4 per hour. In comparison to mobile phones this is similar as one hour of talking is about $ 10[20].

9.5   Summary

This chapter presented the historical and economical development of the case study's main ethnic group "Bakiga" to explore the sociocultural environment and social constraints which come into operation in this case study. I discussed that the history of Uganda's South-West area called "Kigezi" was crucially shaped by the arrival of colonialists. I investigated both on the environmental (pre-)conditions and the problem of overpopulation which resulted from high natural fertility, migration from bordering Rwanda and the attempt of lineage leaders to increase their power by attracting more people. I showed that both the mean "tax-trick" under the colonial and Baganda adminstration and the countryside's overpopulation, which is dated back as early as in the 1930's, lead to a wicked economical situation. As an effect of colonial education, agricultural activities had become unattractive and due to soil erosion crops were (and are still) decreasing. This is confirmed to a certain extent by the fact that household expenditures of people living in Kabale's urban area are four times of those living in agricultural areas. But as both trade was increasingly controlled by Uganda's Asians and agricultural activities remained worrying and unpopular, male workers left the

---

[19] On weekends it was cheaper and amounted for $ 3 per hour

[20] According to the website of the Ugandan service provider MTN (www.mtn.co.ug). Accessed: 5.12. 2006.

countryside to look for better jobs. This tendency to leave the countryside is called "brain drain" and is still evident, which I underlined by statistical figures. As explained earlier, by the arrival of colonialistic education, "educated" people started to consider agricultural jobs as minor ones and in this context I claim that formal education is gaining on importance as influenced by the vision of "brain draining" a Mukiga is convinced that the more he/she is (formally) educated, the better job he/she will get and the better earning he/she can yield.

At the end of the chapter I listed some ICT related key figures on the intervention's district to induce that Internet access is currently quite expensive at least for privates and indicates that ICTs carry the risk to become a trap of debts. This information on the local circumstances within Kabale district will be put into context in the subsequent Chapter 12, which discusses the frame of the intervention in detail.

Chapter 10

Setting of Intervention

The following sections explain the intervention's background and list some facts on the intervention itself. Statistical figures are named and technical details are given about the established computer lab itself. Please note that certain issues are named in this chapter, but are discussed in detail in the subsequent analytical part of this thesis.

10.1 School's Background and Participants' Constellation

The school where the intervention took place is a primary school in Kabale and was founded in 1933 by missionaries. The school offers, according to the Ugandan school system, seven different levels (P1-P7) with one class per school level. In 2002, 287 pupils aged between 6 and 13 years attended this school, which makes a pupil per class ratio of about 40. This is quite low compared to Uganda's benchmark of 75 [Cisler & Yocam, 2003]. As a matter of fact, classes with younger pupils show higher ratios (about 45 in P1) than higher ones (about 35 in P7). Together with 18 members of staff my colleague and I instructed about 300 people in the basics of computing. The teaching staff consists of ten full time teachers and one Headmistress. The remaining staff members are grouped into two secretaries, one purser and four matrons. Actually, talking about the participants' constellation implies presenting statistical figures as well. This would mean that the average staff member is 29-years old and a Mukiga by ethnicity. Though, these figures cannot be taken for granted, as contradicting statements were made. To give an example, one female informant claimed to be 26-years old, although I figured out that she has to be about 40. According to an informant such misleading statements are not uncommon for unmarried women. They pretend a younger age in order to increase the chances of getting married. Another informant pretended to be a Mukiga although her son informed me that she was a Banyarwanda. Though it might have been interesting to correlate such figures to the performed interviews, I leave it with personal statistical figures. In 2003 all members of staff who had participated in 2002 and were still working for the school were interviewed. This was a total of 14, five men and nine women. Out of 14 interviews pupils between the age of 10 and 13 four were male and ten female. As mentioned above an adult interview lasted 41 minutes on average, a pupil's interview took 21 minutes.

10.2   School Infrastructure and Financing

The school offers a boarding possibility for 72 pupils and is usually full. 2002, the boarding school fee was $ 166 per term[1] per child. The maintenance of the school is provided by public means and the school fees have to be additionally paid by the pupils parents. Payments by the government seemed to be unreliable and late. In Summer 2002 teachers' wages were not paid since March 2002. Like in many other schools, pupils' parents contribute to the school's maintenance. Due to the fact that several parents hold highly appreciated (local) occupational positions (lawyers, head of district, physicians, customs officer etc.) the school seems to be an elitist school at least in terms of financial means. In Spring 2003 the school started to construct a new dormitory with a three-year lasting budget-plan of $ 22,000. According to the Headmistress this project is mainly sponsored by pupils' parents and just marginally[2] by the government. Thereby it becomes evident that the parents and its representational institution, the *Parents Teacher Association* (PTA), form a powerful body and drastically influences the school development . Even the whole intervention was contrived by a former member of the PTA who knew Projekt=Uganda through some earlier donor activities. From the very first day the intervention's plans, curricula, and further proceedings were coordinated between representatives of the PTA, the school's administration (Headmistress and Deputy Teacher), who together form the *school management committee* (SMC[3]), my colleague, and me.

Apparently, this school can raise sufficient means to maintain a computer lab, if prioritized and mandatory. For an intervention the financial preconditions seem to be promising, as sufficient financial means to maintain a computer class cannot be taken for granted for such interventions, as Nel reports in his case study on ICT related intervention about sincere financing problems like paying the electricity bill [Nel & Wilkinson, 2001].

10.3   Computer Equipment

The equipment consisted of 13 Pentium 100 with two hard disks with a ghost image of a standard installation on it and one Pentium II 800, fourteen 17" screens plus ten laser printers. Several spare parts were also available. The software was MSWindows 98 with MSOffice, some music files, games and further utility programs. The installation and configuration of the computers was prepared by an Austrian institution which is

---

[1] One Ugandan school year consists of three terms á 14 weeks.

[2] Unfortunately, I could not figure out what "marginally" means.

[3] "*Each school is required to have a SMC, which are charged with providing overall direction to the operation of the school, ensuring that the school has a development plan for ensuring quality education within and outside the classroom, approving the school's annual budget, monitoring the school's finances to ensure that they are properly used, linking the school to the community, promoting harmony among the head teacher and members of the staff, ensuring that teachers and parents do not cause undue psychological stress on pupils or cause them to withdraw from school, and liaising with school foundation bodies on the best way of utilising foundation resources for promoting school objectives and goals*" [[Kiyaga-Nsubuga, 2005], p.14]. The SMCs have also a reporting function to the pupil's parents and to community leaders about financial adn operational status of the school [ibi., p.15]

specialized on the handling of development interventions, but not on ICTs. Therefore the configuration was predetermined and not coordinated to the local needs[4]. Due to power instability and lack of voltage regulators at the beginning, screens burnt and after three months four were broken. In the meanwhile (from 2002 to 2003) the school had organized two new screens to keep the lab running. Nevertheless, when I returned in Summer 2003 ten screens were reported to be broken though the school had provided power stabilizers. Electricity problems are regionally common and cause constantly damages. To give an example, one building reported one blowing bulb per month. Though the screens seemed to be affected by the same damage it was not possible to repair them, as the potentially damaged transformer could not be replaced. Investigations retrieved the results that the donated screens are prone to power instabilities, which underlines the impression that the intervention was not planned very well.

10.4   Intervention's Curriculum

After the arrival in July 2002 the school provided a lockable computer classroom which was guarded during the night by two watchmen. As soon as we arrived, computer lessons were integrated into the curriculum. However, at first, it was not possible to teach practical usage as new screens arrived three weeks late. Therefore, we were focusing on theory in the meanwhile. Figure 10.1 shows relevant theoretical topics and Figure 10.2 the practical ones which were taught in each class throughout the whole first period of 13 weeks. To reduce the pupil per computer ratio classes were split into two groups for practical lessons. Figure 10.2 shows that due to several other activities (e.g. Sports Day, Examination Day) progress differed from group to group (e.g. the second group of P5 covered more word topics than the first one). Whereas pupils were mainly taught during the day within the standard time table, we offered lessons to the teaching staff and non-teaching staff in the early morning from 7 a.m. to 9 a.m. and from 3 p.m. to 9 p.m..

   All in all eight different scheduled lessons[5] were offered each week for the staff. In an initial meeting with the teaching board everyone agreed to participate three times a week in order to allow a collective proceeding within the syllabus. Though, attendance varied a lot and was far from being homogenous throughout the staff. We made no extra arrangements with the matrons, but usually they had some time for free during the early afternoon, which they used about three times a week. Secretaries used computers from the very beginning on a daily basis in their office. Within a short time (about one week) they needed our support just about half an hour a day. Our help was mainly asked for trouble shooting. The advises we gave were noted down pedantically. They used these notes to deal with problematic situations on their own. The workload of a teacher[6] at this primary school means 23 hours of teaching and

---

[4] At least one member of the PTA has profound computer knowledge and could have specified the school's needs.

[5] One lesson lasted between one and three hours, depending on the topic discussed and tasks given.

[6] At this point it has to be mentioned that at this school the pupils have up to 50 lessons á 40 minutes a week!

Topics-theoretical	P1	P2	P3	P4	P5	P6	P7
What is a Computer?		X	X	X	X	X	X
How can we use a computer?		X	X	X	X	X	X
CAR/Computer	X	O	X				
Parts detailed			X	X	X	X	X
Motherboard					X	X	X
CPU				X	X	X	X
Hardware/Software					X	X	X
Connection/Network						X	X
Plugs						X	X
BIOS					X	X	X
OS					X	X	X
Mouse-detailed/usage			X	X	X		X
Internet/History						X	X
Internet/Builtup						X	X
10011-Arbeitsweise							X
Storage devices				X		X	X

Figure 10.1: The touched theoretical topics. [Schneeberger & Sedlacek, 2002].

some additional duties: Looking after boarders in the evening and even sometimes at weekends, giving remedial teaching and stepping in for colleagues who have lost a relative. Especially, the attendance and preparation of funerals makes a teacher absent for several days. In 2002, it happened four times in three months. In addition to the teaching staff, matrons, and secretaries, one member of the PTA was addressed by the PTA and school administration to be in charge of the lab. He had a lot of basic knowledge and in extensive weekend sessions he was instructed on a more hardware oriented level to repair plugs, the LAN, set the proper jumper settings of additional hard disks, etc. He was also supposed to do the computer lessons after our departure. Unfortunately, a few months after our departure in 2002 he stopped his teaching activities (although he got paid by the school), failed to show up at scheduled teaching hours and was finally dismissed by the school management (i.e. the school's administration and PTA) - during the interviews this teacher is named Mr. Rukinga. My colleague and I had intensively instructed another trusted person (the son of a teacher) who had also acquired "deep" computer knowledge. Although, it had been arranged that he will support the school in case Mr. Rukinga is not available, he neglected his agreement and did not take care of the lab. As a consequence to these happenings, the school appointed one of the permanent teachers to become the new lab administrator and offered her to take courses in town which she did. During the second period of the intervention in 2003, I was teaching mainly her and two other colleagues, who had also acquired good computer knowledge. This indicates that the school might have been taught a lesson by the unreliability of the son of the one teacher and Mr. Rukinga. Both had apparently better jobs in town than to teach at the school. Obviously, the school perceived this as a form of a local brain drain and aimed at disseminating computer knowledge on several, socially integrated, teachers to become less dependent on them.

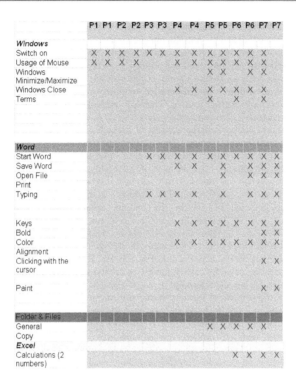

	P1	P1	P2	P2	P3	P3	P4	P4	P5	P5	P6	P6	P7	P7
Windows														
Switch on	X	X	X	X	X	X	X	X	X	X	X	X	X	
Usage of Mouse	X	X	X	X			X	X	X	X	X	X	X	
Windows Minimize/Maximize									X	X			X	X
Windows Close							X	X	X	X	X	X	X	
Terms									X		X		X	
Word														
Start Word					X	X	X	X	X	X	X	X	X	X
Save Word							X	X		X		X	X	X
Open File									X		X	X	X	
Print														
Typing					X	X	X	X		X		X	X	X
Keys							X	X	X	X	X	X	X	X
Bold													X	X
Color							X	X	X	X	X	X	X	X
Alignment														
Clicking with the cursor													X	X
Paint													X	X
Folder & Files														
General									X	X	X	X	X	
Copy														
Excel														
Calculations (2 numbers)											X	X	X	X

Figure 10.2: Practical topics at the Primary School. [Schneeberger & Sedlacek, 2002].

Previous computer knowledge and handling of acquired knowledge

To get hold of previous existing computer knowledge is difficult. To my knowledge some members of staff had taken some private computer lessons in town. Though, those performing good, were not necessarily those who had attended computer lessons in advance. Some of the pupils had a computer at home, though they hardly had access to it. In the analytical chapter (see Chapter 11) I will deduce relevant arguments for this fact.

When the first stay reached its end, the participants asked for certificates which my colleague and I issued. As people were proceeding differently, we listed those topics which we considered that each participant had covered; e.g. MSWord 97, Introduction to MSWindows 98, etc. Our decision was based on immanent testing and notes we took on covered topics. We did not make an explicit test. In a final meeting the certificates were handed out.

Internet

At the beginning, it was intended to connect the school to the Internet which was finally achieved. Though, initially our aim was to participate in a satellite Internet connection of a nearby (100m) educational institution via a wireless local area network (*LAN*). The installation was donor funded as well and equipped with four separate IP-addresses with a satellite uplink. Unfortunately, our efforts to collaborate were in vain, as according to the responsible administrator of this institution, it was not allowed to connect further routers. Therefore we had to use the landline at the PC of the Headmistress's office and established a LAN to the neighboring computer room. In the meanwhile we used the account of a familiar institution which allowed us to use its account in the evening. Thus, "only" the connection fee to Kampala had to be paid. After some theoretical discussions we had established a modem sharing and were using it simultaneously with most of the teaching staff at once. After ten minutes the computer class was almost empty and after half an hour no one out of ten people was still in the computer room. This incidence is discussed among the intercultural phenomena in Chapter 12. During the communication which took place in advance to the second period in 2003 all the communication was done via the same pupil's parent who had initiated the intervention, but not by the school itself. After my return to Kabale in 2003 I found the LAN between the Headmistress's office and computer class room broken (as the rest was and unfortunately Mr. Rukinga had left the place without fixing it). After an additional training I asked the new administrator to reestablish this connection, but she refused to do so, as the Headmistress was fearing that pupils and other teachers could spy out marks and teacher assessments (see Chapter 11). Nevertheless, the PC in the Headmistress's office was still connected to the Internet via the landline. As a consequence I focused my Internet teaching in 2003 on the Headmistress and the computer lab administrators to allow e-mail communication with other schools and with the Projekt=Uganda. Before I left, at least three people were able to connect to the Internet and could send e-mails on their own, which we trained quite intensively. However, no e-mail communication has been undertaken via the school's free e-mail account since my return in 2003. Further thoughts on this fact will be discussed later on.

10.5 Summary

The aim of this chapter was to present the setting of the intervention which took place in 2202 and 2003. I showed some figures on the school and depicted the participants' constellation. I listed a few figures on the school's financial situation and infrastructure. Thereby, I concluded that the school seems to has sufficient financial means to maintain the computer lab on their own. This presentation was followed by a list of transferred computer equipment and the indication of the problem of broken screens. This problem influences the integration of computers into the participants' daily life, which will be discussed later in Chapter 11. I explained also how the intervention's curriculum was implemented, and a short introduction to the way of teaching. Finally, this chapter is closed with the description of the technical situation how the Internet connection was realized.

Chapter 11

Analysis

11.1 Methodology Used

As mentioned above the collected interviews have been discussed in groups of four to five people by the method of objective hermeneutics. This method of interpretation draws upon the stance that the world is structured by meanings. These structures are constituted through language and become materialized in texts (cp. [Wernet, 2006], p.11). The objective-hermeneutical text interpretation aims to reconstruct (latent) structures ([Wernet, 2006], p.15), which are open to scrutiny and do not vary accidentally. This means that passages in the text (so called "sequences") are selected and analyzed - see Section 6.4.2. By carrying out group discussions I aim to retrieve as many different versions of reading as possible. As the participants of the discussions do not have any context knowledge, the scope of versions of reading is widened as much as possible. The (latent) structures are revealed by the application of five principles and within a group discussion each member checks the other to make sure that these principles are thoroughly put into practise. Such a proceeding is a form of quality assurance for the research results, too. Though, as a consequence these principle become complexly intertwined and - in particular because of the applied group discussions - therefore cannot be considered and presented as distinct and separated steps. The principles are:

- Deconstruction of Context [*orig.* "Kontextfreiheit"]

  This principle asks to look for the meaning of the text, regardless of its actual context. Phantom contexts are created to investigate in which situations the text could be well-formed ([Wernet, 2006], p.90). It does not mean that the actual context is neglected at all, but at first, in order not to miss any possible version of reading, the "real" context is consciously disregarded or simply not mentioned. This was provided as I withdrew myself as much as possible from the analytical process and by the fact that the members of the discussion groups did not know anything about the intervention investigated and it's actual progress.

- Literality [*orig.* "Wörtlichkeit"]

  It is important to focus on the de facto produced text and not on the actual intended words; e.g. this asks to discuss slips of tongues ([Wernet, 2006], p.90).

Though, as both interviewer and interviewee were no native speakers grammatical mistakes and linguistic problems got less important. To do justice to this principle smaller phrases were aggregated to longer text elements to assure that the carved out meaning is not mislead by linguistic problems. Nevertheless, as the knowledge of the language was deep, I draw upon the assumption that both speakers are capable of saying what he/she wants to express. This claims for the analysis of occurring variations in dialogic structures.

- Sequentality [*orig.* "Sequenzialität"]

  This principle aims to reconstruct the development process of a textual structure. Thereby, one does not look for interesting snippets, but analyzes one sequence after the other. It is especially important to disobey below text elements, as they did not exist at that point of time, when the currently analyzed sequence emerged. This is important as in the following considerations how the text might be continued are made. These thoughts of extrapolation are compared to the "real produced" text and allows an enhanced reconstruction why the text was produced like it was. Sentence by sentence, each group of text elements is related to previous scopes of interpretation, which are decreasing by each sequence. Consequently, a sequence is not limited to one version. Some valid versions can remain open, which can be closed at later sequence positions ([Wernet, 2006], p.90).

- Extensivity [*orig.* "Extensivität"]

  This means that all elements have to be considered carefully and as many versions of reading as possible have to be evolved. Only if all possible versions are figured out and discussed this principle is adequately treated ([Wernet, 2006], p.91). To do justice to this claim, we agreed to reexamine our versions as soon as we had agreed too quickly on a final one.

- Parsimony [*orig.* "Sparsamkeit"]

  The principle of parsimony implies to exclude versions with too far-reaching contextual assumptions and to take into account only those versions which are really constrained by the text ([Wernet, 2006], p.91). Those readings, which cannot been validated have to be discarded and whenever unreasonable assumptions according to the context were made by the group, I called for this principle.

Apparently, these principles are heavily intertwined and especially in an analytical process undertaken by a group, these principle cannot be separately reconstructed. Therefore, it is not my intention to depict the whole process of analysis and all possible and discarded versions [*orig.* "Lesarten"]. Consequently, I limit my presentation on some revealed (latent) structures of meaning. These are assembled with other assembled sociocultural context information and discussed in Part III.

11.2 Group Analysis

At the beginning of each discussion (all in all we have analyzed four teacher and four pupil interviews) I selected three text passages (sequences) from a chosen[1] interview. Usually, the beginning of the interview is investigated first. To avoid a bias by specific scientific disciplines the constellation of discussion partners has been always heterogeneous and varying: including a professional sociologist, computer scientist, musician theorist, philologist, anthropologist, English teacher, theologian, statistician or whatever. Analyzing social interactions between actors implies identifying emerging actors first. After the identification of possible actors, one can discuss the integration of computers into a participants everyday life. Therefore, I have aimed to elucidate what kind of different actors are prevalent in this study and to focus how " control" and " hierarchic positions" are manifested and implemented in the Ugandan case study. The analysis of this case study deals also with the relationship between the involved actors under the aspects of power and control, the fear towards technology and the nature of relationship. Thereby, the sociocultural environment is marked out and explored. In practical terms this is achieved by two steps which each member of the discussion group performed on his own - starting with the beginning sequence.

1. The first step of analysis was to reconstruct the thinking of the informant. Such as her attitudes, her emotions and (tacit) opinions about the theme-problem "computer".

2. Afterwards the following criteria were applied to shift the discussion to a meta-level. For our discussion this meant to discuss following criteria:

   - Which authorities can be found?
   - What is the role/hierarchic position of the interrogator/interviewee?
   - What does "control" mean?
   - What level of control is to be expected?
   - What/Who is being controlled?
   - Who prosecutes the power and how does it emerge?
   - What is expected from different hierarchies – are these expectations represented explicitly or implicitly?

Each member of the discussion group presents his/her analytical results. Afterwards the revealed attitudes, hierarchical positions, etc. are discussed. These opinions need not necessarily to be conciliated to one final option, as several versions are possible. During this discussion it becomes evident that sometimes it is not possible to determine or agree on one final version and so the research team can leave the door open.

When the second text passage is discussed, it is done in a similar way to the first one, the only difference is that prior discarded possibilities of certain positions or attitudes are not reexamined and not taken into account any longer. Thereby, by a

---

[1]The interviews are chosen by the criteria of maximum contrast ([Oevermann, 2002], p.9).

systematic precluding of impossible constellations, positions are carved out. At the end of each text passage I note the research groups commitment on valid and invalid versions, which will be presented in the upcoming sections. To clarify the analysis I will quote text passages of one pupil interview and one teacher interviews to show how the results were carved out. According to Oevermann, the research can be finished when a saturation level has been reached, which means that no further new results are carved out. He mentions a number of ten to twelve analyzed text passages which might be sufficient to answer complex research questions ([Oevermann, 2002], p.17), whereas for this study twenty text sequences appeared to be sufficient to reach the saturation level. Oevermann draws also upon the assumption that, if one structure is adequately revealed, it will be quickly recognized in other situations as well, which allows me to utilize all the collected material. Consequently, all other interviews will be taken into account in the interpretational part, too. This aims at retrieving estimates on relative frequencies of structural elements' interrelationships. To utilize all interviews the transcribed interviews were assigned to a categorization matrix, which emerged during the group's discussions. In order to gain a better overview relevant aspects are clustered to categories. The main categories are presented in Part III and are named "actors and their roles", the "presentation of a computer" and the "application of a computer". The relevant revealed subcategories are listed as corresponding subsections and can be additionally found in Appendix A.

11.3 Pupil Interview with a 13-year-old Girl

1 **Y:** Are they hard?
2 **R:** R, R, really don't be scared, it's absolutely n, no problem, anything. Y, you have to help
3 me. Yeah?
4 **Y:** Ok.
5 **R:** So, b, b, because , I'd, I'd like to get a feedback, yeah and I am not asking you anything
6 for tests or anything like that it's really for me to help me.
7 **Y:** Ok

Example 11.1: *A beginning sequence*

At the very beginning sequence of this interview the pupil asks: *"Are they hard?"* (Ex11.1Line2). Investigating this sequence without any context it could refer to some tasks to be done or to the ripeness of fruits on the market. When the sequence was extended, the fruit-market option was discarded, and so this means that the girl perceives the situation of the interview as a task. Referring to the not yet asked questions (=tasks), depicts her role as a student. The only thing the girl knew was that she was to participate in an interview. Apparently, she is expecting to be tested. When the interviewer says *"really don't be scared, it's absolutely no problem, anything. You have to help me. Yeah?"* (Ex11.1Line3), he approves this version. This means that during the interviews the interviewee is playing at least two different kind of roles: On the one hand the one of the interviewee and on the other hand the "classical" pupil who is supposed to answer questions in a formal and proper way. Up to that point any other assumptions as e.g. that the interviewer aimed to put the interviewee in a more powerful positions and take up a less powerful were discarded in order to do justice to the principle of parsimony. However, the interviewer clearly states that he wants to retrieve a feedback of her and does not want to test her (or anything like that). Thereby, he underlines that he does not want to take up the role of a teacher.

56 **R:** Can you still remember when you were using it for the first time, the computer.
57 **Y:** Yes.
58 **R:** How *(∗klack∗)* was it?
59 **Y:** Well, you told us to go to "START"
60 **R:** Yeah.
61 **Y:** Then after click there, when we saw that, when the cursor went there, than we saw, click
62 here, than we click through "My documents", Microsoft Word, what, what, what, whaat.
63 **R:** Mhm.
64 **Y:** Then at first I didn't know what to do. Even everything you said I didn't understand it,
65 also ..
66 **R:** Why?
67 **Y:** I couldn't understand anything about computers.
68 **R:** Mhm.
69 **Y:** So, I just go there for the sake of it, because you have said it.
70 **R:** *(∗Laughing∗)*
71 **Y:** Then sometimes when, when you would copy, you would do something that we were
72 forbidden to do,
73 **R:** Mhm.
74 **Y:** just click on anything we find there.
75 **R:** And, and when did you start feeling more comfortable?
76 **Y:** Oh, well, I said
77 **R:** Or didn't you feel uncomfortable? Did you, did you feel uncomfortable?
78 **Y:** At first I wasn't very comfortable.
79 **R:** O, o,o or how did you feel.
80 **Y:** Well, I felt a bit afraid whether I would spoil them and they get viruses or that type.
81 **R:** Mhm. So that's what you had heard about, too.
82 **Y:** Yes. *[∗.∗]* It was like: oh my gosh, how much will I have to pay if I spoil it.
83 **R:** Oh. *(∗small laugh∗) [∗...∗]*
84 **Y:** But then, mhm, it turned a bit easy, and I think when I grow up it might help me really
85 much.

Example 11.2: *Reproducing knowledge*

Looking at Ex11.2 from a functional point of view the informant is reproducing the process of using the computer for the first time like in a test. This is the corresponding sequence to the very general question of *"How was it?"*, which aims to retrieve information about any experience which has happened previously. In her following answer she describes a process how she used a computer (for the first time). She lists every single step (Ex11.2Line60,62-63) and can even remember the rollover text which appears when you move the mouse over the " *START* "-Button in MSWindows. This interactive *"Click here"*-tooltip and the domino-effect like *"what, what, what"* put her in a quite passive role. Thus, we perceive her in the role of a pupil at this part (Ex11.2Line60,62-63), who aims to act in an expected way. As regards content the girl was visibly overextended with the computer. This is not only because she does not know how to use it, but she cannot handle the information which has been given - Ex11.2Line64.

Another interesting fact is that during the whole interview, the interviewer uses "
Mhm." as a sign of giving ear to her words. Except in the mentioned paragraph the
interviewer approves her initial process of reproducing (Ex11.2Line60) with a rather
untypical " *Yeah*" (Ex11.2Line61), which might cause her to remain in her role of a
pupil. Via this approval the interviewer himself shifts from the initial role of an interviewer to an evaluator (in the meaning of a teacher who is interrogating pupils and
asking for knowledge). Ex11.2Line65 opens problems correlated to the understanding of computers and when she underlines in Ex11.2Line70 that she just went there
"*because you have said*" it, the passive and deferring position becomes evident once
more[2]. At this point of time it is unclear whether the "you" refers to the personal
interviewers additional role as an instructor or an impersonal, not named other point
of reference. It is only evident that the instructor takes up a more powerful position
than the interviewee, but if the interviewer is perceived as a member of the group
which was "*forbidden to do*" - Ex11.2Line72 remains uncertain. The prohibition is
present, but the authority of power is unsettled (Ex11.2Line72-73). It is unclear if this
interdiction is valid for both instructor and pupils, as it is possible that the interviewer
is seen as an instructor which is more powerful. If this refers to both it will mean that
the interviewee is an instructor and a member of the community. In Ex11.2Line76 the
interviewer asks rather suggestively, but aims to undo this, by the appliance of the
opposite statement - Ex11.2Line76. Nevertheless, we perceive this as the switching
back from the evaluator's role to the interviewer's one. The fact that there is some
kind of development in the examination of computers is annotated in Ex11.2Line79,
which indicates a temporal process. An interesting issue is found in Ex11.2Line81
when the fear of sanctions is put into foreground. By the phrase like "spoil them"
and "get viruses" the technology gets antropomorphized, though it remains unclear
whether the analogy of a virus is resolved. This offers to elucidate the prevalent image of technology, which is shaped by an apparently adopted "learnt-by-heart" image
of technology (Ex11.2Line85). Actually, the girl uses a computer only, because "*it
might help me*" (Ex11.2Line85). Two possibilities are valid: instructors or KK. This
line induces also that the handling of computers turned easier and implicitly indicates
that the fear got less. Nevertheless, we have to keep the possibility in mind that all
these words could have been said in order to satisfy the instructor, though we perceive
Ex11.2Line81 as if the girl slips into the role of an interviewee.

To identify the actors and their roles this means that the interviewer takes up
the roles of an instructor and of an evaluator as well. The girl is both pupil and
interviewee. Figure 11.1 shows how the roles mute. The numbers marking the arrows
correspond to the line numbers.

If she performs her first steps on a computer (Ex11.2Line69), as she relies on the
person who gives the commands, because of technical skills or of his authoritarian
position is not sure. But evidently, her first usage of the computer was characterized
of a general uneasiness towards the way of using, most likely caused by the fear to
spoil it and, as a consequence, to pay for it. When the first barrier had been broken,
her fear of approaching the technology was overcome and her interest in computers

---

[2]Furthermore, it indicates that she had no concept how to approach a computer on her own.

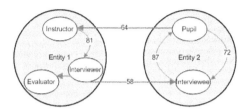

Figure 11.1: Example 11.2: Actors and their shifting roles.

was supported by the expectation that she will benefit from a computer.

On a functional level Example 11.3 deals with the question if there were differences in the way of teaching of the Austrian computer teachers and others. During this interview the interviewer fails to remain in his role of an interrogator several times and switches to the one of an evaluator. This is promoted by the fact that the interviewer actually is both interviewer and teacher of the child. Several times the informant switches from a free and vivid narration to special expressions, which sound as being learned by heart. We interpret these situations as a shift to the pupil's role and consider this as an insecureness towards the whole interview situation - Ex11.3Line317,318.

292 **R:** Wh, what else do you think is the difference between the teaching Oswald and me did
293 and *[*..*]* and others?
294 **Y:** Well, you were the first to teach us, so we were happy.
295 **R:** *(*Laughing*)*
296 **Y:** And *[*..*]* ahm *[*.*]* really *(*emphasized*)[*..*]*, when I said you were teaching us, *[*..*]*
297 it was somehow it, you seemed to made things easy for us, not like, sometimes you have
298 asked "what does this mean", when we go there, but they are just like, it's what I told
299 you to write down. So, there for you, you have to explain what the word means, don't
300 you.
301 **R:** Mhm.
302 **Y:** And *(*cough— singing voice*)* Sometimes we get shouted at and sometime they can kick
303 you ...*(*??*)* *[*too less voice*]*
304 **R:** And, and, and, yeah, and, and what was the difference for example with Mr. Rukinga?
305 **Y:** Oho, *[*...*]* oh well, mhm, I remember one day, we increased our letters on the computer
306 and we were almost dead, he almost cut off our heads. It was like:
307 **R:** Why?
308 **Y:** *(*speech Mr. Rukinga*)* That's not what I told you to do, follow instructions. For us, we
309 had just tried to copy things, it was no problem. *[*..*]* We were scared. Ahm *[*..*]* And
310 by the way, he, some things they write on the board there, they write on the boarder,
311 like first learn them and wait some day. Ok, well, ahm *[*...*]*.
312 **R:** What did you want to say?
313 **Y:** I seem I have forgotten what I wanted to say.
314 **R:** Oh, no, it's ok. Have you wrote on the blackboard something?
315 **Y:** Mhm, well, ok, when you are having lessons, sometimes you have to wr, write down the
316 things.
317 **R:** Mhm.
318 **Y:** And sometimes we leave out the notes, some of them, like click on *[*..*]* what, what,
319 sometimes you leave out those synthesis. So when you are going on a computer you

320   somehow be mislead and will end up like in a fight but then it seams for you, you have to
321   follow your mouse and that's a bit easy, because it's not writing down something you
322   don't know how to do and writing.
323   **R:** But would you remember la, next time?

Example 11.3: *Differences in teaching computers*

It has to be stressed that the teacher's group is divided into the *Austrian teacher* (project implementers) group (AT), the *Ugandan computer staff* (UCS) group and the non-computer-teaching permanent staff. The distinction between the AT-group and UCS-group is made both by the interviewee and the interviewer. The informant addresses the interviewer/instructor with "you" and the other computer teachers with "they" (Ex11.3Line297-301) – this differentiation occurs on the basis of the informant's teaching experiences. The interviewer also differentiates between the UCS and himself with his (Austrian) colleague (Ex11.3Line293). A further characteristic of these two teacher groups can be found in the same passage, when the informant notes the difference in the way of teaching. It is to be noted that the informant demands discovering and practical teaching methods, whereas unto this stage of analysis the UCS is teaching in an authoritarian way, including corporal punishment (Ex11.3Line297-301). So far it seems as if the relationship between the Austrian teachers and pupils and the UCS group and the pupils is different. The differences appear in the way how computers are instructed, whereas the teaching of Mr. Rukinga is presented as a theoretical based education (Ex11.3Line307-311), the ones of the ATs is more practical and easier to the interviewee (Ex11.3Line317-321). Apparently, it is difficult to imagine the instructions of Mr. Rukinga and to internalize the process thereby (Ex11.3Line321), whereas the visual tracking of the mouse is easier than writing instructions down (Ex11.3Line320).

Another interesting point is the depiction of Mr. Rukinga, who stands as a representative of the group of computer teachers. Each time the interviewee speaks with the words of Mr. Rukinga, she uses the I-form when formulating a command, instead of the possible We-form. This gives us the impressions that the rules were presented as made by the teacher himself (Ex11.3Line299-300,310) and not formulated by a community of teachers. This underlines his strong hierarchic position.

589   **Y:** You find you have made a mess, a really big mess, *[*..*]* and well, some parents are strict
590   they find you and and give you up, they beat you,

Example 11.4: *Measurements of sanctions*

These measurements of sanctions and (corporal) punishment are, additionally to the UCS (Ex11.3Line301,302) also performed by the pupil's parents (Ex11.4Line589,590). This clinical presentation indicates the ubiquity of this methods of sanctions. The girl presents the examples as a part of reality and does not consider corporal punishment as something awkward or evil, as otherwise she would have explicated in a different way. If she had mentioned this fact at all. Thinking about reasons why she explained it, one might reason that she wanted to denigrate her teachers. Though I am convinced that a child does not want to cause problems to her parents and that she might also fear further consequences by the teacher. Regarding the results of the participant observation show that corporal punishment is widely accepted and used in this school. Anyway, it is interesting that parents give their children access to computers at home as soon as they are taught at school. Within that interview it remains unknown, if the parents crosscheck their children's knowledge.

39 **R:** So did he teach you then how to use it or a bit, or?
40 **Y:** No, he said, since we haven't yet st, since we haven't yet s*(*??*)*, learning at school, well,
41   when you learn, when you start learning, he'll let me use his.
42 **R:** Aha.
43 **Y:** So now when we started, he has let me use it.

Example 11.5: Computer lessons allow usage at home

Referring to the above lines (Ex11.5) it seems as if the fact of having computer lessons at school is reason enough to enable the children to access computers. This means that the school controls the access of technology in two ways: both at school and at the children's home. It seems that the parents, who contribute almost the whole school budget, completely trust in the school's education and quality.

11.4   Staff Interview with a Woman

The following three sequences are part of an interview with a member of the teaching staff. The first one is once more the introductory sequence of the interview and two further.

1 **R:** Yeah, yeah, so it's better, do you mind saying something?
2 **J:** About What?
3 **R:** No, ok, you know that I would get some feedback about the teaching we did last year,
4   Oswald and me and what you general think about computers.
5 **J:** What I think about computers.
6 **R:** So, yeah, I'll ask you, just let' s keep talking and anything, yeah. So my first question is,
7   when you saw the computer for the first time, how was it? Can you still remember the
8   situation?
9 **J:** Yes
10 **R:** And, *[*..*]* yeah, how was the situation like.
11 **J:** They were nice and lovely, I was excited, I had seen it for the first time, ahm, *[*..*]* and I
12   was eager to touch, and and learn how to use the computer.
13 **R:** mmm.
14 **J:** The people who were here and you were teaching was really enjoyed the teaching.
15 **R:** Oh it's very loud in here *(*background noise*)* Yeah, yeah,
16 **R:** yeah, b, b, but why was it to, why did they look like lovely, because you said they looked
17   lovely and you wanted to touch them. What was so lovely about them?
18 **J:** According to the way they were packed
19 **R:** mm.
20 **J:** It was all those...*(*stammer*)*, they were *(*stammer*)* properly packed, *(*..*)* and when
21   you got them out, the things were... they were look, as if they are very new hm..
22 **R:** *(*response hmm*)*
23 **J:** and indeed they were new because...mm they were *(*stress*)* new to me first of all, *[*.*]*
24   they had a good looking

Example 11.6: Beginning sequence of an interview

The initial question of this sequence indicates that the interviewee is supposed to know about the interview situation. By "*so it's better*" the interviewer is referring to an improvement of "*it*", which can be read in multiple ways. It can refer to the location where the interview took place, to the situation itself, or to the positions of the microphone. In any case it shows an improvement to prior circumstances. Increasing the length of the sequence with the phrase "*do you mind saying something?*", the interviewer aims to start the interview or even to test the microphone. Anyway, the interviewee reacts to this quite courteously formulated question by "About what?" which is answered by "*No. OK, you know that I would get some feedback about [...]*". Thereby, it becomes evident that the interviewer had intended Ex11.6Line1 as the first question and assumes that the setting of the interview is clarified. The "*No. OK.*" makes only sense if it is applied as the marking of a new beginning of the interview situation. We have agreed that it this functions as a rewind. This is underlined by the subsequent statement "*you know*", which refers once more to a prior situation. In Ex11.6Line3 he concludes his statement by two implied questions: "*get some feedback about the teaching we did last year*" and "*what you general think about computers*". Within that statement the interviewer assigns himself as a teacher and defines the used "*we*" as his colleague and himself. From a hierarchical point of view the interviewer aims to avoid a powerful position and chooses terms like "*feedback*", or phrases like "*let's keep talking and anything*" (Ex11.6Line6). Nevertheless, he takes up a leading role and is in charge of the situation when he proceeds by "*my first question is*" (Ex11.6Line6). If power between the interviewer and the interviewee is really that unbalanced at this point of time is uncertain, as up to Ex11.6Line11 (which we perceive as one general attempt of establishing a convenient interview situation), the interviewee's short statements can also be seen as an effort by the interviewee to gain power in order to "force" the interviewer to explicate his questions. Although, consequently (beginning from Ex11.6Line11), the informant answers with longer lasting statements, which might be perceived as an interim outcome of a negotiation process. This statement refers first to Ex11.6Line7 and goes on with her personal statement how she perceived the first encounter with computers: "I was eager to touch, and and learn how to use the computer" (Ex11.6Line12). To "*touch*" offers two versions, either in a haptic or in an explorative way. This options remain open at first, but the annexed "*learn how to use the computer*" at least indicates that the image of technology is that it has to be learnt. Apparently, it cannot be explored by oneself, which closes the second reading of "touch". This is also supported by the presentation of the physical appearance of computers by "*properly packed*" (Ex11.6Line20), "*if they are very new*" (Ex11.6Line21), and "*they had a good looking*" (Ex11.6Line24).

On a content level, Ex11.6Line14 offers that "*the people*" who were taught "*really enjoyed the teaching*". This is interpreted in three ways: she wanted to fulfill the interviewer's expectations, she aims to generate a more explicit question, or she adds these words to make sure that the above questions have been answered sufficiently. Valid to all cases is the fact that she emphasizes her statement by "*really*" and that she is not explicitly member of the group. By "*the people who were here*" she refers to the location or surrounding where the interview took place and as she is encouraged by the interviewer to give a feedback on the teaching she obviously participated in, she latently belongs also to the group of "*people*". If the interviewer's answer in Ex11.6Line15 is due to the mentioned "*background noise*" or insecureness of the interviewer towards the interviewee is unsolved. Consequently, he asks for an explication why she wanted to touch them and why they looked lovely (Ex11.6Line16-17). The central word of the following answers is "*new*". Whereas she begins her reply with a reference to optical and haptic elements (e.g. "*things*" in Ex11.6Line21 or "*good looking*" in Ex11.6Line24) she repeats the newness. Her interest is more based in the physical appearance than in the joy of having access to it. If this was due to her ignorance of computer is assumed. At first "*new*" (Ex11.6Line21) is once more correlated to the external characteristic, but in Ex11.6Line23 she introduces a personal input by "*new to me*" which is initiated by the preceding "*new*" in Ex11.6Line23. If this is meant as a relativization of the forerun "new" or as a further association to this new as she was struggling for words could not be figured out.

This sequence presents the multiple role of both the interviewer and interviewee. It shows also the existence of ongoing negotiation processes during the establishment of interviews and the subjective perception of computers. We carved also out that the way to approach computers is to learn to use it and not to immediately explore/use it at a first encounter.

153 **R:** What do you think was our aim of teaching?
154 **J:** Your aim of teaching..
155 **R:** yaa..
156 **J:** ...*(*softer and matter of fact...*)* of course you wanted to, to use the computers that you
157   brought.
158 **R:** hm..
159 **J:** because of course you know that you were.../*..*/... you wanted to go, hm?
160 **R:** hmm,
161 **J:** so you , you were teaching us to learn how to computers since you have already brought
162   them.
163 **R:** hmm,
164 **J:** mmm /*...*/ you had had wanted to go and leave out and leave behind the computers, so
165   you were teaching us to use those computers /*..*/
166 **R:** and the others in town, they behave different you think
167 **J:** That what *(*sounds unsure of what the question is asking*)*.
168 **R:** in teaching, so they teach you that you have to come back to them.
169 **J:** They teach you such that you can go back to them, if you fail to understand
170 **R:** hmm
171 **J:** you go back to them and pay another money.../*.*/...then for you, you were teaching
172   details, /*..*/ because you were teaching us, ah as if you were teaching someone whom
173   you know,
174 **R:** hmm
175 **J:** ...without any logic behind
176 **R:** hmm..
177 **J:** mmm, you were open
178 **R:** hm,
179 **J:** to us, but for them, they teach you, if you fail to understand you go back and pay
180   another money /*..*/ they are, they are /*..*/ they are working money,
181 **R:** hm
182 **J:** mm

Example 11.7: *Aim of teaching*

In his initial statement the interviewer represents himself as part of a group of teachers, which is congruent to the beginning sequence (Ex11.6) of this interview.

Like in the first sequence (Ex11.6Line5) the interviewee repeats the question, maybe to visualize it or for clarification to avoid misunderstandings (related to language). After the acknowledgement by the interviewer, the informant responds in a factual way (Ex11.7Line156) and identifies the aim in the usage of the computers. In this context she also identifies the interviewer or "the group of teachers" as the bearer of technology. Filling the missing word in Ex11.7Line159 verbs like "*leaving*", "*going*" but also "*teaching*" are readings at first, but after hesitating she closes the last option and explicates that by "*you wanted to go*" (Ex11.7Line159). This statement has two readings, either he/they ("*you*") wanted to go away from their home place to go to the place where they were teaching or they wanted to leave the place were they were teaching. In any option of these the final "*hm*" indicates a uncertainty towards this statement, whereas it is not picked up by the interviewer and leads to the repeated reply in Ex11.7Line161. Thereby, the interviewee identifies herself as a member of taught people, but interestingly puts it into a causal correlation to the bringing of the computers. Thus, it becomes evident that the computer teaching group did only teach them computers, *as* they have brought them. Followingly, the computer teachers' aim on a very general level was to teach computers, but this implies that this was the only purpose of their coming. In Ex11.7Line164 she once more emphasizes the computer teachers' intention of leaving. This closes the option from Ex11.7Line159 that the computer teachers had wanted to leave their home place. She uses three verbal constellations, "*wanted to go*", "*leave out*" and "*leave behind the computers*". Whereas the first two still keep both options valid, later one refers to the local place. This indicates that the first two verbs are also related to the local place, because otherwise, to do justice to a well-formed statement, a local supplement would have been mandatory.

In Ex11.7Line166 the interviewer provides some kind of a break, when he introduces new actors, which are depicted as "*others in town*" who "*behave different*". By the sharp distinction "the others in town" and that he talks about their behaving (and not about their teaching) he opposes this group to the one he belongs to. The term "*behave*" outreaches "*teaching*" and implies activities beyond teaching. At first the interviewee reacts insecurely by a question (Ex11.7Line167), but when the interviewer adds "in teaching" and alleges that "they teach you that you have to come back to them", she seems to understand the interviewer was asking for. "*They*" in Ex11.7Line168 obviously refers to "*the others in town*" which are precised as teachers in the following. If they are computer teachers as well cannot be taken for granted, but that they are also teaching computers can be assumed as the whole inner context deals explicitly with computer teaching and as this group is opposed to the teaching group the interviewer belongs to. By the verbal construct of "have to" the interviewer ascribes the teachers a powerful position who can force those who go there ("*you*" in Ex11.7Line169) to come back again. This is relativized by the informant, when she says that "*you can go back to them, if you fail to understand*". Thereby, it becomes a deliberate decision and exemplarily adds that this is the case "*if you fail to understand*". If there are other reasons to come back remains open, but obviously the failure of managing the subject matters is the major one. By Ex11.7Line171 she depicts the process and tells us that the act of going back is connected with the payment of "*another money*". This phrase ("*another money*") is understood that in town one has to pay for the lessons, though it is not indicated if one has to pay at the place where the interviewer is teaching. After a short break the informant opposes the way of teaching the interviewer did to those of town. She presents the interviewer's (or computer teacher of the local place) way of teaching by "*you were teaching details*" (Ex11.7Line172), "*without any logic behind*" (Ex11.7Line175) and also by describing the way of teaching in a personal manner: "*as if you were teaching someone whom you know*" (Ex11.7Line172-173). If one has to return to lessons if he/she did not manage the subject matters, this means that evidently problems cannot be solved during the teaching sufficiently. The reasons for this can vary as it might be an inappropriate teaching method, an uncertainty of the students how to formulate questions to retrieve clarifying answers, or, as a matter of respect to other attending students, not to disturb the teaching. All these options remain valid when she emphasizes the personal contact, which implies that the other teachers in town did not teach that way. Though, she does not say that the computer teacher group of the interviewer did know the students, but actually puts it in a conditional clause. By this the distant predicate becomes objectified and gives information on an apparently more individual focus of the teaching method the interviewer and his colleagues have applied. This is underlined by "*you were open*" in Ex11.7Line177 which can be completed by "*to questions*" or "*to us*". Putting both in the previous opposition of the teacher groups, which is taken over by the informant means that they were not open to the students or to questions by the students, which is de facto equivalent: If I am not open to the questions of my students I am not open to them in person as well. This gives also information on the given "details" from Ex11.7Line172, which argue for an individual way of teaching by the teacher group the interviewer belongs to. When she ascribes no logic behind the teaching to the teaching group the interviewer belongs to, she claims that "*the others in town*" had one. At first "*logic*" can be understood as a way of thinking or explaining something, which would be perceived as a criticism of the interviewer as teacher. But as the interviewee previously appreciates the way of teaching the term logic becomes negatively connoted and can be understood like "*financial intention*", "*ulterior motives*" or "*formal structures of a way how to proceed*". This makes more sense, especially when she adds "*you were open*" in Ex11.7Line177.

Though, the distinction between the teacher groups was introduced by the interviewer it is taken over by the interviewee and further explicated. Of course, it might be sedulous flattery, but the whole interview situation is rather on an interviewer-interviewee relationship than on a teacher-pupil one, which lead us to discard this option of reading the whole sequence. Of course, the interviewer is the one who introduces the new actor, but retrieves longer narrative statements by the interviewee. In Ex11.7Line179 she repeats the statement of Ex11.7Line171, but after some seconds of hesitation and two attempts of forming a sentence, she finally judges upon this other teacher group and claims that "*they are working money*" (Ex11.7Line180). If this is negative or positive connoted remains unclear. It can be either a simple neutral statement or a criticism as if those people were squeezing money out of the people attending lessons there.

252 **R:** So you are using the computer for private things hm?

253 **J:** hm sometimes.
254 **R:** What did you want to say?
255 **J:** amm like my passwords/*.*/ those are the private things that I put in ..hmm.. but since
256   these ones are publicly used, if something is personal, you just put it in, they may delete
257 **R:** hm so you don't protect it with passwords?
258 **J:** Only passwords
259 **R:** H, How do you mean *(∗stammers... hesitates∗)* please help me.. how do you mean by
260   typing in passwords.
261 **J:** Okay, if I am trying to hide my, my information
262 **R:** yes of course..
263 **J:** ...it's when I use a password.
264 **R:** hm
265 **J:** a password *(∗very softly, almost inaudible∗)*
266 **R:** on the word document.
267 **J:** on my document
268 **R:** Yea, and what are you deleting then?
269 **J:** *(∗Stammers∗)*, I 've told you that when I want to leave it in, I use a password and if I
270   don't want , I delete.
271 **R:** ah.. /*...*/., and , and this is something you discovered on your own, how to use a
272   password?
273 **J:** Yes.
274 **R:** How did you do that?
275 **J:** How to use a password?
276 **R:** Yea. How to use the password.
277 **J:** I was told by someone I didn't...
278 **R:** *(∗interrupts∗)*...ah he showed to you /*.*/
279 **J:** hm... *(∗pause∗)*

Example 11.8: *Usage for private purposes*

Although, the interviewer asks after the usage of computers for private things this sequence is marked by the keyword "privacy". When in Ex11.8Line252 he explicitly asks on this issue, she avoids a definite answer and replies "*sometimes*". This is interpreted as an all in all quite nebulous answer, which provides us with several versions: it can refer to the composition morphological of "*some*" and "*time*" in the meaning of rarely or to the meaning that she uses computer privately if there is time available. Furthermore, it can be understood as an "*whenever it is necessary*" or a blunt confession "*yes, I do*". All of these options have in common that they are not explicitly expressed and are forming a pool of assumptions which might the interviewee have wanted to cook her statement. The interviewer is not satisfied with this answer and asks in a rather harsh way "*What did you want to say?*" (Ex11.8Line254). By that the interviewer requests the informant to define private activities, which were latently acknowledged by "*sometimes*". He takes up a hierarchical position, which is encountered by the interviewee, who is loss at words at first ("*amm*"), but presents an example to do justice to this situation with "*like my passwords*" Ex11.8Line255. Thus, she shifts on a computer-knowledge related level and describes the situation when she uses passwords to protect her personal information. According to her this is necessary, as "*these ones are publicly used*" (Ex11.8Line256), to avoid the deletion (of what?) by someone else. Apparently, there is a further actor ("*They*") who can delete on the computer ("*these ones*"). This could be other users or persons with more privileges (e.g. administrators). During this statement the informant identifies private things as those which are put in by her (*"I put in"*) and generalizes that *"if something is personal, you just put it in"*. *"It"* can infer two readings, either information or password. When the interviewer asks once more and uses *"don't protect it with passwords"* he applies it in the meaning of information and thus one version is closed. According to our perception the following *"Only passwords"* by the interviewee refers once more to the initial question on the usage of computers for private things. The limiting term *"only"* hints for a bad conscience. The interviewer seems to be confused by her answers and definitely not satisfied and consequently states a further question (Ex11.8Line259-260). During the answer in Ex11.8Line261, the previously discussed *"it"* becomes explicitly depicted as *"information"* which is hidden by a password on her document. Interestingly, in Ex11.8Line266 the interviewer is picky again and wants to know if it is a word document she is protecting. Her response makes it clear that she talks about any type of documents, but only about those which belong to her. This leads us to the fact that information is perceived to be owned by someone and that others are excluded from access to it. Information is seen as an instrument of power. Important information is protected/hidden or deleted, which is similar to being destroyed. By any action on information, informational integrity is preserved.

If the interviewer really did not understand her statement in Ex11.8Line255-256 is unclear, but in any case it takes three more questions up to Ex11.8Line270 until he is satisfied with the woman's answers to proceed with the interview on a content basis. In Ex11.8Line271 he investigates the knowledge acquiring process of password protection and is not interested in private things any longer. At first she answers with a short *"Yes"* (Ex11.8Line273), but the interviewer is stubborn and persists on the process of knowledge acquisition itself (*"How did you do that"* in Ex11.8Line274). When the interviewee states that she *"was told by someone"* (Ex11.8Line277), the interviewer interrupts and concludes with *"he showed to you"* (Ex11.8Line278). Evidently, the interviewer has a concrete person in mind, when he notes *"ah, he showed to you"* as otherwise he would not have used the personal pronoun. Although, this was obviously not clear to the informant as she did not say *"I was told by him"*. If the final *"hm"* in Ex11.8Line279 is an acknowledgement or sign of resignation is difficult to guess.

So actually this sequence aimed to clarify if the computer was used for private things. Due to an insecureness if this usage is ok at all, the respondent avoids a binding answer and responds by the depiction of a computer-knowledge related activity. She describes the protection of personal documents by passwords as an example what she is doing if she wants to protect personal information. This need arises as the computers are shared and publicly used and the information on can be deleted. Let us keep it in mind that she omits the fact that her information could be read, but focuses only on the information's destruction. Furthermore, we learn that the knowledge was acquired by the help of someone else, like *"the other teacher in town"* of Ex11.7 or other computer literate people.

11.5 Summary

This chapter depicted the process of analysis in detail. I showed how the method of objective hermeneutics was utilized in group discussions to achieve context free statements. This is from a peculiar importance, as one of my research goals is to elucidate to what extent the multiple role of the researcher as evaluator, donor, teacher, colleague, and interviewer influences the investigation of the computer's integration into everyday life. Furthermore it provides objective explicit statement to the perception of computers within the sociocultural environment. This would be rather impossible to be performed by the researcher himself, but by the aid of text affine and heterogenous discussion groups latent structures are independently carved out. Versions of reading would be prerestricted a priori and indicate a qualitative loss of the study. I presented one pupil and one staff interview to give the reader an understanding of how this method of analysis has been performed. In addition to that I showed how heterogenous the analysis's ways of progress can look like. All in all the analysis of twenty sequences has lead to a multitude of key concepts which are underlying this study. The presentation and discussion of these in combination with other relevant sociocultural context information will be part of the following Part III.

Part III

Empirical Study - Results

Chapter 12

Introduction

The following pages present the analysis' results and carve out key concepts of this thesis. I will depict how a computer is terminologically perceived and what kind of benefits and expectations throughout the application of a computer are prevalent. Furthermore, I will discuss the image of computers and how the factor "time" influences the usage and integration into the daily life. Notes of the participant observation slip into these results and make relevant facts ready for the discussion on the research questions in Chapter 16. According to Oevermann I draw upon all interviews, as once a structure is revealed, it is quickly recognized in other situations as well - see Page 128. The diversity of various text passages helps to demonstrate certain aspects from different point of views. By that the reader's understanding of the situation is enhanced.

Chapter 13

Actors and their Roles

In the previous Chapter I have described the necessity of exploring involved actors first. In these following sections I present the retrieved actors and the roles they take over within the intervention and study. As expected the participants of the intervention can be considered as the core actors of the investigation, but surprisingly some further actors like the pupils' parents or different computer teachers (than the Austrians) as well have an impact on the progress of the intervention. All these actors are related to each other and by analyzing the nature of the relationship it seems as if these actors are subjected to different hierarchical positions. Most likely, these hierarchical positions are not limited to the access of computer, but also extend to other fields of daily life. Therefore, the way how the access to technology is constructed gives information about the society in which it is applied and its social differences.

13.1 Pupils

Though all members of the school were students (pupils and staff) and instructed by the Austrian Teacher group (the interviewer and his colleague), it is a matter of fact that pupils perceive themselves as a separate group (see Chapter 11, Page 129). During the interviews the pupils play two different kinds of role. On the one hand the one of the interviewee and on the other hand the "classical" pupil who is supposed to answer questions in a formal and proper way. E.g. already at the very beginning of our analyzed interview (Chapter 11) the pupil asks: *"Are they hard?"* - In26Line1, referring to questions, which have not been asked yet, which depicts her role as a classical pupil. The girl knew that she was to participate in an interview, but she also expected to be tested. Only as the interviewer says *"really don't be scared, it's absolutely no problem, anything. You have to help me. Yeah?"* - In26Line2 she slowly shifts into the interviewee's role. The informant initializes the interview taking up the role of the pupil, but the longer the interview is proceeding, the more often her answers seem to be the one out of an interviewee's view. Though, in case of uncertainties, pupils immediately return to their trusted and "proper" role of a decent pupil. On Page 128 in Chapter 11 I have shown that during the interview, the pupil constantly changes roles. This happens throughout several interviews. The example below indicates, that as soon as insecurity arises, the male informant which is mentioned below, takes over his well-known role of a pupil. Taking over such a role implies shorter statements and reproduction of formal educational knowledge.

17 **R:** hm *[*..*]*
18 **Rn:** also he told me how, he taught me how to write letters
19 **R:** hm
20 **Rn:** and how to print, to send them, *[*..*]* and to send the letters I have printed
21 **R:** mm
22 **Rn:** and it was very interesting that's why I liked it

Example 13.1: *Pupils are changing their roles*

The unequal relationship teacher-pupil is not only based on the power of judgement upon the pupil's performance, but also by physical power. Corporal punishment is still widely used as a measurement of sanction and sometimes despotically applied[1]. Nevertheless, attempts of the interviewer to avoid this hierarchical relationship (like in 26Line2) were fruitful and longer narrative statements could be achieved[2]. I have figured out that the less the interviewer succeeds in establishing a comfortable interview situation the shorter the interview lasts. As mentioned above a pupil's interview lasts 21 minutes in average, with a minimum at 13 minutes and a maximum at 40 minutes. Still their heterogeneity is not a sign of uselessness. The exact opposite is the fact, manifold interesting and illuminating answers are given in shorter interviews, as such statements can represent less the informants own opinion, but more the public opinion about computers which is evidently regarded as a more valid answer than his/her own. The following two examples emerge from the same informant. Whereas the first example shows the reservation of the child, a 12-year-old girl, the second one states important information about the educative benefit of a computer to her. This is discussed in Page 175.

---

[1]Corporal punishment as an educational method shall not be the issue of this discussion, but the following two examples show, how randomly this "right" was applied.

1. When I gave theory lessons I had to admonish a pupil who was doing some other homework below the bench. The classroom teacher who happend to attend the class immediately slapped her face, though, according to my experience, this is not a general consequence.

2. There is a flag parade and a morning prayer on the compound before classes begin. From time to time it happens that some pupils, who are transported by car are running late. Although it is not welcomed it generally is accepted as a fact without any big fuss. One day a teacher wanted to punish a girl from P7 for being some minutes late and asked her to knee down to get ready for the strokes. Apparently, about 5 mates of the girl considered this unjustifiably and kneed down as well and claimed that they had been late as well. This sign of solidarity prevented the teacher to execute his penalization. This indicates that the punishment did not represent the school's form of measurement and could have been avoided a priori.

[2]It is important to mention that the indicated hierarchical position cannot be avoided at all. As soon as the interviewer knows about his hierarchical position and attempts to change this by consciously handing over power, he consequently applies power and reconstructs the hierarchical relationship once more. Of course, an interview situation is always an unbalanced one and the interviewer is in charge of the interview at any time, but a teacher-pupil interview shows a bigger power difference than an interviewer-interviewee relationship. Apparently, such a change can be achieved by hortative statements and prosecuted empowerment of the interviewee which leads to a more convenient "chat"-atmosphere and offers more space for interpreting the (latent) interviewer's and the interviewee's positions.

63 **R:** aha,...so... *[*..*]* if you, if you had a free wish, let's assume that the computer can do
64   everything you want for you, what would you wish of the computer, how can it support
65   you? *[*..*]*
66 **M:** I don't know
67 **R:** So you have no wishes?
68 **M:** Yes.
69 **R:** ahaa! *(*sounds shocked*)* really?
70 **M:** yes
71 **R:** Okay *(*laughs a bit*)*, when you, *[*..*]* when you hear the word computer yes?
72 **M:** Yes
73 **R:** What comes into your mind *[*.*]*
74 **M:** I don't know
75 **R:** So you don't feel nothing?
76 **M:** Yea

Example 13.2: Short answers

Whereas above example depicts the interviewees scantiness, eight lines later following remark was made, which is discussed in Page 210.

84 **R:** For example, what class work do you mean, because I am not on the school, so its
85   difficult, so I ask you to help me...
86 **M:** English, *(*was interrupted by above part sentence of R*)* maths, and English
87 **R:** And how do you use it there?
88 **M:** In English like letter writing when you make a mistake it will underline the word and
89   you will know that you have made a mistake

Example 13.3: Longer statements

Pupils are granted access by teachers, to be more precise, by the teacher who is currently in the computer room. Children, as soon as they find time, drop by and ask "*Please, may I enter*" - In19Line193 to find out if they are granted access. To which extend and if such pupils are allowed to use computers is discussed later in Section 15.1. Still it is important to point out that the teacher who is in the computer lab and who decides if a child is allowed to enter. In general pupils are allowed to use computers during lessons and under supervision when they are free. A pupil in the computer room without any teacher is usually impossible.

13.2   Staff

13.2.1   Non-Teaching Staff

From a school administration's perspective the staff can be divided into non-teaching staff and teaching staff. In terms of ways of access these groups remain the same, although the group of non-teaching staff (secretaries, matrons and bursar) is divided into small groups[3]. They are split by the fact if they have a realizable chance of using computers. This means that some members of the non-teaching staff do not have access (any longer) to computers, whereas others do:

---

[3] Actually, the non-teaching staff consists of some further groups like cooks, maidens, or watchmen. These groups did not have any encounter with computers at all and have been disregarded throughout this investigation.

Matrons are put at the end of the queue and may use computers only when there are no regular lessons in the computer classroom. However, this implies that the computer room is locked[4]. The following example (Ex13.4) shows the statement of a matron, who is indirectly excluded from computers. By "indirectly" I address the issue of additional power expenses for the school. So access is officially and ideologically granted, but physically and practically hardly claimable. Access to computers becomes extensionalized and is dependant on the physical access to a certain room. Thus, it becomes temporally-spatially controlled by the one who keeps the key.

158 **C:** O.K. *[*...*]* The problem is *[*...*]* mm, the time we want come in, especially during the
159   night. That's when we are most of the time free doing. The night, like the evening at
160   around 8, but other time we are not allowed in *[*...*]*
161 **R:** Mm.
162 **C:** We are not allowed in and *[*..*]* at times, O.K., we, ah, ts, *[*...*]* I am working, then we
163   are *(*??*)*, since after class when the class is locked, it is locked.
164 **R:** Mm.
165 **C:** We have to be economical in power, *[*...*]* mm, power. Otherwise we talk to power use *[**
166   *..*]* it is not there. We saw it, *[*...*]* I, not good you using them during the night. Just
167   being economical with power.

Example 13.4: *Matrons are exclused from access*

What is striking in Ex13.4 and Ex13.5 are the diverging reasons why the matrons are excluded from the computer access. Whereas the SMC seems to have reasoned the denied access in expenses for power (Ex13.4Line164), the teacher informant (see Ex13.5) names the SMC's opinion that computers are unimportant for the matron's profession. Consequently, this distinction is made upon their profession and reasoned as a benefit for the school. Which reason is more important was not made clear. As mentioned above electricity in Uganda is limited to town areas. The cost for electricity poses problems to schools like the following informant states, "*electricity b,b by passes but the school can't manage I think*" (In4Line428). This coincides with findings of [Nel & Wilkinson, 2001] who reports maintenance problems of ICT laboratories like paying the electricity bill. From the teacher's sequence one realizes that the informant herself would welcome the matrons qualification, as she thinks that the matrons could support her by e.g. typing their exams. When the interviewer asks whether the teacher had discussed this issue or not, she notes that due to her hierarchical position she has avoided to discuss it - see Ex13.5Line733. This underlines the strong hierarchical position of the SMC, as the informant has been teaching at the school for more than

---

[4]In terms of power this means as follows: The person who is in charge of the key to the computer class room is the one who finally enables someone to use the equipment. Usually, the keys are placed in the Headmistress's office. Only the Deputy Teacher and the Headmistress have access (once more with a key) to it. This means, that if a teacher wants to use a computer on his/her own (during the day the computer class room is occupied by classes or staffed by the administrator anyway) in the evening, first either he/she has to go to the houses of the Deputy Teacher or the Headmistress (which are located on the compound!). Afterwards one of these two persons accompanies the person who wants to get access to the computer lab to the Headmistress's office. This procedure is repeated when the user intends to leave the room. In the case of both, the Deputy Teacher and the Headmistress are not available in the evening, the computer room remains locked. During the day the computer teacher is the carrier of the key.

25 years and is the longest-serving staff member. In reference to power I state that by denying the matrons access to the computer, the teachers preserve their powerful position and are not endangered by a knowledge accumulation from the matrons. Regarding the school's perspective it can also indicate that the school aims at preventing the matrons from leaving the school to get a better job by their enhanced computer skills[5]. This can be avoided or at least prolonged by taking above mentioned measurements. Of course, they could take private lessons in town as well, but as a matter of fact it is more cumbersome to attend lessons in town (it means a walk of about 20 minutes), to pay for its usage and the teachers than to use the local computer room for free - see Section 15.4.1.

698   **R:** but matrons they have been stopped isn't it?
699   **C:** mm
700   **R:** that they use
701   **C:** they were stopped by the Head Mistress [*.*] not the Head Mistress necessarily but they
702   said that the computers came purposely for the teachers and the children they never
703   included
704   **R:** they came for the school
705   **C:** for the school and matrons are not serving inside the classroom
706   **R:** mmhm
707   **C:** mm [*..*]
708   **R:** and so ...but why are they prevented?
709   **C:** I don't know, it depends on the s, administration they may not see the main point of
710   them learning (*2 sec. Inaudible*)..
711   **R:** mmhm, nor the importance
712   **C:** nor the importance
713   **R:** mm
714   **C:** mm they see that the main part of importance is on the teachers and pupils
715   **R:** mmhm
716   **C:** I think that's why but there are some matrons who already know the computer so if they
717   knew and added on, it would help them in case teachers were very busy
718   **R:** mm
719   **C:** they would help us
720   **R:** mm of course
721   **C:** because most of the time they are they don't have a lot of work, so if they knew and it is
722   like setting exams they would help to set it
723   **R:** mmhm
724   **C:** already put them on, in the books and do them for us while we are
725   **R:** Yea
726   **C:** you have got really, doing the marking or supervising
727   **R:** Yes, of course, really, it's a good point, have you ever discussed it?
728   **C:** Oh yea so it would be the division of labour for me I would have loved them also to learn
729   them, the matrons.
730   **R:** have you ever discussed it?
731   **C:** mm
732   **R:** have you ever discussed this topic?
733   **C:** but I am not supposed to discuss, to discuss it, it is not part of my duty
734   **R:** aagh

---

[5] This implies that computer skills enable someone to get a better job, which will be discussed later on in Section 14.3.

735 **C:** I am just a classroom teacher
736 **R:** mmhm

Example 13.5: Matrons are not allowed to access computers

The actor **"secretaries"** has a special status anyway, as each entity has a computer of her own disposal. Nevertheless, they assume or know[6] that they are obviously not supposed to use a computer for private use. In Ex13.6Line45 she denies her private usage first, but admits it in Ex13.6Line49 by playing down her usage as "*maybe sometimes*".

44 **R:** So are you using a computer also for private things?
45 **A:** No.
46 **R:** Never?
47 **A:** Nmhmh.
48 **R:** You don't write letters to your friends. Sometimes with the computer.
49 **A:** Maybe sometimes.
50 **R:** Mhm
51 **A:** I can write a letter to my friend, but not in *[*..*]* not, not everyday.
52 **R:** Mhm.
53 **A:** Yeah.

Example 13.6: A secretary about private usage

The **bursar**'s position turns out to be special. Albeit the bursar did not attend lessons during the first period in 2002, she emerges as an actor, when an informant explains that the bursar had intended to pass over work to the secretaries. In fact she is, due to her profession supposed to learn computers which she apparently started after the Headmistress had told her - Ex13.10Line964. During the intervention her role is minor related to computers in general, but the way she aimed at passing over her work helps us to illuminate the function of a computer as an instrument of power.

13.2.2 Teaching Staff

An entity of the actor "teaching staff" is coined by the general allowance to use the computer whenever one needs it. Statements like the following example evolve this assumption. From her statement (Ex13.7Line16-20) the accessibility of the computers clearly emerges. The teacher uses self-explaining phrases like "*own use for a long time*" and "*any time*", which can only be true, if she had been able to retrieve the key to unlock the computer room in the evening.

1 **R:** *(*both laughing*)* Hello, Hello − Yes
2 **H:** Is this a microphone? *(*laughs*)*
3 **R:** Yes
4 **H:** The smallest I've ever seen.
5 **R:** Yea *[*.*]* so it, it doesn't belong to me *[*..*]* aha *[*..*]* yes as you know I'd like to get
6   some feedback
7 **H:** hm,
8 **R:** about my computer teaching, because in the very near future I'm going to do some
9   computer projects in other countries and in other schools

---

[6]It could not be surmised whether agreements in this respect were made.

10 **H:** hmm
11 **R:** ... most likely I hope and therefore, yeah, I'd like to ask you some questions, respectively,
12   let's simply talk about computers, so what I am interest in is/*..*/ if you know, if you
13   still know /*.*/ the situation when you saw computers for the first time
14 **H:** hm /*..*/
15 **R:** How was it, the situation, /*..*/
16 **H:** (*hesitant*) I, I, because it wasn't for the first time mainly but it was the first
17   experience when I had it for my own use for a long time, any time I wanted to so I felt
18   like being there all the time for the first few couple of weeks like a month and then later I
19   got used and the the first morale that I had for the first couple of weeks kept increasing
20   with time. /*..*/
21 **R:** And can you still remember when you saw it for the very first time /*....*/
22 **H:** Yea, I, I can /*..*/, actually I can remember. I was somehow confused not as confused
23   but /*..*/ I, I when I looked at, for example the keyboard, I related it to the keyboard of
24   a typewriter, alright? So I know definitely there was something to do with typing but I
25   didn't know anything beyond that apart from the, the keyboard that seemed to be
26   related to the keyboard of a typewriter, because I had seen a typewriter sometime.
27 **R:** And, the mouse
28 **H:** mm..I hadn't, the the rest of the parts of the computer I had never seen. Actually, the
29   keyboard I had never seen but I, when I first saw it for the first time I just related it to,
30   to a typewriter

Example 13.7: *Unrestricted access for teachers*

Searching for equivalent examples retrieves a more precise definition of usage, as apparently the usage of the computer lab for **private** tasks is not taken for granted by the teachers. To give an example, one informant explains that he was obtaining permission by the Headmistress (as an entity of the SMC within his physical reach) to use the computers for the homework in his private study course that he was attending at an advanced education institution - Ex13.8 points this out. More information on the private usage can be found in Section 15.1.

318 **R:** How are you u, using the computer now? For purposes?
319 **A:** The, these days?
320 **R:** Yeah.
321 **A:** For examination purposes.
322 **R:** Mhm.
323 **A:** For, for letter writing, /*.*/ for, as I have told you for my study cause I asked the school
324   they give me that privilege.
325 **R:** Mm.
326 **A:** If I want to use I'm free

Example 13.8: *Teacher asks for private usage*

The issue of using computers for private purposes is not only indicated by the example mentioned above, but also by sidestepping answers according to the interviewer's question if the teacher uses the school's computers privately. As shown in Chapter 11, Ex11.8 a direct response to the question about her private usage is avoided and a work around established. (Though, this is not limited to the teaching staff and applies also to the secretaries.) Ex13.6 shows this clearly. Hierarchically speaking, the teaching staff is placed below the SMC, whom they are asking for access - be it to use the computers for private issues or to get the key to unlock the door. As shown in Ex11.3 teachers are more powerful than pupils, as they create their own rules. They can punish them physically without consequences and are able to deny them access to the computer laboratory.

13.3   School Management Committee

Another actor who influences the daily business of the school and therefore also the computer lab is the SMC, which decides not only about the (financial) ongoing expenses at the school (see Page 119), but also about the access to the school's computers. It was shown above (see Ex13.5Line701,709) that the SMC is obviously responsible for the fact that matrons are not allowed to use computers any longer. It consists of representatives of the pupil's parents, the Headmistress and the Deputy Teacher.

13.3.1   Headmistress & Deputy Teacher

Investigating the physical access to computers shows that these two entities have access to the Headmistress office and consequently to the keys of the computer room: the Headmistress and the Deputy Teacher. Apparently, they are the most powerful people permanently living on the compound[7]. They schedule the staff work and decide whether someone is allowed to leave the school temporarily to prepare funerals or weddings. Their responsibility as members of the SMC is not only to provide the quality of teaching, but also to guarantee that employees like the bursar do their jobs.

The Headmistress presents herself clearly as superior to all others on the compound when she protects the new computer administrator and tells the other teachers to study computers - see Ex13.9. A further example is when she, as a member of the SMC decides upon the computer access of the matrons and that the bursar should not pass over work to the secretaries any longer. Later Examples (Ex13.9 and Ex13.10) show that the incapacity of using computers can be instrumentalized by people to pass on work to those who have the capability. Thereby, nescience is covered and utilized to gain power over the others. Nevertheless, the Headmistress is in charge of the situation and tells the others "*Learn, then you can't now tell ...*" - Ex13.9Line917 or "*Please, do your work, [...] produce it yourself*" which was evidently fruitful, as the bursar is learning computers from then on - Ex13.10Line943. This underlines her exceptional position which, of course becomes evident e.g. in staff meetings or by the fact that she is the only person who has a driving license on the compound and does the school's weekly shopping in town. Due to her official obligations she teaches less which makes her different to other teachers as well.

915 **A:** So a person commented and said: "Ehh, Juliet's work, Juliet's exam is neater a, a,a, you
916   know to handle it", I said: "I told you earlier"
917 **R:** Yeah.
918 **A:** *(*laughing*)* Learn, then you can't now tell Juliet print our work she has other things to
919   do.

Example 13.9: *Headmistress and new computer administrator*

---

[7]Except the cook, secretaries and the bursar all the employess live in houses on the compound.

This example shows that the Headmistress orders the bursar to carry out her duties more and to focus on learning computers than on handing on work to the secretaries and being unsatisfied with the product afterwards:

924 **A:** So some of the things they will have to learn, anyway where, eheh, either they keep on
925   suffering or they learn in order to /*.*/ bring out their work or /*.*/ even not waste a lot
926   of time, ahm, another example is Patricia, the bursar, the, our bursar,
927 **R:** .. buzzer?
928 **A:** the cashier, ..
929 **R:** Ah, yeah. Yeah, yeah.
930 **A:** Ah, sometimes like at the end of the term, we have various meeting, meeting, various
931   comittees, the management and.
932 **R:** mhm.
933 **A:** And she has some of the things to print ou, to, to, print out and you know to bring for,
934   for presentation.
935 **R:** Mhm.
936 **A:** So you think, whether last term or the other term, she gave the secretaries "type this
937   work in this way and this .." and then when they made some mistake she was: "
938   wraupqupqeup" I told her: "Please, do your work, .."
939 **R:** Mhm.
940 **A:** ..produce it yourself .. *(*small laugh*)* .. the computers are here. So the *(*small laughter*
941   *again*)* ..
942 **R:** And did she?
943 **A:** Eh, yeah, I, she is now serious, i think she, she utilized, Caroline seems, Caroline is
944   giving her some lessons, ... I said you should learn now how to produce work, this is the
945   work to be presented, ... Mhm.

Example 13.10: *Bursar asks secretaries for help*

In teacher interviews her superior hierarchical position also becomes evident and shows that the superior hierarchical position of the SMC is perceived in the same way by the teachers, too - see Ex13.5Line731f.

The position of the Deputy Teacher is more difficult to grasp and I understand his position more as an "in-between" for the Headmistress's position and the other teachers. In addition to his general load of teaching (like the other teaching staff) the Deputy Teacher is responsible for organizing the timetable, maintaining the school library and managing school related events like the term sports day or the school's participation in diverse competitions. The Deputy Teacher has to be considered as more powerful than his colleagues, as he is participating in the meetings of the SMC and has access to the Headmistress office (the "key").

13.3.2   Pupil's Parents

By applying the criteria of analysis it is evident that in addition to the obvious participants (members of the school) there is a further one: the pupil's parents, who can be seen as an entity of society and a representation of its rules.

When the informant is asked whether she was allowed to use the computer at home, she denies it, but notes that as soon as she was taught at school she gained access to computers at home - see Ex13.11.

37 **R:** So did he teach you then how to use it or a bit, or?

38 **Y:** No, he said, since we haven't yet st, since we haven't yet s(*??*), learning at school, well,

39   when you learn, when you start learning, he'll let me use his.

40 **R:** Aha.

41 **Y:** So now when we started, he has let me use it.

Example 13.11: *Parents are granting access*

Referring to the above lines (Ex13.11) it seems as if the fact of having computer lessons at school is reason enough to enable the children to have access to computers. This means that the school controls the access to technology in two ways: both at school and at the children's home. At first it seems that the parents rely on the school's quality and trust in the school's education. However, there are indications that some parents had to be convinced by showing them in practical terms that the children are capable of using computers. Ex13.12 below shows that some children were not allowed to use computers at home because of the fact that they were learning it at school. They had to prove their knowledge by showing it practically. Therefore, the usage at home is controlled by the parents, but enabled by the school. The main fear of the parents is that their children "*may cause a fault*" or "*loose some of their files*" - In26Line118. According to the child, who varies in including and excluding herself from the mentioned pupil group (she is shifting from "*we*" (Ex13.12Line114) to "*they*" (Ex13.12Line119) and back to "*we*" (Ex13.12Line131)), some class mates attempted to convince their parents, which is perceived as hard (Ex13.12Line130).

112 **R:** Hm. Hm. I see. So do you have, ah, the possibility to use a computer at home? /*...*/

113   When you are at your parents' place?

114 **P:** There is a small /*..*/ possibility that you can use computer at home, because most

115   parents cannot accept us to use computers because they know we don't know how to use

116   them. And we may cause problems.

117 **R:** They know or they think?

118 **P:** They think. They think that you don't know how to use them and maybe you may cause

119   a fault, /*..*/ you may loose some of their files and different things.

120 **R:** So, have you ever tried to convince them?

121 **P:** Yes. No, because some of us have convinced them now they, they can use the computer.

122   Some parents have now, have now, offered maybe, shall, shall, now shall we have better

123   thinking of computer. Now they accept us. When, when we have nothing to do like any

124   homework accept us to use their computers.

125 **R:** Hm.

126 **P:** Like, we can also play games, we can write letters, we can now maybe send emails. They

127   can easily get for us an email, email. Yes. e−mailaddress.

128 **R:** And your parents think so as well?

129 **P:** Yeah.

130 **R:** So was it difficult to convince them or?

131 **P:** No, I, it was kind of, because kind of to convince, hard to convince them. But now they

132   understand because, we'd, we, we tried out some things like maybe opening Windows,

133   like Microsoft Excel, maybe writing a short letter when not shutting down the computer

134   so maybe it's /*.*/ it was easier to convince them that way than speaking to them.

Example 13.12: *Parents want to see the capability*

Investigating the way how parents were convinced shows that practical presented activities on the computer, which "*was easier to convince them that way than speaking to them"* (Ex13.12Line131f). These activities are representational forms of computer knowledge - starting programs, opening and closing windows is enough knowledge assurance to overcome the fear of the parents that their children could spoil something - Ex11.2Line81. If this is due to the parent's own lack of knowledge is not definite. Consequently, the pupils were sufficiently self-confident that they are capable of using the technology, as otherwise they would not have made the attempt. This is emphasized when informants indirectly admit that they did not dare to ask, yet. This might be correlated to the fact that pupils are afraid of measurements of sanctions and (corporal) punishment, which are, in addition to the local computer teachers (LCTs) (Ex11.2Line11,12), also performed by the pupil's parents (Ex11.4).

588 **Y:** You find you have made a mess, a really big mess, *[*..*]* and well, some parents are strict
589   they find you and and give you up, they beat you,

Example 13.13: *Parents punishing pupils*

Obviously, the way parents decide on "how" and "when" their children access computers is not deliberate, but bound to preconditions. This is represented by phrases like *"when we have nothing to do like any homework"* - Ex13.12Line124, which prioritizes school related education to the private dedication to computers. Statements like "[I can use the computer] *Not every time because my dad is also doing something"* - In17Line198, or *"Because much times my grandma is on the computer doing some work"* - In23Line142 emphasize this impression. The last two statements indicate an access privilege to the one who is using it for work. All in all, the decision about the certain point of time of computer access, the allocated time and how computers are used at home is still placed within the power range of the parents. They define the purpose of usage and when pupils are allowed to use computers. The usage sometimes coincides with some additional computer teaching by the parents - see Ex13.14. Being educated by parents means that this is a measurement of control as well, because as long as one teaches the other one, he/she is in charge of what activities are performed on the computer. The following examples point this out:

155 **R:** So at home you don't use it for games is it?
156 **A:** At times we do, other times we don't, but when dad is teaching us mainly, it's it's
157   learning not playing games.

Example 13.14: *Parents are teaching as well*

142 **S:** Because much times my grandma is on the computer doing some work
143 **R:** Mhm.
144 **S:** So I don't get chance.
145 **R:** Mhm.
146 **S:** I don't get time.
147 **R:** But she is teaching you sometimes. Or?
148 **S:** Mhm. She teaches me sometimes and then she tells me to go on and do what she told me.

Example 13.15: *Relatives are teaching as well*

Unfortunately, no information on how much time the children are using a computer to do work or to play games could be found. Those programs, which are evidently preferred and commonly used by the pupils are discussed in Section 15.1.

13.4   Computer Teachers

If it is an unknown teacher in a town or a relative of a pupil, repetitive learning experiences with other (computer) teachers emerge. These other teachers can be relatives of the informant, the school's computer teachers, or unnamed people of learning institutions where some informants attend computer lessons.

13.4.1   Relatives as Teachers

As indicated one group of LCTs can be grouped as the informant's relatives, e.g. the informant's grandmother, father, mother, or uncle. The extent of how far relatives contribute to the transfer of computer knowledge is not retrievable, but remains a fact. In Ex13.14 the pupil lists her father as a teacher and in Ex13.15 the grandmother.

13.4.2   Teachers in Town

During the interviews it turns out that some members of staff attend lessons both at the place the interviewer was teaching and at a certain place in town where other computer teachers are teaching.

   Some of them wanted to be prepared for the school's project, whereas others did not mention any reason at all. Differences in the way of teaching are discussed later on (see Page 227). For now, it is enough to show that there are other computer teachers in Kabale:

35 **J:** So, ah I just ignored it that, ah maybe the computer is not yet come I, I, I will go
36   somewhere and, and learn it. So, in the year 2000 I went down in town. 99. Someone
37   start t, to teach how you can open it, how you can shut it and, and for the cursor it was
38   also difficult to get something with the mouse, to get the cursor there.

   Example 13.16: *A secretary is taking lessons in town*

13.4.3   School's Computer Staff

With the departure of the AT in October 2002, the school hired an extra computer teacher to take care of the lab and continue with the teaching. The first teacher was Mr. Rukinga, ("*when you left there was this man Rukinga*" - In5Line309) who was member of the PTA as well. He has a good basic computer knowledge[8] and has been extensively trained by the ATs. After the departure of the ATs he was employed by the school, but failed to hold his lessons several times and was consequently dismissed after a few months. The reason why he did not perform his obligations seems to be some kind of local brain drain, which is discussed in Section 15.4.2. Consequently, the SMC decided to allow[9] young teachers, who had already been teaching on the school previously, to enrich their (previous) computer knowledge by some extra lessons in town in order to take care of the computer lab. The SMC reduced their "normal" teaching hours in favor of computer lessons. In addition to these "official" school

---

[8] As good basic computer knowledge I consider e.g. the capability to mount a disc-drive on his own.
[9] To my knowledge this measurement aimed to get less depended on one person, as after the departure of Mr. Rukinga the school had problems to maintain the lab up and running.

computer administrators, a familiar British Parish had sent an English teacher, who took over some computer teaching as well. This is shown in Ex13.10, where she is identified as the computer teacher of the bursar. In terms of access to the computers, the school's computer staff is considered as something more powerful than the common teacher, as during the day they are usually in charge of the key to the computer classroom.

13.4.4   Austrian Computer Teachers

A priori it was not apparent that the *Austrian Computer Teachers* (ATs) represent an own actor. During the analysis it became evident that they represented an autonomous actor with several roles, not only because this separated role is introduced by the ATs themselves (Ex13.17), but is also identified by the other teaching staff and pupils.

112 **R:** What do you think is the difference between the lessons you got in town and *[*..*]* of
113   ours?
114 **H:** The main difference is that mainly because those people were paid, I, I had to pay for
115   those lessons and these were free anyway so that makes a difference on its own. And
116   another thing is that because there were few computers and there were many of us, so we
117   would be allocated a certain period of time so if your time would be over even if you
118   hadn't got what you wanted to get you would leave and someone else gets in but I think
119   here I was free to ask whatever I wanted and use it for the time I wanted to use it and
120   being around with you. But that was like a school, you go there, time goes you go but
121   here we would be around with you anytime we feel like coming and asking even when it
122   was not time for lessons we would be able to do it so I think it was also a good part of it
123   and a difference *[*.*]* and it's like there were some facilities there that we would not use
124   like the printing *[*..*]* they would, actually the printing would be done theoretically they
125   would tell you do this command this and put like this and you would end eld *(*??*)* not
126   even see where the printer is working from because the printer would be somewhere in a
127   room, they would just be connected so they would simply tell you when you are going to
128   print you do this like this like this theoretically but with this you would say "follow the
129   following procedures and then print", you print on your own and see what you have done
130   has produced something practically. *[*.*]* So I think it was also a difference.

Example 13.17: *Other computer teachers emerge as actors*

Apparently, the interviewer distances himself from the above mentioned computer teacher groups by requesting the informant to differ between the lessons she attended in town and of the group he belongs to. Doing so he refers to the teachers in town - Ex13.17Line118. The interviewee's implicit agreement underlines that there were further computer teachers in town and are distinguishable to the ATs. These differences are pointed out in several further interviews and are discussed on Page 221.

As deduced in Ex11.2 pupils are also distinguishing between the ATs and the school's computer staff, at least by their way of teaching. If the pupils perceive the ATs as more powerful than other school teachers could not be determined but what we could figure out are several different roles which are ascribed to the ATs. These roles are the focus of the following sections.

13.5   Interviewer and his Roles

13.5.1   Interviewer

The typical role an interviewer takes up during an interview is one which implies a quite powerful position in relation to the interviewee, as he/she is in charge of the interview situation. He/She is the one who decides about the starting and ending point of time of the conversation. He/She is the one who defines the progress of the interview and it is up to him to decide whether a question has been answered sufficiently. According to the methodology of this study (as discussed in Chapter 6) the interviewer is supposed to achieve a rather domination-free sphere during the interview in order to retrieve as much narrative statements as possible. Whenever the interviewee responds freely and e.g. criticizes the interviewer in his role as a teacher, this indicates that the interviewer in fact is perceived as a less hierarchical person at this point of time. Please note that a teacher-pupil relationship is strongly hierarchical and poses a difficult relational frame to achieve narrative statements[10]. As mentioned above the relationship between interviewer and interviewee is hierarchically constructed as well, but due to the fact that both are taking up their roles during an interview these roles are constantly re-negotiated, but reconstructed as well. Therefore, the teacher-pupil relationship is considered to be more desirable than a teacher-pupil relationship. The optimum situation would mean a trustful relationship between two friends who have a "free chat" about certain topics.

During the interviews of this study such a relationship occurred only to a limited extent, as several times the interviewer was perceived in different roles instead as the one of an interviewer, which he obviously had previously taken over. Ex13.7 shows how the interviewer attempts to establish an egalitarian interview situation. The first four lines of this sequence deals with the annotation of the used microphone. When in Ex13.7Line4 the interviewee notes that it is the smallest one she has ever seen, the interviewer considers himself obliged to understate this fact and emphasizes that the microphone does not belong to him. Actually, as a matter of fact the microphone is noticed, which means that both are aware that this interview is recorded, as the interviewee compares the microphone to others, and not necessarily that the interviewer is superior. This might not have been necessary at all as the informant is familiar with the usage of a microphone. After some seconds of hesitation the interviewer continues with his attempt of specifying the question, which lasts until Ex13.7Line15. Within these lines he aims at reasoning why he wants to get a "*feedback*" (Ex13.7Line6) and "*talk about computers*" (Ex13.7Line12) and aims at achieving domination-free sphere. As indicated above a domination-free sphere is de facto not possible during an interview due to the fact that the interviewer is supposed to conduct the interview and to direct it. Nevertheless, by understating his powerful position and ascribing the interviewee as more powerful by phrases like "*it's really for me to help me*" (Ex11.1Line6) or "*so everything is ok*" (In13Line346) he reduces restraints to talk freely.

The following sections will outline what kind of other roles are ascribed to the interviewer as he takes over.

---

[10]I have shown this several times above. Please see Ex13.2 or Ex11.1.

13.5.2   Computer Teacher

One apparent and dominant role during the whole research is the role of a computer teacher. With his activities as a computer teacher the interviewer was ascribed and in fact took over the role of a computer teacher, together with his AT colleague. He ascribes this role also to himself and identifies himself as a member of the ATs during the interviews. This is acknowledged and approved by the pupils (see Ex11.2) and teachers (see Page 221).

This role seems to be quite deeply ascribed to him, as during the interviews, the informants perceive him several times in the role of a teacher though he had explicitly aimed at avoiding this with statements like: "*... and it's of course, it's no examination or test or something, it is just for me to get a feedback [...] because I'd like to improve my teaching ..*" - In12Line8. As discussed in Section 13.1 and shown in Ex11.2 the role of a teacher is a priori assumed by the interviewees and rather predominant. However, both roles are constantly exchanged. As insinuated in Chapter 11, this changing of roles is also influenced by a general uncertainty towards an interview situation in general. Statements like "[The interview was hard], *because it was my first interview*" - In25Line205 indicate this assumption. Memos of the participant observations show that some teachers were watching the interviewer doing other interviews in the computer lab although they are usually not in there[11].

The following example explicates how the changing of roles has happened:

174   **R:** Mhm, when you are thinking about the computer lessons, what did you like especially?
175   **P:** Hm, I like the subject, the computer.
176   **R:** Yea,.. or, or, yeah, what subject, or just in general anything
177   **P:** I like Microsoft Excel.
178   **R:** Mhm.
179   **P:** Ah there, we, I think we only studied it recently, like last week though we were making
180   cards bringing them out, it was cool so you would like, you would choose anything from
181   the table or at whatev, whatever it was and *(*clears throat*)* you would make a card of
182   your own choice and I thought that maybe someone had a dis, a diskette, they would add
183   in more and you even make a much better card than what you have made.
184   **R:** And about the way of teaching?
185   **P:** the way of teaching, it's okay though they teach because like they, they tell you to, you,
186   you, switch on, they no longer tell us to switch or do anything you just enter, switch on
187   the computer get the subject you were learning and then aah you switch on then, you
188   open were you were, the subject you were on then like if you get stuck, like in aah *[*..*]*
189   in reducing on the size of the picture or something like sort of the, something like that
190   you ask a teacher and they come and help you but the rest of them are easy *[*.*]* you
191   understand from the first lesson you get, then second lesson in just a step out.

---

[11]Most of the interviews were done in the computer classroom to allow associations related to the computer. It was also aimed to conduct the interviews in a well-known/trusted atmosphere. However, to be in the computer classroom did not necessarily imply to feel comfortable. As several teachers do not use the computer lab regularly like some colleagues it was obviously more inconvenient to them than to the regular users. I conclude this from the fact, because those interviews I was taking with adults outside the computer-classroom were those with the longest duration. Almost all interviews above the average of 44 minutes were not conducted in the computer class. The only interview which was not taken outside the computer lab and which is above average as well is the one with the new computer administrator. From a methodological point of view this underlines the approach to conduct interviews in a trusted environment.

192 **R:** Mhm, *[*.*]* what do you think is so special about computers?
193 **P:** *[*...*]* there is *(*whispers*)* nothing special about computers, I don't *(*laughs*)*
194 **R:** Yeah, it's, *(*seems to be smiling*)* aah, what would you do, if you had too teach
195 computers. What do you think would be the most important topics, subjects to teach
196 about?
197 **P:** mm *[*..*]*, mm *[*.................*]* pardon me your question?

Example 13.18: *Interviewer changes roles*

At the beginning of this sequence the interviewer asks in the role of an interviewer what the interviewee especially likes about the computer lessons. The replied "subject", "*the computer*" is vague at first (Ex13.18Line175), as it could mean the physical object of a computer or the fact that lessons are given. Nevertheless, she does not directly reply to the question of Ex13.18Line174 about the computer lessons themselves, but defines the preferred subject as "*Microsoft Excel*". Previously, the interviewer has asked for clarification ("*what subject*"). Thereby, he takes up a powerful position and asks the girl to specify, though he relativizes this by "*or just in general anything*" (Ex13.18Line177). Consequently, her mentioning of and specification of "*Microsoft Excel*" is done from a pupils perspective, as she immediately shifts in her narration to "*making cards*", which is apparently more interesting.

In addition to that phrases such as "*own choice*", "*whatever*", or "*anything*" represent personal responsibility and the "*bringing out of the cards*" is perceived as an evidence of accomplishment. Thus, knowledge is presented in a processual and reconstructive way, which is continued in Ex13.18Line186-192, as well. When the interviewee begins with the processual representation, she switches from a personal perspective to indirect addressing (Ex13.18Line182-184, Ex13.18Line186-192). Interestingly, she does not identify the interviewer as her computer teacher at that point of time, but perceives him as an interviewer.

The AT-pupil relationship is dominated by the different rights of access to technology, which also shows the different hierarchic levels. The words "*Then sometimes when, when you would copy, you would do something that we were forbidden to do,*" indicate that the "*we*" group is different from the interrogator who did things beyond an area of power. This incites to consider the teacher group superior to the children. At this point of the interview the "*we*" term could still be considered both as the pupils and as the group of teachers and pupils, but in (Ex11.2Line11-12,18-21) the pronoun "*we*" is used for the pupils and "*they*" for the *local computer teachers* (LCT). This shows that the ATs have more power when using the computer than the pupils.

Hierarchically speaking, when the ATs[12] are perceived in this specific role they were not only powerful because they controlled the physical key to the computer lab. Being the school's computer experts they participated in SMC meetings to advise how the computer lab should be established. The ATs were also discussing future necessary steps, which the school is supposed to take. As the school realized some of their inputs, they apparently influenced school political decisions. Their powerful position is also found in the fact that their teaching methods were not criticized during their first stay, though they were proceeding too fast and had a syllabus, which is negatively connoted

---

[12]From now on, as my colleague and I were working as teachers at the school, I refer to us as ATs and avoid shifting between the other role names.

- see Section 15.4.1.

13.5.3 Evaluator

The role of an evaluator and a teacher are related to each other, as both are evaluating something. Especially in interviews with adults these roles are mixed up and inseparable. As the interviewer was a (computer) teacher of the school's teaching staff as well, one cannot define whether the interviewees behaved according to their internalized norm of an interviewee or of a student.

I have shown that by asking about the differences to other computer teachers or by asking questions about the usage of computers for private things, the interviewer becomes an evaluator - see Ex11.8. This occurs several times, but is also inevitable when one is asking about differences between e.g. the ways of teaching. Another interview issue which placed the interviewer apparently in the role of an evaluator was (again especially during adult interviews), when he asked about the private usage of the school computers. As Ex13.6 apparently depicts that the interviewee feels examined, and consequently avoids a direct response. She denies her private usage at first, but contradicts this previous statement later on.

As shown in Chapter 11 the interviewer fails to remain in his role as an interrogator and switches to the one of an evaluator several times .

13.5.4 Donor and Representatives of Donors

Any intervention induces new resources in an existing environment, be it personal or materialistic ones. The investigated intervention provided both: two computer teachers for the period of three months and the equipment of a computer lab. Apparently, as the appearance of both temporally coincided, the ATs are sometimes perceived as those who bought and brought the computers - see Ex13.19.

556 **L:** But you are the guys who brought these computers.
557 **R:** We brought, yeah, but we didn't buy.
558 **L:** You didn't buy?
559 **R:** No.
560 **L:** I though you were the one who bought them.
561 **R:** No.

Example 13.19: *ATs are perceived as donors*

Most teachers know that this is not true, but when the ATs were attending meetings of the SMC about the school's strategy according the computer lab and when they were holding workshops about the future of the computer lab, they acted in fact in the name of the donor. Between the first intervention period in 2002 and the one in 2003 the e-mail correspondence about further material equipment was also lead between Mr. Dekanya[13] and the ATs and not between the school and Ilse Kroisenbacher of the Projekt=Uganda. Even during the interviews the interviewer was confronted as a member of ATs with this role, as Ex13.20 shows.

---

[13] He is the local manager of an international NGO and has arranged the whole intervention.

A: Yes, but, but what about the donors. Do they still have interest in us?
R: Well, they have interest but there is no money left for now. Even for this project their was hardly any money left. So, I, I got this as a sponsorship, my flight here from, from the government and the spare parts we only, could, we had to cut down the amount of spare parts we could offer like the scanner, because we had, the, the donor, she, she got, she got a bill of a company who br, who was delivering last year and they were giving her a bill one year later, /*..*/ of one million.

Example 13.20: *ATs are representing the donors*

From the above example it becomes obvious that the interviewer freely takes over the position of a representative of the donor. He represents the Projekt=Uganda and is well informed about the financial situation. This shows at least a good informed status, but not necessarily that he is a factual representative of the donor. However, in correlation with the fact that as long as the ATs were teaching at the school the matrons had access to the computer, highlight that the AT had a quite powerful position. This could not be explained by other roles. This position is emphasized, when concerns about an existing LAN (due to privacy see Page 191) were not mentioned in the first period. Furthermore, there was one incident, which clearly describes the hierarchical relationship between the school's teacher, pupils and ATs best: When the computer room was equipped for the first time the ATs asked the teachers to help them in order to practice the "plugging in" of the components. A few weeks later the hedgerow which was bordering to the computer lab caught fire, which caused the ATs to evacuate the computer room. By the help of the children who were in the lab at that point of time the equipment was saved. The fire was put out and fortunately the room was not damaged. Therefore, the ATs intended to re-equip the room immediately. As a form of reward they wanted to do so with the children. After the action had started teachers chased away the children and claimed for their privilege to do the re-equip the room. The ATs insisted to do this work with the children as a reward for their help. Furthermore they argued that the teachers had previously equipped the computer lab and could not gain more knowledge in plugging in parts anyway. These arguments were somehow not plausible to the teachers, but within a few sentences the ATs **"made their decision clear"**[14] and the computer room was re-installed with the help of the children. This meant that the ATs were, hierarchically speaking, above the other school staff and pupils. Furthermore, we realize once more that the teachers consider themselves in a prioritized position to children. Considering the whole context the ATs aimed at treating them all equally, but in this certain situation they favored the pupils. This indicates that the ATs empowered the pupils according to their understanding how children are supposed to be educated. Apparently, this is contradicting to the local way.

---

[14] *orig.* "Die Lehrerkörperschaft protestierte lautstark, weil sie in der Einrichtung des Computerraumes ein Monopol ihrerseits sehen/sahen. Lehrer sind in der gesellschaftlichen Ordnung gegenüber Kindern „höherrangig" . Da ich deren Proteste sehr entschlossen entgegnete, setzte ich meine Bestrebungen doch durch." [TD|p.4] (*itmo* see in *thick description* (TD), Page 4)

13.5.5   Colleague

The role of a colleague, though a priori assumed and desired, does not emerge. From a terminological view this term turned up just once in an interview. Even in that case it was applied by the interviewer himself. No other member of the teaching staff, which actually had been colleagues to a certain extent, refers elsewhere to the ATs as colleagues. This coincides with other observations, which show that when the ATs were *invited* to participate in certain activities, e.g. join the school's field day, other teachers were *obliged* to do so. Let us now turn to the example of when the interviewer discusses with an informant the reason why he did not make any formal tests and presents himself as a colleague:

678 **R:** But then the problem is, because of my authority and out of the hierarchic position,
679   because I am a colleague so I don't want ..

Example 13.21: *Interviewer refers himself as a colleague*

This Example (Ex13.21) shows that the interviewer himself has difficulties with the ascribed authority and his self-perception as a colleague. This situation is contradicting to him and can be seen as a result of role conflicts, because as an "equal" colleague he would not have had doubts to discuss such issues in public. Only, when he is aware of his other, more powerful roles, as a teacher, evaluator or donor might be, this conflict arises. Any teacher-student relationship implies a mandatory strong hierarchical relationship which cannot be avoided, as one is actively aiming at conveying knowledge. The relationship between the ATs and the local teaching staff is evidently more coined by a teacher-student relationship as the one of a colleague. Another informant mentions that he did not dare to say anything about his desire to be tested, as *"you can't ask for an exam"* - In27Line740.

13.5.6   Researcher

This role, did not really emerge to a significant extent as the main reason for making interviews was based on improving teaching for other interventions and not for academic research. Nevertheless, this role is immanently involved and has influenced the progress of the interviews to a certain extent which will be discussed in Chapter 16.

13.6   Public Opinion

One actor stands for the public opinion about computers, which is shaped by media, retold experiences, and cultural leaders, such as Festo Karwemera. Such entities influence computer introduction, which the following examples show.

543 **R:** do you know if there is a Rukiga word for computer?
544 **C:** aaah yea
545 **R:** mmhm
546 **C:** 'Kalimagezi'
547 **R:** mmhm
548 **C:** mm, and there the Bakiga who hear it on the radio before we didn't know it so we got it
549   from the radio
550 **R:** aaah

551   **C:** from that man who teaches Rukiga

Example 13.22: Radio as opinion leader

1206   **H:** But I knew it was a Rukiga word. Because, I came to know that I knew a computer is
1207   kalimagezi. From the way I always hear it on the radio.

Example 13.23: Radio as an information source

Concluding from the above example that the term "*kalimagezi*" for computer was promoted via radio and shaped the adoption by the local population. When one looks at the example on Page 207 from In19Line190ff we find a further indication that due to lacking personal experience one refers to the opinion of others. Thus, their importance becomes evident and indicates that the public opinion encourages people to use computers by ascribing benefits to it ("*makes work easy*", "*helps you to earn money*") - see Section 14.3. Another form of transferring information is the exchange of experiences. Evidently, the exchanging and retelling of computer experiences contributes to the local concept of a computer. Experiences are adapted to own wishes and consequently slightly falsified. This seems to have lead to generalized statements about computers. Although, these benefits are not necessarily experienced by the people themselves, which is discussed later on - see Section 14.3. Another issue is that the public opinion has an impact on is the conveyed opinion on how the technology should be approached (see Section 15.4.1).

13.7   Educational and Political Institutions

The ways how institutions and the Ugandan government influence the adoption of computers are different, as they do not promote computers and their characteristics abstractly, but setup more time-related preferences and restrictions. When one informant explains that in his vocational training he is required "*to have some work done on the computer*" - In27Line238, the institution where he attends the course actually encourages him to learn computers, if he wants to pass -"*you cannot submit that work which is not computerized*" - In27Line246. Such courses are usually held in semester breaks and so the course participant has about three months to do his exercises. According to the informant outsourcing the work is theoretically possible ("*going out for to look for people to do my work*" - In29Line330), but due to high production costs[15] this option is discarded and the possibility to use the school's equipment utilized.

Regarding implications induced by governmental constraints reveals that the nationwide curriculum and exams for primary schools have involved the shortening of regular computer lessons for the pupils. The school has no scope to de-emphasize certain subjects in favour of computer lessons, as e.g. the marks reached in the nationwide exam is decisive for the admittance to secondary schools. During the first period it became apparent that teachers avoided to victimize "their" lessons in favour of computer lessons, as the feared failing their teaching aims. Consequently, the Deputy Teacher replaced some Science and Social Studies (SST) lessons with computer instruction. This was also just a temporary measurement and did not continue after the ATs departure.

---

[15] The example homework the informant mentions amounts to $ 42.

If a school does not teach on behalf of the government's curriculum, it implies a cut down or even loss of financial support. As a consequence to above reasons, computer lessons are considered as "*an extra curricula activity*" - In1Line244, which is not tested nationwide and has to be neglected in favor of those subject which are tested. A more flexible model for innovative schools should be possible. Regarding the multitude of lessons for the pupils and the high workload of a teacher makes it evident that inducing computer lessons in addition to the basic workload is problematic.

237   **A:** Because of the government curricula. We have had to cut down the computer *(*laughing*
238   **)* lessons ..
239   **R:** Mhm.
240   **A:** .. for the children. Because these, you know the Ugandan system is such that, because
241   there are certain *(*??*)* things that are going to be examined and on some days that are
242   going to be examined.

Example 13.24: *Governments curricula influence computer lessons*

13.8   Intervention Broker

The man who arranged the intervention, Mr. Dekanya is also an actor in this socio-cultural environment. He has levered the whole intervention and therefore his opinion is apparently valuated highly. Evidently, he has taken up the role of the local donor's representative, when he acted as a communication broker between the intervention's pause from 2002 until 2003 between the school and the ATs, because he has obviously influenced the matrons' level of access to the computers, as well. By the following example this assumption arises, though it could not be discovered to whom the subject "*they*" refers to - Ex13.25Line739. Apparently, the pronoun "*they*" means either "*teachers*" or the "*pupil's parents*", but it cannot be settled.

737   **C:** they I think they went to the Head Mistress and I think they said that they are for the
738   pupils and teachers
739   **R:** mmhm
740   **C:** that when they were coming from Austria the main purpose w, was, were, were, was
741   mainly from Mr. Dekanya was for the pupils to learn them not the matrons so they were
742   ignored in that way

Example 13.25: *Mr. Dekanya emerges as a powerful actor*

Another example for his powerful position is the fact that the ATs "had to repair"[16] some computers in his office and to teach some members of his staff. This issue becomes a sign of power in combination with the fact that the ATs were previously rebuked by the Headmistress that their duty is to work for this school and not someone else, when they had installed a computer lab in a neighboring school[17]. Thus, the ATs were

---

[16] Actually, he asked the ATs to help him, but as he was the one who arranged the whole intervention and the one who picked the ATs up in Kampala and transported them to Kabale, the ATs felt themselves obliged to meet his inquiry.

[17] A missionary from Ohio had donated the equipment for a 15 PC strong computer lab, but did not provide the school with skilled personal to install the computers. The Headmaster of this school asked the ATs for their help and in one afternoon they installed the computer class room. After their return, the Headmistress immediately signified them that she expects them to work for the school and not for anybody else.

seen as a resource dedicated to the school without any exceptions. To give a further example; When some members of the Computer Club of the neighboring university were dropping by to ask some questions, the ATs were informed by the Headmistress that they were not supposed to teach other people other than members of the school. The reason why Mr. Dekanya was not obliged by this rule seems to be because the school owed him something and therefore, granted him access to the resource ATs. Obviously, the Headmistress did not want to burden the relationship between her and Mr. Dekanya. If these decisions were made upon her own personal opinion or in consultation with the SMC is unknown.

Chapter 14

The Object "Computer" - Ways of Presentation

14.1 Computer Names

14.1.1 Kalimagezi

Searching the interviews for used names for computers, allows to get closer to the peoples' perception and concepts of this technology. During the interviews I have asked the interviewees about their nicknames for computers and if they know about translations of the term "computer" into their mother tongue. Answers proved to be quite different, and whereas some people are not aware of a translation for computers, (which indicates that it has not infiltrated the environment yet), some others claim that the Rukiga term for computer is "*Kalimagezi*", which is translated as "*something full of knowledge*". Investigations have shown that this translation is actually a term which has been introduced via the neighboring district Buganda and is a Bugandan expression. The local cultural expert Festo Karwemera claims that he has introduced the Rukiga word "*kanyabwengye*" in his local radio program to achieve a better integration of computers in the area of Kabale.

1203 **H:** For example now you asked me another name, a name for the computer in Rukiga I told
1204 you kalimagezi. But Kalimagezi is an Ruganda word.
1205 **R:** Yeah.
1206 **H:** But I knew it was a Rukiga word. Because, I came to know that I knew a computer is
1207 kalimagezi. From the way I always hear it on the radio.
1208 **R:** Yeah, but, uah, I, that *(∗background noises − paper∗)* I think the, I don't know if I find
1209 it now. *[∗..∗]* Kanyabwengye. *(∗?∗)*
1210 **H:** Kania?
1211 **R:** Kanyabwengye.
1212 **H:** Where is it? *[∗..∗]* Mhm. *(∗silent∗)* kaniavengi. It's kanyabwengye
1213 **R:** But, yeah.
1214 **H:** Because this is the translations of someone.
1215 **R:** You can not invent, I think because if the people know kalimagezi, it's kalimagezi. In
1216 Austria computer is a computer.
1217 **H:** Kanyabwengye. This is another name which has been adapted by, *[∗.∗]* by some people
1218 who have been translating also who have been trying to translate.

1219 **R:** Mhm.
1220 **H:** As a Muganda translated computer to kalimagezi. Although some few Bakiga have been
1221 also trying to translate it, calling it kanyabwengya.
1222 **R:** Yeah, of course.
1223 **H:** Mhm. And Kanyabwengya it makes sense in Rukiga.
1224 **R:** (*laughing*)
1225 **H:** Kanyabwengya it means "something which produces knowledge". That's the legged (*?*)
1226 translation of kanyabwengya.
1227 **R:** Mhm.
1228 **H:** "Something which produces knowledge". That's what it means by the way (*mumbling*)

Example 14.1: *Informants explain "kalimagezi"*

According to this example (Ex14.1) the interviewee is unfamiliar of the Rukiga
term, but can reconstruct its composition morphological meaning, which shows that
it is close to the Muganda word "kalimagezi". Though, "kanyabwengye" seems to be
the official Rukiga term [Karwemera, 2006], it is a matter of fact that "kalimagezi" is
considered as the "real" Rukiga term by several interviewees. Only above mentioned
interviewee is aware that it is a Buganda term in fact which is utilized in the area of
Kabale. If one wants to find out the real name for "computer" these statements appear
to be contradicting at a first sight. However, obviously both terms are not yet applied
at all. A further indication for the unfamiliarity with the term "kalimagezi" arises,
when informants claim that this term "[...] *is even created for people who already
have an idea*" - In1Ex1Line508 and consequently is not appropriable to "explain" a
computer, as the following two examples (Ex14.2, Ex14.3) emphasize:

490 **A:** .. because even if I want to somebody deep in the village and you say "ebyoma
491 kalimagezi" then he will not have a image of a computer in the head.
492 **R:** He won't or he will?
493 **A:** He won't. Somebody in the village, ..
494 **R:** Yeah.
495 **A:** ... who has not heard of a computer or seen something like it ..
496 **R:** Mhm.
497 **A:** .. and I say "ebyoma" that's a machine, "ebyoma kalimagezi" he will not have the image
498 of a computer. He will think of a machine and then think of what knowledge is that
499 maybe he will think of it's the, the, the, the machinery that creates other things that
500 makes other things and all that.

Example 14.2: *Kalimagezi is not commonly used*

424 **R:** And how do you call it? Do you have a nick name?
425 **C:** A nick name?
426 **R:** Yeah.
427 **C:** That's it. That's "kalimagezi". I'm telling you!
428 **R:** And you use it?
429 **C:** Yes.
430 **R:** If you talk to other people?
431 **C:** O.K.! Oh, look. O.K. Yeah, you say, "ebyoma − ha", "kalimagezi" Yes, to people who
432 know, who know the computers. If you say "kalimagezi" they immediately say, they
433 mean that you are meaning, they mean that you are meaning a computer. [*...*] Mhm.
434 **R:** And to others?
435 **C:** Others who know, who knows it you say computers.

436 **R:** Mm.
437 **C:** But we don't use that "kalimagezi" very much.

Example 14.3: Explanation of kalimagezi

Another issue arising in this context, especially in correlation to the above mentioned examples, is: Though "kalimagezi" is usually presented in an isolated manner, it is actually an attribution to the word *"ebyoma"*. When the informant uses the constellation *"Raphael is kalimagezi"* in Ex14.2Line477 it is undoubtedly used as an adjective to the subject. The example above shows this explicitly. Thereby, it becomes contradicting to those statements, in which the adjective is already detached from its subject and where it becomes substantiated - Ex14.4.

1020 **R:** Mhm.
1021 **A:** Then, I would, would show them this is the kalimagezi, you have ever heard of, if some of
1022   you have never seen it.
1023 **R:** Mm.

Example 14.4: Something IS kalimagezi

In consequence to Ex14.2, the term kalimagezi is obviously derived from the phrase *"ebyoma kalimagezi"*. This reveals the most prevalent image and analogy of a computer: *"ebyoma"*. In Taylor's dictionary this Kiga term is translated with *"iron things; printing; machines; apparatus"* [Taylor, 1969]. Concluding from this definition, the original indigenous name for computer is originally derived from the image of a machine, which is validated and discussed in Section 14.2 in detail.

14.1.2 Nicknames

According to my understanding, a nickname, be it an abbreviation of a proper name or the replacement for an actual name, indicates a trustful and rather private relationship between actors and/or objects. In this case-study nicknames rarely arise in the context of computers.

One example on how nicknames can contribute to the improvement of the relationship between actor and computer is listed below (Ex14.5). It shows that the informant nicknamed "his" computer for identification by using his own name. Though, his named statement *"because most of the programmes where there"* - Ex14.5Line989 shall be considered more as an explication than a real reason, because all computers have the same programs with the same file structure and are not different at all. This statement only makes sense if the informant wrongly applies the term "programmes" actually referring to the documents he has saved on that specific PC[1]. The interviewer aims at clarifying the informant's quote. This evolved the answer that the informant preferred a specific computer to become familiar. This is perceived as an attempt to gain safety and also stability in the handling of computers. The wish of increasing secureness means that there has to be - at least a latent - fear to approach and use the object involved. By making it to one's own through constant usage and naming, computers become more trustworthy and a fictive part of oneself.

---

[1] The LAN broke soon after the departure of the ATs and as Mr. Rukinga left as well, it was not fixed, until the second period in 2003. Therfore, the files have to be saved locally.

978   **R:** How? What nickname you give to your computer?
979   **A:** My?
980   **R:** Yes.
981   **A:** I think I had put there my name.
982   **R:** mm.
983   **A:** So it was called according to my name. *(∗mumbling∗)*
984   **R:** So, where, so whenever you are in a computer class you are always using the same?
985   **A:** That is what I, that is why I, I nicknamed it.
986   **R:** mm.
987   **A:** Yes. *(∗laughing∗)*
988   **R:** Why?
989   **A:** For identification, because most of the programmes where there and it was easy for me I
990   was used to it.
991   **R:** Mm.
992   **A:** Mm.
993   **R:** So, that you get somehow a bit familiar?
994   **A:** Yes. You get somehow familiar. I don't want to sit anywhere but on my computer.

Example 14.5: *Nicknames are used to get familiar*

Whereas only one more adult uses the abbreviation "*PC*" (In8Line1142) no further nicknames could be revealed[2]. Among children the abbreviation "*Comp*" (In18Line274) and "*Comps*" (In22Line294, In23Line238) occurred. Nicknames as substitutions for computers arose twice, first as "*Dexter*" (In20Line267) and a second time as an "*A-DO-IT-ALL*" (In26Line486). "Dexter" refers to the a 8-year-old protagonist called "Dexter" of the cartoon Dexter's laboratory. This boy is a scientific genius and by the application of his highly advanced equipment of his laboratory he achieves astonishing things, e.g. like saving the whole world or mutating himself to masses of protoplasm. Investigating on the second nickname leaves a similar impression. The self-explaining term "A-DO-IT-ALL" clearly presents the underlying expectation towards a computer, namely to be capable of doing everything. Though, both nicknames take over very general statements about their characteristics, attributes such as universal and powerful can be ascribed to technology. This coincides with detected analogies which are presented below.

14.2   Concepts & Analogies

Looking at the context of the concept "computer" is how it is understood and used in the interviews, reveals several analogies for computers. These can be grouped into three major concepts:

- Computer is equated or similar to tangible "things" (e.g. machine)

- Computer is like a human being

- Computer, Internet and Development

---

[2]Although the informant of Ex14.3 claims to use "kalimagezi", this cannot be considered as a nickname but as a translation.

All these analogies cannot be considered isolated and not every interviewee refers to exactly one analogy or concept but informants use different of these concepts. The following example depicts how several analogies can coexist and are unified in one "thing": the computer. Ex14.6 offers the chance to present the terminological diversity and to discuss latent expectations towards a computer, which will be done in Section 14.3:

242 **R:** Mhm. And, yeah, ok, ok. When you think just about the name computer, what comes
243   into your mind? *[*..*]* Can you name me five
244 **P:** ...five things?
245 **R:** Yeah, y, y, y, y, it's not about real things, it can be subjects, issues, anything. Just what
246   comes into your mind.
247 **P:** when I talk about computer
248 **R:** Yeah, just associating, yeah *(*referring to above P statement*)* to computers, to the word
249   computers.
250 **P:** Something that knows everything,
251 **R:** Mhm.
252 **P:** something that you can do anything on, aah *[*...*]* something doesn't need any control
253   over it, ahh.. *(*clears throat*)* doesn't it. *[*.*]* It has a colour and,
254 **R:** Yeah, yeah, yeah, it's ok. Go on.
255 **P:** something that, *[*...*]* its almost like its almost like a TV
256 **R:** Mhm.
257 **P:** Mhm. Only that one *[*.....*]* something like that can make me world famous,
258 **R:** Pardon?
259 **P:** something like a Nokia, if I know computer, if I know what okay on what you are asking
260   me like, like I am the only one in Kabale who knows what computer is, so it makes me
261   famous

Example 14.6: *Analogies of a computer*

From the above mentioned example (Ex14.6) the general "*thing*"-like physical characteristics that are ascribed by the female informant to the computer leap to one's eye. Answers like "*It has a color*" and "*it's almost like a TV*", is "*something like a Nokia*" and "*it makes me famous*" present different kind of analogies and expectations. Interestingly, the girl does not use the word computer, but applies the word "something" instead, by which her replies become more ominous. These answers offer also the possibility of discussion of the image of technology. A computer is presented as "*something that knows everything*" (Ex14.6Line250) which relates more to a human being than to a machine, "*something that you can do anything on*" (Ex14.6Line252), which does not need to be controlled (though she seems to be uncertain on this feature as this is relativized by her post-placed "*doesn't it*" and clearing of throat) (Ex14.6Line252), and can even "*make her famous*" (Ex14.6Line257). In short: the computer is consider as a very powerful thing with human characteristics.

Drawing upon the previous discussion thread on the indigenous presentation of a computer I conclude from the quote "thing" that a computer is associated with the concept of a machine. This will now be discussed.

14.2.1   A Computer as a Machine

Apparently, the key concept from which the informants have started to approach computers[3] has been that of a machine - see Ex14.7 and Ex14.8.

253 **R:** So, ah, if you hear the word computer [*..*]
254 **B:** Mhm.
255 **R:** what comes into your mind
256 **B:** if I hear of a computer?
257 **R:** hm, what comes into your mind?
258 **B:** aah, aah, what comes in my mind is the machine and the work it does.

Example 14.7: *A Computer is a machine and does work*

865 **A:** We shall, in the, in the future we shall not have, we shall not have many people working
866 but the machine, the computer will be doing most of the work.

Example 14.8: *Computers promote unemployment*

Over and over informants call a computer a machine, even when they do not use the term "machine" explicitly, but ascribe values to a computer to such an extended degree that there is no doubt about an internalized "machine"-concept. To give an example: Following informant talks about a computer in the way someone talks about a machine. A computer is presented as "something" which "produces work" and which has to be "fed by someone" - Ex14.9.

15 **B:** in other words I had been hearing people say computers, computers and I didn't know
16 what it was
17 **R:** What did they tell about computer?
18 **B:** Computer?
19 **R:** What did they tell about?
20 **B:** I, I know it was, it produced some kind of work
21 **R:** Mhm.
22 **B:** may be processed by somebody who feeds it.

Example 14.9: *Something that processes things*

Analyzing the memos of the participant observation underlines the "machine"-concept of a computer. The situation: The computers of the computer laboratory were automatically switched off by the operating system, but had to be switched on. In the first weeks of computer lessons it was very difficult for the students to prevent themselves from pushing the switching on button once more after the operating system had shut down. A typical functionality of a machine is switching it on to use it and then off after its usage. Apparently this process of usage is internalized by the students and as computers are connoted as machines, they pressed the switch again.

27 **R:** Do you remember what you were thinking about that so far.
28 **N:** At first I thought, when I saw the screen first I thought it was a television but to my
29 surprise it was something else, I had never heard of, it was just something new to me but
30 , I got used to it as, as we continued explaining more to me about computer.

Example 14.10: *Screen is like a TV*

---

[3]The expectations how the technology is supposed to be approached will be discussed in Section 15.4.1.

Other interviewees presenting analogies are more accurate and refer to certain technical products and use for example the analogy of a TV. Some of them are very clear on the hardware concept[4] of a computer. By being on these separations some correlate the screen to the TV - see Ex14.10, whereas others, that are not acknowledged with the hardware concept compare the whole computer to a TV. One might claim that this general comparison could be a verbal impreciseness, but indications like in Ex14.11 raise doubts if the hardware concept of computers has been internalized by some informants. The informant talks about the fear that her information might be lost, if a screen blows because of high voltage[5].

317 **R:** So for what purposes y, because you are just talking about purposes are you using a
318   computer now?
319 **A:** I am using it for a, basically now, personally it is for [*...*], writing letters, setting exams
320   , a [*...*] yeahhh...meeting minutes, the, the, the, the writing out minutes for meetings (*
321   slow speed*) mhm, but keeping like files, files maybe like [*..*] can I say, i,i information
322   on staff, word on my teachers, it's not yet [*..*] fed in the computer, say I now want to
323   see the the information about the late L (* ?*) and it is still manual... (*laughing out lout
324   *)
325 **R:** Why?
326 **A:** It's still manual. I think just because, as I said we have [*...*] not gone, [*..*] not got
327   ourselves used to it to say "I have got to keep my information" and this, but second there
328   are there are fears that now, I don't know whether it, it affects it that now if I put on
329   information there and these are confidential things, and then over set and switch on and
330   then it it blows and that's gone.

Example 14.11: *Information gets lost by blowing screens*

Even though, the informants usually explicitly note that they do not consider a computer to be equal to a TV, they still refer to them as similar. Nevertheless, they compare it to known technologies and statements like "*I felt it was like a TV*" - In5Ex1Line23 or "*a computer has the shape of a television*" - In6Line212, which indicates a longing to compare the new subjects to something familiar. Differences are pointed out and answers like "*It's like a TV, but it's not a TV*" are common. Actually, the TV does not serve as an analogy, but is commonly used as a starting point to differentiate the computer later on, as it was a basis. According to that computers are perceived as something "*almost like a TV*", but as "*they are more talented*" (I9Line59), the concept of a TV is not sufficient to explain a computer any longer. It seems as if it appears difficult to express these differences in detail, because most of the interviewees stick to vague statements, but the following example will illuminate this to a certain extent and will show how the computer becomes an actor. This will be discussed in detail in Section 14.2.2.

270 **Y:** Well, they seemed easy [*..*] to deal with. And exactly the good thing with the
271   computers is when you click on a word it shows you exactly what to do.
272 **R:** Mhm.
273 **Y:** And all the time.When you tell us to go to Microsoft Word and we are supposed to type.
274   Take with there what type from, the Files, the Insert, Tables and all that. So, ok exactly

---

[4]By hardware concept I understand that a computer consists of three major components: input devices as keyboard or mouse are, central processing unit as the tower, which is the heart of a computer, and output devices (screen, printer)

[5]This happened several times, as the screens could not resist power instability.

275   you are able to know where you are going, by then when it's like a television set. There
276   is only /*..*/ a remote control that sometimes you don't know how to use it until a grown
277   up teaches you, even sometimes they forget, but at computers they lead really you
278   properly.
279   **R:** Pardon, they?
280   **Y:** Lead you, /*.*/ to the right thing.

Example 14.12: *A computer leads you to the right thing*

According to the informant in opposition to the TV a computer *"leads you"* -
Ex14.12Line277 to do the right activities, which indicates that the computer knows
what you want to do on it. The informant indicates that a TV is obviously more
difficult to use than a computer, as the effect of the remote control is not visible,
whereas a computer "asks one" what to do. The reason why other informants could
not describe the differences between TV and computer in a more detailed manner
might be that their computer-machine analogy is obviously not sufficient to explain
the "work" a computer does. If the functionality of a computer is substantiated, like
to the work of a typewriter, the informants are able to present specific differences.
The following example (Ex14.13) shows that the keyboard is perceived similar to the
typewriter.

45   **J:** okay, according to, ./*.*/ to the part with the keyboard *(*cheerful*)*..it may look like a
46   type writer..
47   **R:** hmm
48   **J:** I, so at first I thought that it's a type writer then when they *(*voices in the background*)*
49   started it /*.*/ it started differently..*(*giggle*)*..haa it was not a typewriter. So after then
50   I came to know that it is a computer..

Example 14.13: *Typewriter analogy*

This described relationship is also marked by the mentioning of differences, but
whereas the differences of the screen-TV similarity are not mentioned explicitly (the
comparisons build upon the physical appearance), those between keyboard-typewriter
are. The advantages of a keyboard in opposition to a typewriter are placed by phrases
like *"it requires less energy*[6]*"* in Ex14.16 or *"like a typewriter but more nice than
cyclostyling"* - In5Line139. At first it seems as the informants compare the basis
of the physical appearance of a keyboard to a typewriter, but it turns out that the
informants compare the results achieved by a printer (as a whole system) to the ones
of a typewriter.

137   **R:** So what do you like about computers? What do you like /*....*/
138   **B:** ah... *(*very hesitant*)*...what should I say about the computers../*..*/ in fact I like their
139   neat work so when you produce work its like not that of a typewriter. At times it is very
140   neat in fact when it comes out the work on it there is ink.. it is clearly ready and unlike

---

[6]Referring to the thick description of the first period of the intervention, one of the first insights in
the usage of the computer was that the typing on a keyboard requires less physical effort. Especially
affected people were those who knew how to use the typewriter. Another interesting contextual issue
is that the secretaries were wondering why the computer needed an additional device to bring the
letters to paper, when it was already on the screen. This shows that the secretaries were not familiar
with the fact that, in general, any information on a computer is first validated by one's eyes via screen,
before one can print it.

141   the typed one now... *(∗Yawning very audibly∗)*.. When they do cyclostyling

Example 14.14: *Product by a computer is nicer than cyclostyling*

Benefits like the quick reproduction of stored information are also important as "t*he neat work it produces*".

480   **C:** The work which the ma, which the computer too. Like if you are using a *[∗...∗]* a
481   typewriter. The work which would you, during a day the computer can make it within
482   one hour.

Example 14.15: *Computer-typewriter comparison*

35   **A:** These are the simple things. And .. *(∗cracking door − talks to secretary 3 sec∗)* and even
36   making work easier and neat, whereas the typewriter, the typewriter it requires energy
37   and that the computer .. it makes work easier.

Example 14.16: *Typewriter requires energy*

The importance of the neatness of work is discussed in extra on Page 208. Concluding from the above mentioned statements one the prevalent concept of a computer seems to be strongly related to an "improved typewriter".

In some interviews the issue arises how the informant would introduce a computer to the people in the rural area. In one case the informant draws upon the analogy to a radio. Doing so, it becomes apparent that she once more limits the explanation to the one of the typical characteristics of a machine, as she would show those people to switch it on like a radio - Ex14.17.

406   **C:** the machine because you can't just go to the machine when you don't know how it works
407   its like switching on a radio if you do not know to switch on a radio can the radio give
408   you the news?

Example 14.17: *Switch on like a radio*

When the informant compares the driving of a car to the usage of a computer it is an analogy on the meta-level: Clearly said, how to approach computers. Apparently, this informant, as the only person on the compound who knows how to drive a car, uses the analogy that you can only become a good driver, when you drive a lot. According to her you can only learn computers if you use a computer quite often. This shows that her perception of approaching a complex technology is obviously more based on the practical usage, than on a theoretical approach. Nevertheless, her presented goal of driving a car, or learning about computers, is to "master" it. Even though she had started with "man ...", which most likely should have been finished by "manage" - Ex14.18Line382. Though, we do not know in fact. This indicates that apparently the proper way to approach technology is to master it, which will be discussed in detail in section 15.4.1.

381   **A:** Then, maybe they are special because they, I *[∗.∗]* what I found, its, I was telling
382   someone, I think computers are like driving, you can not man...you can not master
383   driving unless you are in a vehicle often.
384   **R:** Mhm.
385   **A:** And most of the things like in my experience on driving these are small things you, you
386   learn them by through experience.

387 **R:** mhm.
388 **A:** And I think I have found it out with the computer.

Example 14.18: *Using a computer is like driving a car*

14.2.2   A Computer is Human Like

Whereas the above mentioned concepts refer basically to physical things some concepts draw upon human beings and informants consider the computer as "*something that knows everything*" (Ex14.6Line250). Whereas this example ascribes human characteristics to a computer, some informants put it more directly as Ex14.19 depicts:

392 **A:** The, the more you sit at it and say now I am going to do with it, you find you have
393   learnt something else that you had not planned for or that you had not aimed at. And /*
394   ..*/ they are, I wonder, they are like sometimes like a person. (*laughing*)

Example 14.19: *Computer is like a person*

The above depicted example shows that one human-like characteristic of a computer is the one of teaching someone new things. A computer "*tells you*" what to do and prevents you from making mistakes. A computer is educative and can even act like a teacher.

932 **A:** Yes. So, and sometimes a computer teaches much better than an individual studying in
933   the front of
934 **R:** what means, what are you imagining?
935 **A:** Like eh, if it wants to, /*..*/ if you want say if it is teaching about physical features.
936 **R:** Mm.
937 **A:** I don't know, are you a students what did you learn?
938 **R:** Computer sciences and business economics. (*silent*)
939 **A:** Say, if want show you the let me now go to economics.
940 **R:** Mm.
941 **A:** If it wants to show you the, the graph
942 **R:** Mm.
943 **A:** of developing and developed and developed countries.
944 **R:** Mm.
945 **A:** It can use bars. You know what I mean?

Example 14.20: *Computer is like a person*

According to above mentioned informant a computer actively helps studying certain topics. In some passages and especially when the interviewer refers to a strange behavior of the computer in the interviews, the computers are represented as an actor which advise the user what to do next.

153 **R:** Do you think that they behave sometimes strange?
154 **B:** the computers,
155 **R:** hm,
156 **B:** Yes at times they behave, they behaved strange like, like, /*..*/ yaa, they give you like
157   information like, they give you like a warning
158 **R:** Mhm.
159 **B:** that's very strange, you are working on something that tells you something is in danger
160 **R:** hm

161 **B:** When actually like my heart can't tell me that we you are going to hit a sto, hit a stone
162 or a snake.

Example 14.21: Computer behaves strange

The analogy of this example (Ex14.21) might be induced by the interviewer himself, when he ascribes the computer a human like behavior. Nevertheless, it is taken up by the interviewee when referring to the fact that computers could look into the future and give you warnings about something dangerous. Phrases like the computer "*tells you*" - Ex14.21Line158 emphasize his perception and together with his annexed comparison it becomes evident that this "way of acting" is unexplainable to him. In this case the informants' attempt to compare the behavior of the computer with the one of a human being is not successful. On a latent level the computer becomes something very clever which can look into the future, something of which a human being is not capable of.

Apparently, all these examples have in common that the underlying concept of a computer goes beyond the understanding of a direct cause-effect worldview. Interactive dialogues (such as pop-ups) emerge on the screen are as an effect if something third is done. To give an example: If someone closes a window[7] without saving (though he/she has changed his document), he/she gets an error message. This refers to the fact that the user did not save the document, but not about his activity of closing the window. This "reaction" is indirect and apparently not transparent. The concept is too complex to the user and the "computers actions" are not explainable with existing explanatory models. According to this perception of computers, the user is guided by the computer, (as in what to do next) and not informed about the inconsistencies in his process of work. Following example (Ex14.22) describes that this does not apply to all informants, when the informant states:

87 **R:** Do you think that sometimes the computer is a bit strange to you? Behaves strange?
88 **H:** Yea /*.*/ like now, /*..*/ like you, you command it to do something and it refuses and
89 tells you something else and it can't do it and you really feel like, how can a machine
90 refuse to do what a person has commanded it to do so sometimes it would behave in a
91 strange way, but certainly I would understand that I am the one who may be made a, a
92 wrong command and which does not fit it

Example 14.22: Informant knows the concept of a computer

In opposition to the previous example (Ex14.21), this informant (Ex14.22) explains that it is the persons who commands the computer what to do and not the other way around. Although she admits that this seems confusing and strange to her. The following lines by the same interviewee, illuminate the background of this extraordinary concept of a computer. Obviously, our informant from Ex14.21 is just one out of several people, who broadcast the perception that a computer "*is able to read even the thoughts of a person*" (Ex14.23Line218) and thereby emphasizes/reconstructs the term "kalimagezi", and by doing so makes a computer even more ominous.

211 **R:** So what were you expecting from the computer before you were using it for the first time
212 ?
213 **H:** Hm from what I used to hear people say what I used to hear people say, they would tell
214 you a computer can tell you, they ..to actually used to say it a computer can speak like

---

[7]No physical one, but in the operating system!

215 people so I expected something like what now,...*(*softly*)*.. so they would be like you can
216 tell a computer to do this and it, it tells you something else like you tell it you want to
217 do something and it asks you "are you sure you" want to do this so I was imaging in my
218 head what kind of thing is this that is able to read even the thoughts of a person so I
219 expected something so sophisticated and all that but I actually came to know that it was
220 what you do to it, what you, you tell the computer to do that it does, it's, it's you who
221 commands it
222 **R:** hm
223 **H:** So unless you command it, it's not going to do what, to do anything but from the start I
224 had expected something so sophisticated as and so knowledgeable, as knowledgeable as a
225 human being *[*..*]*

Example 14.23: *People say that they are human like*

As a consequence to this mutual interaction, a computer becomes something very clever and seems to be even cleverer than a human being. In combination with other difficult understandable "unexplainable" rumors, for example to be able to communicate with people in America or read the latest news of America (see Section 14.3) this image is strengthened as it remains abstract and cannot explored by oneself. Another fact which contributes to this powerful concept is that people claim that everybody can talk with a computer in ones own language. One informant argues that his brother participated in an NGO intervention, where the functionality of a computer was adapted to an interactive medical guidebook. This was translated in his brother's mother tongue, which we assume that he had actually intended to say in Ex14.24Line830. However, he goes beyond the local language of his brother and corrects himself that a computer can talk "*in all languages*" (Ex14.24Line830).

813 **A:** To educate. To educate the masses
814 **R:** Mm.
815 **A:** Say about to *[*..*]* these days I am told there are, there are computers which have come
816 and can be set *[*.*]* so much so that *[*.*]* they can talk.
817 **R:** Mm.
818 **A:** Have you come across such computers?
819 **R:** No.
820 **A:** They, they I am told they have come.
821 **R:** Mm.
822 **A:** So, in this case I would use them to educate the people from the villages.
823 **R:** Mhm.
824 **A:** So again is the rather about *[*.*]* courses of diseases. How they can be eliminated. Ah. *[*
825 ...*]* This, eh *[*..*]* ah, feeding.
826 **R:** Mm.
827 **A:** about feeding the, their children, balanced diet, about to *[*..*]* about *[*..*]* civil wars,
828 why do they fight amongst them, civil wars, such things and other one I would use it.
829 **R:** So, who told you about this *[*.*]* computer?
830 **A:** My brother. *[*.*]* That you can set it. It talks even the language, in all languages.

Example 14.24: *Computer speaks all languages*

When a computer is presented as such a clever instrument, its real functionality becomes more and more abstracted from its realizable and factual possibilities and evolves to statements like the following ones, when the informant claims, that a computer makes you free:

288 **R:** Ok. Ahm *[*....*]* When you hear the word computer, just the word, what comes into
289 your mind. Anything is possible.
290 **S:** Me I know that computer are the ones, *[*.*]* Ok, they told us, they told me that
291 computers do everything, when you have a computer you are free, they do everything you
292 want.

Example 14.25: *A computer sets one free*

According to this concepts a computer is in no means a (language) barrier, but can
free you. Since these "qualities" of the computer are reported and not experienced by
the informants, it thereby remains ominous. This also indicates that a person that
knows how to use a computer becomes very powerful as well, as he/she can utilize
"everything". This will be discussed in the following sections.

In general, during the interviews, there are hardly any questions directed to the
interviewer, but if there are arising some then they are either about the Internet or
on a meta-level about the person who has made a computer. Be it pupils: "*Who was
the first person to make a computer?*" (In14Line460) or adults, apparently it becomes
interesting to know who the "creator" is (Ex14.26). Even the Headmistress, who was
one of the fastest and best performing students, who is used to hold speeches, and who
is very good at repartee, stammers and is at a loss, as the following example depicts:

548 **A:** Eh... So *[*.*]* I think *[*.*]* it's a great, it's a great achievement, *[*.*]* and at times when,
549 ahm, computers though I am not often usually there, I usually ask myself, whoever
550 created a computer, I mean, what did he have in mind, does he have the same brain ... *(*
551 laughing*)*
552 **R:** Yeah. That's what I am asking myself, too.
553 **A:** Does he have the same brain, I mean, we do have?
554 **R:** Mhm.
555 **A:** What, what, what, prompted him ...
556 **R:** laughing
557 **A:** .. I mean to create such a thing and to have so much in it. *(*enphasized*)*
558 **R:** Mhm.
559 **A:** Eish.. ahm, I am, I am at a loss. *(*laughing*)*. Eh.

Example 14.26: *Who has created a computer?*

This example shows that the possible person behind this "clever machine" becomes
powerful, important and interesting.

14.2.3   Computer, Internet and Development

In some explications about computers, interviewees try to explain computers objectively, by actually presenting their own explanatory model on how they perceive them.
The following example derives from the question asked by the interviewer on how the
interviewee would explain a computer to someone else. This shows how the interviewee
tries to describe a computer, but ends up describing how she perceives the Internet[8].
It becomes manifest that the informant explains the comparison to a TV and that it
might not be sufficient to explain a computer. The informant prefers the comparison
to a radio. The analogy she uses builds upon the idea that the information broadcast

---

[8]In a prior text sequence, the interviewee talks about the Internet.

on the radio is emitted by a remote station. This implies that the information on the radio or on TV is not created within the device itself. It is produced externally and is then transmitted to the receiver. She implies that a computer cannot be utilized to create information locally without access to a network. In Ex14.28Line244 she points out that she does not know if there is a central station for computers and if this is the case, where this station for computers is located. This is understood in the way as follows: When the informant strays off her intention of presenting explanation of a computer, she presents her own idea (explanatory model) how the information on a computer (=screen) turns up. Actually, she presents her concept of the Internet, and in Ex14.27Line117 she notes her eagerness to learn about how the Internet works. According to her, a computer represents a device which retrieves information from a distant station. This concept is disillusioned by the interviewer in Ex14.28Line249-251. Ex14.27 indicates that her explanatory model of a computer is a mixture of different aspects. Obviously, the concept of computer coincides with the one of the Internet ("*I hear, I just hear*" - Ex14.27Line117). The possibility to use a computer to send messages becomes one of its main "known" benefits. Due to rare experiences among the interviewees, this is closely related to the prevalent expectations of a computer, which are discussed in Section 14.3.

117 **C:** Mm. And some interesting things I hear, I just hear But I've no ever done them like the
118 *[∗...∗]* How this, *[∗...∗]* what? *[∗..∗]* the Internet, I just hear how good it is.
119 **R:** Yeah!
120 **C:** but I've never seen it! *[∗.∗]* Mm. *[∗...∗]* I would, I would, O.K. I am eager just to know
121 how it works. Which is, we, we shall, u, I found very interesting also. As I was, as I was
122 told.
123 **R:** The way it works.
124 **C:** Yes, the way it works. Mm.
125 **R:** so, you were expecting the computer too?
126 **C:** To make that, to make it. Yes! Because as I was told I knew to, to have done it. As if we
127 had gone deeper ...
128 **R:** To done what?
129 **C:** ...to, to how the Internet work. Yes. *[∗...∗]* Mm. *[∗..∗]*

Example 14.27: *Hear about the Internet*

211 **C:** Like when you have never seen it, becomes strange to use it. Ah! How comes this appears
212 on the screen or what? Like someone who has never seen a TV. *[∗..∗]* You can't just tell
213 her...
214 **R:** Mm.
215 **C:** ... anything concerning with a TV or *[∗...∗]* ah, a DC at least we have ever seen a
216 television. You would say now, this has a shape of a what?
217 **R:** Mm.
218 **C:** Of a television. So, I've never seen a television. And when you see things appearing in
219 the screen. You say: "I've never seen that television." Now, even these things happening
220 therefore *[∗.∗]* it could always be the same and you would see it not strange.
221 **R:** Mm.
222 **C:** Mm. Cause, we've, we've ever seen things appearing. Pictures appearing on the screen,
223 on the TV, but there are some people who have never seen such and when see it becomes
224 very strange, they wonder where these co, things are coming from. Mhm. You need *(∗??*
225 *∗)*
226 **R:** *(∗laughing∗)* And how do you explain it to them?

227 **C:** You would use them, you would use maybe like the radio. The Radio, when the radio you
228   just hear words, voice and you wonder where it is coming from ...
229 **R:** Mm.
230 **C:** So, as you, as you go deeper you find that there are somewhere else which brings the
231   sound and you don't see the words.
232 **R:** Mm.
233 **C:** And somewhere which, which come by air.
234 **R:** Mm.
235 **C:** For example, someone has the radio you have U, Uganda Tele, Rad, Radio Uganda in
236   Kampala, what we hear those news from here. When we are here, we, we are not there.
237 **R:** Mm.
238 **C:** So, I would think this is the same as or the telephone, even the telephone. You wonder
239   you only have a phone there is now wire which connected to this phone, but you wonder
240   how comes that you hear someone talking from ...
241 **R:** (*silent laughing*)
242 **C:** ... Nairobi, or where.
243 **R:** Mm.
244 **C:** Mm. So, I would take it as the same. Maybe, I don't know whether there is a station
245   where there is their station, for, [*...*] for computers, which connect worldwide
246   computers.
247 **R:** Yes. There is a big network.
248 **C:** O.K. The network. I think there could be the same to the computers.
249 **R:** Mm. [*...*] Well, this one in fact, has already stored a lot of information inside ...
250 **C:** Yes, exactly.
251 **R:** ... which it can show to you.
252 **C:** O.K. Mm. [*...*]

Example 14.28: *How to explain a computer*

Above example (Ex14.28) shows/depicts another technological device which is used by informants to compare the Internet (as a "part" of a computer): the mobile phone. When one boy explains (In6Line133f) that sending an e-mail is the same as sending a message with the mobile phone, the analogy of a computer with the mobile phone is quite clear. A mobile phone, as shown in Chapter 7 Uganda exhibits the sixfold number of cellular subscribers compared to the number of Internet subscribers, is not only used to show functional similarities to a computer, but also validated against it. In the following example (Ex14.29) the interviewee criticizes that the mobile phone of her parents gets blocked when there are no credits on it. This makes a mobile phone unreliable, compared to the Internet connection of a computer which always works. Whereas the credits for a mobile phone have to be bought in advance, Internet costs are paid at the end of a month. Her statement also shows the possibility of using a computer as an alternative to a mobile phone for communication purposes and her wish to be able to communicate whenever she wants to.

446 **R:** That's great. [*...*] So [*..*] do you think computer is important for the Ugandan people
447   ?
448 **Y:** Very important.
449 **R:** Why?
450 **Y:** (*hem*) Some people, ok, well, ahm, [*.*] like a telephone and a computer, I know a
451   computer might be a bit more expensive,
452 **R:** Mhm.

453 **Y:** but then, *[*.*]* a telephone when it rings it too much, it gets blocked,
454 **R:** Mhm.
455 **Y:** Ours, usually it gets blocked.
456 **R:** Mhm.
457 **Y:** So when you think of communicating and communicating. You, like somebody is sick in
458 the house. Now you look at teacher or like, those looking at money when they are hungry
459 . They like, ha, gosh, I need something really to help me which is there all the time and I
460 know I can rely on it. *[*..*]* Ahm. *[*...*]* Ui *[*..*]* Well. *[*..*]* I wish I had computer on
461 my own. I would like to win a price of a computer.

Example 14.29: *Computer and a mobile phone*

The comparison to a mobile phone also emphasizes the high-tech characteristic of computers. In the example below Ex14.30 the informant points out that the Ugandan industry is not capable of producing computers on its own and states that the production of computers is related to "developed countries".

905 **A:** I think it, it has to do with developed countries
906 **R:** Mhm.
907 **A:** because for us, well, as far as making, making a computer is concerned Yeah.
908 **R:** Mm.
909 **A:** for a, for a developing and undeveloped countries. It will take us a long time to
910 manufacture computers. That's what I mean. *(*laughing*)* Because, because they require
911 a high technology. *[*..*]* Eh, like *[*..*]* the developing countries ah, it will take us a long
912 time. And you hear they have done this, they have done that. Ah! When we have not
913 even learnt how to make something like this one. *(*showing mobile*)*

Example 14.30: *Computer as a product of developed countries*

The implications of this relation is discussed in detail in Section 14.3, because it seems (Ex14.31Line144) that if a country cannot produce computers internally, is a restriction in itself. "*Showing to know computers*" or "*factual computer knowledge*"[9] implies being regarded as developed and being part of the modern community (even as a single person) - see Ex14.31Line135,144. When the PTA chairman explains in his opening speech on the occasion of the welcome-dinner of the ATs in 2002 the school's expectations, he made it clear that the school's aim is to get connected to the Internet. He perorates by the words „*And then* [if the school is connected to the Internet] *... we are part of the greater world*" (TD|p.1), which evidently show the idea that to be connected to the Internet means to be connected to the modern/greater world. Ex14.32 clearly depicts that this has not been reality in the local environment[10], when the interviews were made. The local environment is seen as a small world and countries like Austria, U.K., U.S. ((In5Line297f, In12Line364) are considered as parts of the "*bigger world*" and as modern and developed countries. To be modern is obviously

---

[9] The informant restricts herself to the verb "talk about" computers. However, the technical terms "files" and "menues" are attributed to the verb "talking". This could mean further that, by the usage of verbal concepts, one becomes developed. If someone is considered as developed because he/she is in possession of computer knowledge, or represents development through verbal concepts, is unkown, even if there are indications (see Ex14.46Line297) that most benefits will be gained by factual computer knowledge.

[10] If the local environment is understood as the school, the district, or the compound could not be retrieved.

desired, since an interviewee states: "*We said: Ah, we are good to have a computer. For us we are going to be like Austria.*" - In12Line357.

134 **R:** Mhm. So what do you in general like about computers?
135 **A:** What do I like computers? *(*mumbling*)* It brings you to the modern world, to the
136   modern world, then as I said there are a lot of things that can be done with a computer,
137   it has, it has made our work easier as far as exams, writing letters, though I would like to
138   have some other ahm *[*..*]* some other elements or parts of it that we don't have. For
139   instance if we had things like e—mail, to make my work as I communicate to different
140   people much easier. I know I am lazy in communicating, hand writing letters, but if we
141   had maybe e—mail it would be excellent for me.
142 **R:** Mhm.
143 **A:** As far as communication, sending messages, communicating with people i i *[*.*]* is
144   concerned. And *[*.*]* the work, I mean, when one is talking of computers, files and
145   menues, then you are in the modern community, you are, you are part.. *(*laughing*)*

Example 14.31: *Become developed through computers*

424 **R:** Can you help me. What do you mean by globalization?
425 **A:** Globalization, you are, yyy, you, you, you have, you have a wide knowledge, you, I mean,
426   you are part of the bigger world *[*..*]* by that is connected to maybe to the
427   modernization, globalization you are part of the bigger world in a moment if you have e—
428   mail you communicate with someone very far and ..
429 **R:** Mhm. So you mean, bigger world is ..
430 **A:** We are in a small world here.
431 **R:** In Uganda.
432 **A:** No even if it is here, we are, when there is no computer, we are in our own world, but
433   with computer I can be able to communicate with someone *[*..*]* within a, or I can find
434   information that concernes another part of the world.

Example 14.32: *Globalization*

These examples illuminate the background of a very dominant expectation towards a computer's functionality: to be able to communicate: for example by e-mailing, which is known as "*the quickest method of information*". These expectations towards computers will be discussed in the following section.

14.3   Expectations of a Computer

Deriving from the stance that the above mentioned expectations are based on its underlying concepts and analogies associated with computers, I want to stress that expectations, concepts and analogies must not necessarily be regarded as distinct from each other. The opposite is true: all of them mutually reconstruct each other and are changed by the users' experiences and interaction. They are an expression of culture and represent only a snapshot at the certain point of time the interviews were taken.

14.3.1   Increased income by ...

... the usage of the Internet and communication and e-mail.

Over and over informants underline the importance of sending e-mails. They wish to send an e-mail to someone else or to learn about how the Internet works. Interestingly, users hardly specify the expected benefits of communication. Most informants claim that their wish is to communicate with other people, because "*like maybe someone in America can send a e-mail to someone here in Kabale.*" - In19Line102. Only two pupils define certain applications of communication. One informant states: "*if you can want you can send an e-mail and you get a more better computer*" - In14Line258. This statement explicitly induces the expectation correlated to the possibility of communication: to benefit financially. The second informant explains that his expectation is to be supported in his job, while working as a designer in a worldwide network. Even though, this interviewee has a precise concept on how to benefit from communication. It is once more motivated financially.

Obviously, other interviewees cannot place the benefit this clear, since they do not mention the benefit of communication at all, even when they want to use a computer to do so. The following passage exemplarily presents the idea of an informant that relies more on the available information from the Internet as a benefit in itself.

270 **R:** So, you think a computer means a *[∗...∗]* possibility to earn money? *[∗...∗]* Or, or did I ge
271 ..., get you wrong?
272 **C:** Mm. *[∗...∗]* O.K. It has so many, *[∗...∗]* so many duties to perform maybe. When I say.
273 It is not necessary that it should, should only, *[∗...∗]* give you money but now you have it
274 to, to children, learning from it. Not only children, even old people, like for example if it
275 was an office people come and ask you, *[∗...∗]* would come and ask you too. And now you
276 do for, what, what are you to, to get money, what I mean! Like someone say: "Now I
277 have my work to be done." That's when you get money. But if it can just work on other
278 purposes like want to know worldwide news, as I have already to you.

Example 14.33: *Vague perception of benefit of communication*

This example shows the vague imagination about the purpose of reading the worldwide news[11] (Ex14.33Line276-280) and the possibility of using the computer as a service offer. This expected benefit is discussed in the following section.

... a better job through computer knowledge and making business.

During the analysis it emerged that the major expected benefit of knowing how to use a computer is to earn (more) money in an improved job. This expectation emerges not only among pupils, but also among the interviewed adults. Informants hope to improve their living standard by the knowledge of computer and set high expectations on it.

Pupils. According to this example (Ex14.34), the expectations towards a computer are set quite high, in means of providing a job as for example a secretary which would help earn money - Ex14.34.

97 **P:** It, its benefits to most people and it, it earns a, a life end a good, a good, a nicer
98 standard.

---

[11] The Internet is represented either by providing the user with worldwide news, a not real benefit, or the download of movies and music. To give an example: One pupil, which also merges the concept of computer and Internet (In27Line405f) explains in this context: "*So, but they can find the music there and some movies, like, some people don't have Yes - TV, but the movies we watch at home, they watch them on the computer.*" - In27Line411-412. This way of usage is discussed in Section 15.1.1.

⁹⁹ **R:** Hm.
¹⁰⁰ **P:** Yes. Its quiet. It helps others, it may earn income like as I said people can be employed
¹⁰¹   as secretaries if they know computer and it can connect different areas together *(∗the bell*
¹⁰²   *in background∗)* like may be someone in America can send a e−mail to someone here in
¹⁰³   Kabale.

Example 14.34: Improve living circumstances

Whereas this girl ascribes the benefit to "*most people*" critical voices are found among the teachers who fear an increase of unemployment through the adoption of computers: "*you know there will come a time when everything will be done by a computer and manual labour will be less*" - In8Line284 or "*So most of the work which should done by people, it's done by the computer, so people have no work to do.*" - In9Line368-370. This is interesting not only because these statements inform not only about the teacher's fear of having even more unemployed people, but also that their local environment and country will change and that this development could go in this direction: they believe computers will result in people loosing their jobs. However, we cannot claim that this is connoted negatively, as these teachers are people who expect a personal benefit from it. It is not the aim of the research to discuss the effect on computers in the working environment of Uganda, although it would be very interesting to look at.

Despite these negative expectations (by the teachers), children expect in contrary to profit from computer knowledge and claim like in Ex14.35:

²⁵⁸ **R:** Mhm. *[∗..∗]* So how do you think can a computer help you in your future. *[∗.∗]* Can it
²⁵⁹   help you?
²⁶⁰ **S:** Yes. You can earn a living.
²⁶¹ **R:** Pardon?
²⁶² **S:** You can earn a living. When you, when you are a com *(∗interrupts∗)* a teacher who
²⁶³   teaches computer or when you are a computerialist.
²⁶⁴ **R:** Mhm.
²⁶⁵ **S:** You can earn a living.
²⁶⁶ **R:** To earn a living.

Example 14.35: Earn a living

Being a computer expert enables to increase ones income, but some informants focus more on the acquisition of certificates than on knowledge. As formal education is highly validated in Uganda, certificates are popular and also expected. This certificate has the function of approving in computer knowledge. Memos of the participant observation indicate a longing for certificates, which were discussed above in the Chapter 10. Certificates are seen as an evidence for formal education and emerge in pupils' and adults' interviews. Nevertheless, certificates currently do not seem to be a quality assurance for employers at all, as one informant states:

³⁰¹ **H:** that's how they start, and they end up actually finishing a whole course a year someone
³⁰² has a certificate, but when you give him a computer. *(∗R: laughing∗)* To connect the wire
³⁰³ just write a simple letter to Raphael say: how was your day

Example 14.36: Certificates for the future

Consequently, having a certificate does not imply real competence in using a computer. Of course, computer knowledge standards are difficult to set up and enforced, but international accepted computer certificates (like ECDL) seem to be necessary and advisable. When the ATs were asked by the SMC to provide the participants with certificates, the ATs drew upon individually chosen quality criteria and drafted their own certificate with their own levels like "Introduction to Windows" or "Introduction to MS Excel". Thus, this contributed to a vastness of certificates. Thinking about the fact that several schools participate in similar interventions indicates that such variations in certificates are going to spring up like mushrooms. This multitudinousness of certificates is counterproductive to the achievement of an standard- educational level and should be avoided in future interventions. This is important in relation to the criticized quality of current LCTs and their teaching methods - see Section 15.4.1.

106 **R:** And what made you so excited when you saw it for the first time?
107 **P:** We thought that the school maybe, yeah, we thought maybe they were /∗...∗/ It ma,
108   made us, it made us excited, I thought maybe we can send emails to our, our friends, we
109   would get certificates in future maybe and maybe played a new jive (∗???∗) in your
110   computer /∗.∗/ earlier, we learnt computer earlier and it even developed the school kind
111   of.

Example 14.37: *Certificates for the future*

Ex14.37: The excitement about getting a computer certificate played a significant role during the first encounter with the computer. Some benefits for the future, though due to the interviewee's mumbling not clearly audible, are expected. These benefits are not placed directly[12] and the informant focuses more on indirect benefits which provide for a more promising future. According to her the advantage of learning computer in an earlier stage of education than others and certificates are highly valued[13]. Obviously, the benefit achieved through a certificate is more valorized than the gain of competence acquired through the learning of computer. A certificate legitimizes to apply for a better job and consequently is highly valued.

Adults. This fact is not only acknowledged among the pupils but is evidently also something important for adults, when the following informant states, that she was expecting to acquire a certificate:

132 **R:** Ok, no problem. Ahm /∗..∗/ what did you expect of the computer, when you were using
133   it, or when, when you started using it? /∗..∗/ Did you /∗..∗/
134 **S:** Mhm. /∗..∗/ I was expecting a certificate,

Example 14.38: *Informants are expecting certificates*

Previous statement points out once more the matron's focus on formal education instead of the acquisition of computer knowledge. This puts the actual learning process in the rear. Thus, knowledge is represented by formal qualifications.

453 **B:** Their money is given according to the number of children
454 **R:** mhm, oh yea
455 **X:** but I think if such schools are given computers they can have even more children

---

[12] As discussed above the sending of e-mail is not a direct benefit.

[13] Besides, the above example indicates that the school "*was developed by computers*", which underlines the power of computers, because it appears as if the computer actively developed the school which implies a gain of status. This will be discussed in Section 14.3.3.

456 **B:** Yea, by attracting the other side
457 **X:** Yea, by attracting the other side
458 **R:** Yea I think so too [*....*] because the I have heard even because KPS has a computer
459   classroom it attracts others
460 **X:** yes, so many people now want to bring children to KPS
461 **R:** hm
462 **X:** because there is some computer lessons
463 **R:** Because the people think that when their, when their child knows about computer then
464   it will have better chances in the future
465 **X:** yes even for us we are adults, when you are looking for a job and you have a certificate in
466   computer, then you have better chances of getting into a good school [*..*] because if in
467   some schools they have computers but they have nobody to train their children so if you
468   have some good knowledge about computer you stand a better chance than any other
469   who does not have computer skills

Example 14.39: *Computer improves status of school*

This example (Ex14.39) does not only show the expected gain of status (see Section 14.3.3), but also the fact that the pupil's parents are expecting improved chances for their children, with the gain of computer knowledge, as discussed previously. Furthermore, he states that acquired computer knowledge as a teacher (Mr. X is a teacher) can increase your chances as a teacher when you plan to apply at a "*good school*" - Ex14.39Line466. To my understanding, a "*good school*" is marked by increased financial efforts from the parents. This implies employing better qualified teachers as to assure well educated children. In conclusion to his statement Ex14.39Line466-468 it seems as if ICTs in LDCs are less a matter of facilities, but more of qualified staff. This suits perfectly to the request of the Headmaster of the neighboring school to also install a computer room in his school. This emphasizes to discuss rather strategies to convey the knowledge, than developing new products.

In this text passage it becomes obvious that the importance of a certificate is not strictly limited to the pupils. This underlines once more the dominance of formal education in the local environment and its deep rootedness in everyday's life.

The expectation of an increased income by the usage of a computer as a secretary is present as well. This benefit of usage will be discussed later on (see Section 15.1), but analysis has proven that to the increase in earned money is associated with a confined typing activity.

Following examples document this statement - Ex14.40 and Ex14.41:

283 **A:** I know it can make business I know it can do some documents for me and keep other
284   things, other letters, I say let me say, it can earn money for me.
285 **R:** It can?
286 **A:** mo .. earn money for me.

Example 14.40: *Earn money by producing documents*

Whereas secretaries remain within their job perspectives, other adults, similar to some pupils (see Ex14.35) identify teaching computers as a source of income as well. From the dialogue below we learn that if the precondition of having an own computer is fulfilled, a computer can be used as a mean of knowledge transfer in order to earn money. At a close look, two latent issues arise: Apparently, the informant does not need to be a professional teacher in order to teach computers. It seems as if the bare ownership of a computer legitimizes to instruct computers. Although, if she intends to acquire a certificate anyway is unsolved. The question arises if the bare ownership of a computer legitimizes a person to the extent that formal qualification is neglected. These reading versions remain open. The second issue refers to the way of teaching: if, ownership of a computer is a precondition, this means that computers can obviously not be conveyed by theoretical literature, but at least to utilize it as an object of illustration. Apart from teaching computers the interviewee's second source of earning would be related to office work, such as typing letters, making business or weddings cards - Ex14.41Line340. The records of the thick description emphasize this when one reads: "*Secretaries stopped by already on the second day to write their application for the next year on the computer. This was their first intended purpose, they aimed at using a computer for*"[14].

332 **J:** ... If I can manage to get my personal computer ...
333 **R:** Mm.
334 **J:** ... I can at least teach some people.
335 **R:** Mm.
336 **J:** Mm. If I have mine /*.*/ in future. That's what I am been thinkin'. But if, if I don't have
337 a job, but I haven't, I have a computer. I can at least do something to get some money to
338 , to use.
339 **R:** Mhm.
340 **J:** Mhm. /*...*/ That's what I'm, I have been thinkin'. To have a computer and maybe a
341 photocopy.
342 **R:** Mm.
343 **J:** So that I can employ myself.

Example 14.41: *Get a computer to employ oneself*

However, this opinion is not restricted to teachers and secretaries, but applies also to other members of staff. One matron states:

264 **C:** Like, if I would own a office, an office, if I could own, own an office!
265 **R:** Yeah.
266 **C:** And use it to maintain my family.
267 **R:** O.K.
268 **C:** *(laughing*)* ah, if I own an office and have one, then I would work on it and give people
269 to bring work and I do it for them and earn money.

Example 14.42: *Own an office and earn money*

The informant notes the wish to own an office in order to maintain the family and to earn money. When harking back to the possible fear of the SMC that the matrons might leave the school after they have gained computer qualification, the fear is nourished by the above mentioned statement (Ex14.42). If the school does not want to loose her, as a staff member, it seems, as if there is a necessity to deny this informant the access to computers. Doing so, power is prosecuted by the school over the matrons. According to my understanding, people who have already access to a computer, will go on excluding others from utilizing computers, as they might quit their job and

---

[14] *orig.*: "Bereits am zweiten Tag kamen die Sekretärinnen der Schule vorbei, um ihre Bewerbung für das nächste Jahr zu schreiben. Dies war sozusagen der erste konkrete Verwendungszweck, für den sie den Computer verwenden wollten." (TD|p.4)

apply for a better job, most likely in town. As computers are dependent on electricity, working on computers means to work at places where electricity is rather permanently available and financeable. Consequently, people knowing computers will accumulate at places where there is electricity and therefore computers, whereas others in rural areas are excluded in the meanwhile. Not only to my conviction this strengthens the hierarchical positions of those who have access to computers and those that do not which widens the gap between these groups. One informant states clearly that he thinks that computers are not appropriate for Uganda at all, since too many people live in rural areas. He claims that "*the others [rural people] will keep ignorant, because there are those with electricity that have, will benefit*" (In7Line371). This implies computer knowledgeable people who preserve their status and prestige - see Section 14.3.3 - so that the digital divide emerges.

As shown above, pupils and adults expect computers to improve their living situation by increasing their income. Their preferred activities range from typing letter, making business/wedding cards to teaching computers to others. To put it the other way around: Their expectations are not focused on selling products or applying for a job via the Internet. Their expectations on how they could profit through communication are high, although very vague. This is not surprising, as hardly anyone of the interviewees knows how to send e-mails and therefore presenting certain steps how the living standard can be increased becomes difficult.

14.3.2   To Control Things and Gain Power

Another formulated expectation of a computer is to control things and to gain power. However, the first example (Ex14.43) is based on a fictive wish of the informant, what a computer could do for the informant[15] and represents its latent expected benefits. By aiming at controlling technological products like refrigerator or mobile phones (Ex14.43Line174) he attains more power.

169   **R:** Y... Yeah. Let's say the, the computer asks you : "What can I do for you? You have a
170   free wish" What would you say?
171   **B:** To help me study and finish build a house.
172   **R:** Mhm.
173   **B:** Putting all the things. [*...*] Mhm. And control the, even controlling the interios.
174   **R:** Mhm.
175   **B:** Like coming, enter your number, the refrigerator, the cellular phone.

Example 14.43: *To control things*

Whereas the above mentioned informant restricts himself to haptic and local things the following examples go beyond this concept. It represents the concept the interviewee had in mind, when he used the computer for the first time. In his explication he is not limited to physical things within his environment, but expands his range from the controlling of airplanes[16] to controlling the whole world (Ex14.44Line1035). Evidently, the informant wants to gain power by aggregating more information, which

---

[15]In this question, the interviewer implicitly reconstructs the concept of an "omnipotent" computer.

[16]In fact it would be interesting to discuss the symbol of an airplane in this context, but it goes beyond the scope of this thesis.

is not only latently presented by the concept of controlling things through a computer, but also explicitly, when the interviewer asks for it (Ex14.44Line1043-1045). Thereby, the informant criticizes the lack of information, he is obviously suffering from (Ex14.44Line1046) and reasons that he would feel more powerful, if he was informed (Ex14.44Lin1048). In a previous sequence he has indicated what "to be informed" means: "So, with a computer I feel once I am connected to the Internet I can surf and get informed [...] the latest news. With a computer I feel I can communicate with people who are far from me" (In9Line996-999) The access to latest news[17] and thus information becomes an instrument of power. This illuminates the reason why he states in the course of the interview that he wants to learn how to use the Internet so badly (in In9Line1289).

1027 **H:** For the first time I knew /*..*/ with a computer, I knew, for the first time I knew when I
1028   have it I would see the whole world on the screen maybe.
1029 **R:** Mhm.
1030 **H:** What is happening? Everywhere in the world? At least that's what I knew with the
1031   computer. Though I never tried to find out. It was for the first time for me I knew once I
1032   have a computer I can see the whole world. And what is happening in every corner of
1033   the world.
1034 **R:** Mhm.
1035 **H:** I don't need, to, to know anything about the world when I have a computer. That's what
1036   I, /*..*/ I need to operate and.
1037 **R:** Mhm. (*silent*) So, so does it mean that ...
1038 **H:** I, I would even control the jets, the planes, the what, I coul, I knew I would control
1039   everything in the world with a computer.
1040 **R:** (*laughing*)
1041 **H:** That's what I knew on a computer on my first time.
1042 **R:** So, made you feel powerful. Is it?
1043 **H:** Mhm.
1044 **R:** Or not really?
1045 **H:** A little bit powerful of course, if you can get not much of the things taking place in the
1046   world.
1047 **R:** Yeah, of course. Yeah, of course.
1048 **H:** Therefore I feel a bit powerful. Because, Yeah, you would be informed to some extent. (*
1049   laughing*) To change your at least your idea(*l?*) a little bit. (*mumbling*)

Example 14.44: *Computer makes you powerful*

Since computers are related to the modern world and to the fact that they are a gain in power, the modern world as well is seen as powerful.

14.3.3   Increase Status by Computer (Knowledge)

Deriving from the above stances I conclude to claim that the power gained by the access to computer and its knowledge implies also an improved social status. In fact there are

---

[17]This relation to the Internet emerges several times, when interviewees say: "*if you are connected to Internet, you can listen to news in different areas of the world*" - In22Line307 or "*because Internet has so many things, you get the latest news.*" - In14Line 254. These statements are not explicated further and so it seems as if the purpose of the Internet is restricted to worldwide news. At least it is the example which comes into the interviewees' minds. This also shows the vague and abstract concept of the Internet. The current usage of the Internet is discussed in Section 15.1.2.

indications which nourish this assumption and syllogism. Be it for the school or for a person in private the usage of a computer implies the expectation of a gained status and promises increased incomes. The interviewees distinguish between an **expected** benefit for the school and a private benefit. Therefore, this will be depicted separately. Looking at the example (Ex14.6) shows that, according to the girl, by acquiring the knowledge of computer she is made famous. This points out an increase of her own status attained through computer knowledge. To show her knowledge she applies the analogy with a Nokia. Such analogies underline the status of verbal concepts and indicates that the gain of status can be achieved by the application of such concepts. Implicitly, this is also achieved by a "Nokia" - 14.6Line259 - on which the analogy is built upon[18].

Whereas the characteristics of a computer are amendable to everyone and presented impersonal ("*you can do anything on*" in Ex14.6Line252), she ascribes the potential effect of becoming famous to herself in person. Consequently, this answer is given in "I"-form. It would be interesting to investigate what the concept of becoming famous implies and indicates, but as it is ascribed to her in person we assume that it is positively connoted[19].

Above paragraphs have dealt mainly with the personal and private gain of status through the computer users, although throughout the analysis an expected gain of status for the whole school is expected as well. I have indicated this already in Ex14.37. In example (Ex14.39) it becomes evident that the school hopes to attract more children, which - due to UPE - implies more financial support as well. If this in fact becomes true cannot be made definite, since from year 2002 to 2003 the numbers of pupils dropped from 272 to 255[20]. Though, concluding from above discussion about the powerful representation of a computer it sounds reasonable. Nevertheless, the staff of the school and the SMC expects a gain in status as the teacher in Ex14.45 points out that the "*certain status*" the school has is due to the computer lab. In the course of the same interview, the woman explains why the computer lab, to be more precise the learned computer knowledge, increases (or has increased) the status of the school: "*when you tell them a Primary 1 child studies computer they are like eeh. ... my goodness ...*" (Ex14.46Line297) which shows that the environment looks up to such an institution. However, she indicates that the child has also to be able to use a computer which indicates that a factual computer knowledge is necessary and not only the theoretical access. As a consequence to acquired computer knowledge she expects that many people consider the child as above. (Ex14.46Line296). Consequently, the status of the school is based on the children who are attending it. This increased status influences the reputation of a school's teacher.

256   **H:** Actually this is a helpful project and whoever is given the project feels proud about it
257   like most schools. Actually now look at KPS as a standard school because of the
258   computer project and like even the children who study here, every time you ask them
259   what is special about KPS, they tell you we have a computer lab, who ever they have to

---

[18]It would be interesting to discuss if there are similarities between mobile phones and computers, but this is not within the scope of this reasearch to put the focus on.

[19]This is supported by another interviewee who states: "*So I would wished I would be one of the [* ..*] one of the popular people in the, in the world*" - In18Line219.

[20]This is explained by the informant as a nearby nursery school has started a school of its own.

₂₆₀ write to, they tell them we have a computer thing. So, one is that it, it, it gives a certain
₂₆₁ place, a certain status that is most peop, places in, actually in Uganda haven't got
₂₆₂ especially in primary schools, even secondary schools, even tertiary institutions most of
₂₆₃ them don't, don't have and if they have its not a project like this, its 1.

Example 14.45: Computer improves status of school

₂₉₅ **H:** you have to pay money for it and all that. so being in a school for example a primary 1
₂₉₆ child leaving a school and they can tell this child for example to go in a certain office and
₂₉₇ do certain type of work like typing and printing this work and this child is able to do it,
₂₉₈ I think they look at this child as, as above so many people who have not yet got to this
₂₉₉ facility. So that's why people look at and when you tell them a Primary 1 child studies
₃₀₀ computer they are like eeh... my goodnees .... (*whispers so she is not audible*)... so that'
₃₀₁ s why it gives the school... (*not clear*)... actually a very different status apart from...

Example 14.46: Computer improves the status of a pupil

In terms of hierarchical relationships this underlines that a teacher coming from a technological oriented culture and having computer knowledge is seen as a powerful person. These aspects heavily correlate with Cushner & Brislin who claim that (male) technological personal is placed in a quite high hierarchic position in less technological oriented cultures from the very beginning. To my understanding this means that for a person in this position, it is not possible to take up the role of a colleague which has not been carved out in this analysis.

14.3.4 Privacy and Ownership

During the analysis in Chapter 11 it turned out that the "need" to protect own files has arisen, and that the new computer administrator has been looking for a tactic to accommodate this demand (Ex11.8). I have shown that information is regarded as something which can be dedicated for a restricted private use and should not naturally be accessible to the public. Ex14.47 points out that the school computer users are evidently aware about the fact that one should not read other people's files.

₁₁₅ **R:** Why do you like typing or, what kind of typing do you like?
₁₁₆ **Y:** Ok, well, like typing there "dear what, dear friend", putting in, ok, pings and then, ahm,
₁₁₇ when you go to save your letters, them open them again. You know, sometimes we read
₁₁₈ other people's when we are not supposed to, at least we know that.

Example 14.47: Reading other people's files

Although, she admits that she (and by using the pronoun "*we*" in Ex14.47Line118, she implies others as well) does so. Having access to information of others increases his/her power. But shielding information from others and having exclusive rights to access it, is the other way around and also increases ones power. Therefore, the access to information and consequently to a computer can be seen as a negotiation of power and social interaction and computers are utilized as an instrument of power. The knowledge how to protect or hide ones information has been transferred between 2002 and 2003 and arises in conjunction with the goal to own a computer.

411 **R:** a wish and the computer can fulfill any wish, what kind of wish would you ask the
412   computer to fulfill for you?
413 **I:** *(*laughing*)* I think, *[*...*]* the *[*.*]* the wish would be *[*....*]* well if these computers
414   were like, if I had my personal computer
415 **R:** Mhm.
416 **I:** I wish, I could *[*.*]* well, store all the information of my family
417 **R:** Mhm.
418 **I:** Mhm. And I wish I could, you know, handle it *[*.*]* at my place do work on it, even at
419   night and ensure that all my work is done on the computer.

Example 14.48: *Store information of family*

This example (Ex14.48) depicts that if the informant owned a computer he would
use primarily for storing private information. This is also correlated with the benefit
of having access to a computer whenever the informant wants to - "*even at night*"
(Ex14.48Line418). In addition to that it becomes important that the usage is done
at his own "*place*" (Ex14.48Line418). This appears as an attempt to control the
technology and by importing it to ones private place, the computer penetrates the
home of the informant. A home is a private space which is completely controlled by
oneself or immediate family. The host decides who is allowed to enter and whom
to share his/her private area with, at least temporarily[21]. With the wish to bring a
computer to ones home, it becomes a permanent part of oneself, which is also indicated
through the wish for owning a computer - this will be explained below. By bringing
a computer to his/her home, the person has to allocate a space for the device in ones
private area and consequently the will to protect it arises.

The following text passage shows a similar situation of a different informant who
presents her wish to control personal information and not to share it any longer -
Ex14.49. The girl also indicates the unrestricted access to it ("*no one can come and
interrupt you*" - Ex14.49Line143). This aims at guaranteeing that files are exclusively
accessed by oneself and also to have a timely unlimited access to it.

139 **R:** what could it be, what kind of wish?
140 **I:** I don't really know but I would *[*..*]* hm, *[*.*]* yea, I just wish to have one so that myself
141   which I own myself not like my parents, or something, or my friends me alone came it
142   will be better for me when I can as in there, I can like keep your own private things
143   where no one can come and interrupt you if you have your own personal one. That's the
144   only wish I would have for having a computer.
145 **R:** Where the, is it *[*.*]* so, ah, please correct if I understood it in the wrong way, is to keep
146   your files private.
147 **I:** yeah
148 **R:** So that you can store information and no one other has access to it?
149 **I:** If it's all that all private of course let's say my family members, my closest friends, yea
150   they can have access to it but *[*.*]* there is some other private news which they can't
151   have access to but most of it won't be there normal, e−mails to your friends and what
152   that stuff.

Example 14.49: *Owning a computer*

---

[21]This decision which room is considered to be accessible to the guest is made by the host. He/She
is in control of the situation and leads the guest to him/her adequate room. Depending on the
relationship and closeness of the guest it can vary.

When the LAN between the Headmistress's office and the computer lab was disrupted the ATs asked the new computer administrator to repair it. After some time she told the ATs that the Headmistress did not want to have her office computer connected to the others, as she feared that other people might read her official files on teachers and pupils, or even have access to exams. In that case official files are in fact private information of the Headmistress. If she cannot guarantee that an unauthorized person might read these files, she looses control over her information. This is something she clearly aims at to avoid. Therefore she hindered the repair of the LAN in favor of the protection of these files.

On the surface the combination of information protection and the demand for an own computer does not appear to be a necessary consequence. However, discussions revealed that these things seem to be the reason of motivation for the desire to possess goods. Keeping information private and preventing to share those, indicates the longing for identity. Controlling information within ones reach means to distance oneself from others and to create an own space. The user who wants to be exclusively in charge of information apparently wishes to possess goods (just as information is a good). Consequently, when these wishes emerge in the interviews, it approves as a more individualistic tendency than a collectivistic one. Not everything is shared and part of the community. Historically, this individualistic tendency is considered as an eurocentristic emergence from the Age of Enlightenment (as mentioned in Section 2.2.2). The identity of the user himself/herself is more important than sharing information. This identity is constructed by the exclusive access and control of information.

However, the wish to adopt a computer as one's own takes the same line. As shown above, the desire to possess a computer physically, becomes explicitly evident in the interviews. In consequence to "owning" it, a computer becomes part of oneself and implies that the owner ascribes the positive characteristics of the subject to oneself. As shown above, these characteristics are e.g. to be developed and help by gaining an improved status, which seems to be obtainable by the permanent access to computers. Obviously, a computer seems to promise and to fulfill these expectations. A computer emphasizes the desired feeling of "being developed". This shows that computers are not value-free and fortify the specific desire of possessing goods. Evidently, the characteristic of the product "computer" implies demanding for more. This might be regarded as a rather consumer-orientated attitude which is specific to technologically orientated cultures. Though, considering the fact that the wish to own a computer and to use it for private purposes only, was explicitly raised by those who mention the most possible uses of a computer[22]. However, such an attitude becomes prevalent in the local cultural environment of Kabale. Apparently, the more I know about computer, the more I want to control it, be it virtually, by controlling the information, or physically by owning a computer itself.

---

[22] This indicates a certain time of using computers. In combination with the participant observation, the informants offering their wish to possess a computer, are also those which performed quite good. Although, to limit the wish of ownership on usage time and performance would be over-simplified. Previous computer knowledge and the way of using a computer during the ATs absence are undocumented and are contributing factors.

Chapter 15

The Object "Computer" - Application

15.1 Usage in Everyday Life

The usage of computers in everyday life is apparently influenced by some factors which handicap the informants. In the following sections I will present those factors which emerged during the analysis of the interviews.

To use a computer means that one has to have the possibility to physically access it, to have the know-how to utilize it, and to have the time. The first precondition implies that either one owns a computer or is allowed to use a computer which is owned by someone else. Both require financial resources to gain access to it, except when you are in an institution which grants you access to it, or it belongs to a person who is close to you. Beside this *"physical access"*-precondition, the second precondition of mandatory computer knowledge requires also to have access to human resources which can transfer the knowledge. This *"knowledge"*-precondition is usually correlated to expenses as well, except when one has a relative, friend or employer who enables this knowledge transfer, either as a relative or friend in person or by providing a computer teacher. It was revealed that previous factors are constrained by the resource of available *"time"* to use a computer, which represents the third factor. In the course of this study different combinations have been met, which are discussed for the involved participants subsequently. I will present both main user groups, pupils and staff separately. Concluding from above presented actors and their relationships; three locations arose where the participants of this intervention used computers. Pupils access computers at their parents place and at school, whereas the staff (mainly teachers) at public places like the Post Office and also at the school as well.

15.1.1 Pupils

This section deals mainly with pupils, their possibilities to use computers and the ways they use them. To illustrate their usage the three key barriers *"physical access"*, *"knowledge"*, and *"time"* are discussed.

Physical Access

Usage at Home or Parent's Office. To have physical access at home or at the parent's office requires that the parents own or have unrestricted access to a computer. Concluding from the interviews several parents have a computer at home or in their office (Ex13.12Line121). As mentioned in Chapter 10 the parents of the school pupils are affluent. Most parents work as civil servants[1] which implies more work on computers than in manual labour. Consequently, I consider these parents as computer affine and due to the fact that more money is available to them[2] it becomes "easier" to them to afford a computer. Some pupils have access to computers at their parents' office. Looking for financial constraints for using computers makes it evident that the running costs of electricity seems to be one. When one pupil claims that "*My parents do not like me to use that much electricity.*" - In27Line128, this indicates that electricity plays a financial role in daily life and is not a matter of course. Another factor influencing the usage of computers at the parents place is that parents fear that their child may cause damage to the computer - Ex11.2. Spare parts and skilled people are rare., therefore, if a computer or a screen breaks it takes weeks or even months to get the computer fixed or to find the necessary spare parts[3]. This issues arises in interviews as well, when one informant claims:

505   **R:** So do you think that therefore they sometimes not... Do they fit in the Ugandan
506   environment?
507   **I:** Well, they are suit, suitable, they are suit, suiting in, but they are still having a problem,
508   we are still having a problem handling them and maintain them, because some peopl,
509   some of our people have not acquired the knowledge and skills of repair.
510   **R:** Mhm.
511   **I:** Repairing these computers. And in future you find according to this lack of knowledge, we
512   have a looots a looots of skills, computers and spares
513   **R:** Mhm.
514   **I:** And not using them. Because we failed to get some people to repair it.

Example 15.1: *Spare parts and knowledge are not available*

Whether Internet fees pose a financial problem remains unsolved, but due to its high costs (see Chapter 10) it has to be assumed that it is used economically. However, as discussed in detail in Section 13.1, it seems as if a pupil is allowed to use a computer, if he/she has approved his/her knowledge. Although, some are not allowed at all, as

---

[1] One teacher claims that "*And the fact that our school has children of people who are mainly civil servants*" - In5Line501.

[2] To use computers they have to be placed in urban areas, which indicates an income above the average.

[3] As a matter of fact the school had organized some technicians to repair the screens. They transported them to Kampala and did not report to the Headmistress for months. When the intervention's second period started, the Headmistress asked me to have a look at them. In Kampala the mechanic told the ATs that they had taken a look at them, but could not repair them and advised me to buy new ones. When the ATs got a false report on broken parts, they got suspicious of this analysis and examined one broken screen by themselves. Finally, after replacing visibly damaged condensers and asking an Austrian expert on screens, they noticed that in fact the screens were not repairable. It also turned out that these screens were easily affected by voltage fluctuation. As a consequence they reported this to the Headmistress and recommended to buy new, but different ones. There is evidence that other ones might not break as well, as this seems to be an error of this certain production series.

their day is filled with school tasks: "[...] *at home I didn't get to really enough access to it 'cause I had to read at time I was coming to P.7"* - In21Line109.

Usage at School. The usage at school is bound to an additional financial contribution by the parents for maintaining the lab. Expenditures for e.g. electricity are aimed to be covered by this additional fee.

Knowledge Transfer

Pupils acquire their computer knowledge about computers either from relatives (e.g. grandmother, uncle, parents) or at school from computer teachers during lessons. Though, not expressed explicitly in the interviews, but in reference to my field notes, pupils help each other in playing games[4] or when "*they were stuck*"[5]. According to my observations other teachers did not interfere in the children's games. If this is due to their own lack of experience is not definite, but just an assumption.

Time

Usage at Home or Parent's Office. The pupils' usage time is generally allocated by their parents[6] and depends on the purpose of their intended usage. In general, pupils have a high workload and report that if they are allowed to use computers, they may use a computer mainly for doing their homework or writing letters. A typical statement of how the pupil uses the computer at home is: "*To type some work. [* ...*] Mainly to do some work. [* .*] Or if I am free, I play games*" - In4Line103 or "*when dad is teaching us mainly, it's it's learning not playing games*" - In3Line158. The prevalent opinion is business before pleasure. Children are mainly supposed to carry out their duties (to do homework). Although, some pupils support their parents in their work by e.g. "*making business cards*" - In14Line103 or typing letters. Thereby, computers are implicitly presented as something serious (Ex15.2), something for business and office work ("*the people who need computers are the people who work as officers*" In7Line534), although games would have been actually preferred by the children. In the course of the interviews the pupils preference for games was tremendous. ALL (!) fourteen pupils liked to play games and preferred it (together with listening to music) to other usages of a computer.

156 **R:** So at home you don't use it for games is it?
157 **A:** At times we do, other times we don't, but when dad is teaching us mainly, it's it's
158   learning not playing games.

Example 15.2: *Usage at home means to study computer*

---

[4] Explaining to the other one what to do.

[5] "Excuse me, I am stuck" is the most common phrase to express that one does not know how to proceed on the computer. Whereas this rarely happened for applications within the operating system, games running DOS-mode crashed several times and were quite instable.

[6] As the above example shows it can be close relatives as well.

Usage at School   At school they may use computers during lunch time, those who are borders on Saturday afternoons or between the afternoon lessons and evening tea (this is a time slot of around 45 minutes). This usage heavily depends on the responsible teacher's computer skills and time, (which will be discussed below in Section 15.1.2), if no teacher is in the computer class it will be locked anyway, or if the teacher does not want (or dare) to use a computer by him/herself (e.g. Saturday afternoon) the situation is the same. The usage time for boarders is also rather stingy, as even if they do not have lessons, they have to meet other duties like cleaning their room, or doing homework. Pupils are fond of the computer lab and it makes the impression as if they use "every minute" of their free time to go into the lab. This is not only a result of the participant observation, but even in the interviews the run of pupils on the computer lab is pointed out, when one interviewee claims "*in the evening I don't get my own time, because there are some people here and all of them are playing, are playing games and some are decorating, so I don't get, my own free time of coming*" - In17Line186-188. A pupil-computer ratio of three to one for a period of 45 minutes indicates that each child has 15 minutes. Obviously, this is not enough to improve and develop their own computer skills. This is indicated by an informant who claims that this limited time of usage is even insufficient to train and reconstruct learnt things - Ex15.3.

215 **R:** ah so you have to /*...*/ Yeah. Do you want to say something about computers?
216 **N:** aah, they are very nice and I wish I could have my own time of coming here and like
217   playing and typing my own things and also learning, ah, very much about them so that I
218   can type very quickly and I, I was very excited when I first saw them and I thought that
219   I could also get my own time until P.7 but it came that we don't usually have computer
220   in P.5 so /*.*/ and in my free time in the evening there are boys here and little girls and
221   its time to evening tea and I become late so sometimes I don't get the chance of coming
222   here so like now when I got the chance, I try many things and sometimes what I, I call, I
223   have to call somebody to put it for me, because I have forgotten some things

Example 15.3: *Own time to practice*

During the lessons pupils concentrate mainly on issues like typing or drawing, but are also allowed to play games as soon they have finished their tasks within the computer lessons. As indicated above during lunch-break the pupils prefer to use games, be it for example Super Mario, Prince of Persia, Test Drive, FreeCell, or Solitaire. However, usage time during lessons is short as well and with the reduction of running computer units (breaking screens) the pupil-computer ratio increases.

Ways of Usage

General.   Generally speaking, the represented usage of computers by pupils is implicitly always marked with the permanent latent fear of spoiling the equipment, having to pay for it and being punished - Ex11.3. Most likely this fear is emerging to an extra-ordinary extent in this intervention as several screens of this school blew up during the lessons without any external impact, which proved the technology's unreliability, which is discussed in Section 15.2.2. Parents control the access to PCs at home, whereas pupils would prefer to play games instead of typing. However, they are supposed to use a computer to work. Nevertheless, the fascination with "computer" seems to be strong enough as to desire use at every opportunity[7] to play games or listen to music. To give you an example: When the interviewer asks the interviewee what she generally dislikes about computers she states: "*Me, as they are not a lot of interesting, they are boring for me so I prefer the music and the games*" - In21Line31. The presented benefits of typing, listening to music and playing games are discussed in the Section 15.2.1.

Excursus - Games. Interestingly, when the pupils are naming games in the course of the interview all but one interviewee lists games of the school, which indicates that they are not playing games at home and do not have the opportunity/connections to get new games, respectively. According to the participant observation their level of playing games has improved within one year, but as e.g. the top player reached only the third stage of Super Mario, I take this as a further indication that they use a computer to play games at home rarely. This limited usage is underlined by only one boy who lists different games than those available on the school's computer. Besides he is the most successful in playing games, the only one who has used the Internet and also the most well coordinated of all children: He is the only one who uses one hand to use the cursor to play games whereas the others use both hands or even asked a friend to help them to play the game. To give an example: for a car driving game, one pupil presses the arrow-up key permanently to accelerate the car, whereas the other uses his/her left forefinger to steer left by pressing down the left-arrow and the right forefinger to steer the car to the right. I made this observation to a subject of discussion when I interviewed the above mentioned boy, as he was the only one who used one hand. He states: "*That's funny I also wonder, but I think..[*...*]. I think that's why, because if you are on computer you, they use either one finger [*...*] so typing like this, that's funny, at least I can all of them with the both hands.*" - In14Line289f. To my understanding he correlates playing a game to typing. If this is in fact the reason for the different way of playing games, I am not sure, but at least it indicates that there is a difference, which is obviously based on a time intensive dedication to computers. An exploration of this issue would be interesting, but is considered to be beyond the scope of this work. Anyway obviously, he has not passed over his games to his school mates. This might be as other games are "useless", when the children have only limited time to play games at home anyway.

398 **L:** because Micrografx you can make your words very different and you can make decoration
399   in the side different, [*.*] with different sizes and with different decorations. Yeah I think
400   that's what I expected in a computer but some of my expecations I thought that you
401   just come on a computer, go to Internet Explorer, you find there a, you find there In,
402   Internet, and you tell it what you need. But though you have Internet, I went to Internet,
403   my, for my first time, I didn't know how to, how use it , I didn't know what the purpose
404   was. And so I just got confused , I tried to download movies, I couldn't get the
405   downloading programs, but we downloaded them from our friend and he,
406 **J:** Keep quiet and you sit. (*to others*)
407 **L:** after sometime we tried to download some movies, but it was slow.

---

[7] As soon as the bell rings indicating a break, children sprint to the computer lab and ask to use the computer.

₄₀₈ **R:** Mhm.
₄₀₉ **L:** it was slow downloading, it would take, but I get everything down from Internet, but the
₄₁₀   downloading was slow.
₄₁₁ **R:** Yeah of course.
₄₁₂ **L:** And so I think if there is a better program there they can release, I think that would
₄₁₃   fulfill my expectations.

Example 15.4: Internet too slow

This boy is also the key-user of the pupils in reference to the Internet. Whereas other pupils stick to the vague representation of expected benefits like earning money by communicating, he has aimed at utilizing the Internet for a purpose he had heard about: to download movies - Ex15.4. However, his expectations, to get everything from the Internet were not met. His experiences are based on a landline connection and to download the estimated size of 600mb of a movie takes about 34 hours[8]. Looking at the costs involved this means \$102 per movie[9]. As a consequence the Internet cannot be an option to download movies - Ex15.4Line413 and he stopped using it. This becomes evident in correlation to a prior statement when he explicates that his wish how a computer could support him in his future, would be to get connected to the Internet - In14Line251-258. Merging these statements shows that the issue of financing prevents him from using the Internet.

15.1.2   Staff

Physical Access

Whereas the usage of computers in public institutions entails the payment of fees (Ex11.7), I have previously discussed how the access at school is constructed under the aspects of power. From a financial point of view the school can afford maintenance costs as electricity and printing are, as they are covered by the extra fee of the pupil's parents. As members of staff they do not have to pay extra, but as several screens are damaged, less facilities are available. If money for new screens can be raised by the school is unsure, but in my opinion the school, seems to be economically able to afford at least one screen per term. A preventing factor, why the school might not be very encouraged to buy new screens, is, that they are blowing up - see Ex15.5. During the second stay in 2003, I emphasized that this is apparently the vulnerability of this productions series to voltage fluctuation. If sufficient high voltage regulators had been used, this could have been avoided.

₅₅₁ **R:** Yeah. If you perhaps start now buying one screen per term for example.
₅₅₂ **A:** But the worry is we might buy those screens and they also blow.

Example 15.5: Fear that new screens might blow up as well

---

[8]This is based on an optimized download ratio of 5kb/s. Landlines proved to be unreliable, which means that several download attempts have to be made. Even if the server support resumes, 34 hours download time represents a best case scenario.

[9]This calculation is based on 34 hours multiplied by \$ 3 per hour connection fee.

To my understanding this is the reason, why the school aims at repairing these screens and acquiring the knowledge to repair them:

522 **R:** Yeah, but how do you think can we, *[∗..∗]* can we take, ah, can we prevent this
523 **I:** It's one organizing workshops for some teachers or some skilled people and being taught
524   how to do these repairs
525 **R:** Mhm.
526 **I:** There should some workshops training, a teacher is person who is trained
527 **R:** Mhm.
528 **I:** Make a workshop, training right from *[∗.∗]* every part *[∗.∗]* trained for instance like *(∗*
529   *getting sheet∗)* this is *[∗.∗]* like you know *[∗.∗]* they talk of a computer it's like like what
530   like a simple thing to you.

Example 15.6: *Knowledge to repair the screens*

On the date when the interview was made, the forlornness of repairing the screens was not yet evident. Therefore, it could not be pointed out during the interviews that the screens were not repairable. If Internet fees pose a financial barrier for the school cannot be said and will be discussed later on, as it might not be a financial problem why the school seems not to use the Internet at the present time.

Knowledge Transfer

Staff members learn computing either from paid teachers in town or by the school's computer teachers for free.

Time

Besides the financial expenses it is also the distance to get to the place, if one wants to use a computer in town. The workload of a teacher is high and an additional walk of at least forty minutes return. Obviously, the usage of computers at the compound implies several advantages to the ones in town, which is discussed in Section 15.4.1.

Regarding the spent time of using computers at school, many informants admit that they do not use a computer currently and refer to their main problem as lack of time due to their workload. The following example illustrates the whole time-restricted situation perfectly:

70 **A:** but but as far as my staff is concerned for my I think they don't practice a lot but they
71   can make marksheets, some drawings, like which I can't.
72 **R:** So why do you think they don't practice a lot?
73 **A:** Ah..they have a lot of work here. Different activities.
74 **R:** Mhm.
75 **A:** The different activities that .. that we are involved in.
76 **R:** Mhm.
77 **A:** Sometimes hinders them.
78 **R:** Mhm.
79 **A:** And some of them, of course, are involved in some other, those who live here, we are
80   giving exam after exam. others of course They are involved in marking, they are involved
81   in private studies, they are trying to get, they are gone for course. Those who have

82 certificates are now teaching and the same time studying. The free time they get is their
83 free time they go to their own private work. Write their assignments. They have got time
84 limits. So I think that is the trick *(*?*)* the trick then.
85 **R:** Mhm. So ...
86 **A:** I think even having these girls here who are specifi producing the work, as somehow
87 hindered them in typing. Because when we have a exam they write it with their hand
88 and hand it in. But if there was no secretaries and it was like everybody has produces his
89 own work, marksheets unlist they have done it ...
90 **R:** but they don't do it on their own, do they?
91 **A:** Sometimes.
92 **A:** but when Caroline came in again, she somehow spoilt them.
93 **R:** They
94 **A:** It seems she does some work for them somehow she was supposed to teaching them some
95 elements.
96 **R:** Mhm.
97 **A:** but as I have said others would enter and it would somehow disorganize her and maybe
98 she found it easier again to ..
99 **R:** of course it's easier for her.
100 **A:** hahahaha *(*laughing*)*
101 **R:** because teaching means to take your time and to explain it ten times.
102 **A:** but previously they were doing it.
103 **R:** mhm.
104 **A:** On their own.
105 **R:** oh.
106 **A:** *(*interrupting*)* the marksheets but they have stopped just now.
107 **A:** Mhm. Some time back. In fact there was a time, I think it was last year, third term,they
108 produced marksheets for themselves.
109 **R:** Mhm.
110 **A:** Mhm. But it was hectic. *(*laughing*)*
111 **R:** Yeah. Butt...
112 **A:** They almost slept in the computer room.

Example 15.7: *Usage of computers*

This example shows that if the teachers have the chance to outsource their computer work, they do so. The informant mentions two ways how teachers try to pass on their work: either to hand it over to the secretaries whose job it is to type exams or marksheets is or to Caroline, the new temporary female English teacher from the U.K. parish. When Caroline arrived, she used the computer lab in the evening as well to write her e-mails to send them later on via the post office. Obviously, if a teacher who has computer related problems at the same time she was in the computer lab, he/she asks her. Actually, she was supposed to teach computers (Ex15.7Line94), as she had computer skills. Instead of explaining the problem or solution to them, she did the work for them (Ex15.7Line94). As a consequence it seems as if the teachers were outsourcing the making of marksheets to Caroline. Therefore, there is no necessity for them to use or learn about computers any longer. According to the informant they might be able to use it for their main purposes (Ex15.7Line70), if they really needed it. Critics might claim that this represents only the view of the teachers, but when I was talking to Caroline, she told me that the teachers did not go to the computer lab

very often. She explained that when she arrived in December 2002[10], teachers did go
to the computer lab, but later on they lost interest. This was her individual way of
perceiving things. Evidently, it was not a matter of interest, but of saving time and
outsourcing tasks to her.

Nevertheless, informants claim that they wanted to use computers if they had time:

*"I am expecting maybe to learn more about it in case we get time and maybe teach
us to keep teaching us"* - In5Line110, or

"it is difficult to learn it when you are teaching because you don't have time for it"
- In7Line21, or

"the nature of work [...*] School timetable is the one which brings in the problem"*
- In27Line310.

Repeatedly, the argument of non computer use is the teacher's high workload,
which makes it difficult for them to attend extra computer lessons. However, it could
be argued if this is really based on a teacher's high workload or just a mechanism to
justify ones fear in approaching the technology (see Section 15.4.1). As a matter of fact
the day of a teacher is marked by a general availability in the school from 8 a.m. to
6 p.m.. Within this timeframe teachers are busy. Though, lessons were offered before
(7 a.m.) and after these hours (up to 9 p.m.). The following example underlines that
the daily work of a teacher is perceived as a heavy burden[11] - Ex15.8Line192:

159   **R:** And why do you think did we decide on our proceeding, on that proceeding?
160   **I:** Ahm,[*..*] I think along that study, along that arrangement, [*.*] somehow teachers did
161   not cope, I mean did not have enough time.
162   **R:** Mhm.
163   **I:** That's why sometimes they would tell us to come at a certain time.
164   **R:** Mhm.
165   **I:** and in all, one or two come.
166   **R:** Mhm.
167   **I:** Others would not come. One day two people come, another one, two, three, like that.
168   **R:** Mhm.
169   **I:** So I think time did not permit, not allow the teachers to continue in the learning and
170   teaching of the computer.
171   **R:** How do you think we could support it?
172   **I:** Ahm, there what I, I would imagin, ahm, [*.....*] what I would imagine was to [*..*]
173   arrange any time, for example, when you think teachers are all available.
174   **R:** Mhm.
175   **I:** There could be a time sit.
176   **R:** Mhm.
177   **I:** For the teachers when they are not involved in teaching
178   **R:** Mhm.
179   **I:** So that we say. Monday, Tuesday, like how we had started during the time you arranged
180   to go. At least the time was enough for teachers to come in and learning, because there
181   was no teaching. So I would, I would say to improve that we need to arrange a special
182   week, we say this week is for teachers only.
183   **R:** Mhm.
184   **I:** And we allow teachers to sit like how children sit and on that lessons, after lessons they

---

[10] This correlates with the "end of the third term", which is mentioned in Ex15.7Line107.

[11] As I have discussed in Chapter 10 some teachers succeed in attending the lessons more regularly
than others.

185   go for lunch and after lunch they come back and that
186 **R:** Mhm.
187 **I:** Or they just come half day. Like that. Otherwise if you mix it with these teaching hours,
188   we are here, you have always *(*??*)* seen us, we are very busy.
189 **R:** Yeah.
190 **I:** One item, after one other item, after one item.
191 **R:** Yeah, I know.
192 **I:** Mhm, so it was, it wasn't convenient for teachers, the time wasn't convenient.

Example 15.8: Computer lessons for teachers

The informant aims at explaining the unsteady attendance of the interviewees during the intervention's first period in 2002 and suggests to arrange the lessons on a date, when all teachers are available (Ex15.8Line172-173). In his subsequent suggestion (Ex15.8Line177-182) he refers to an five day course we had arranged in the school holiday after the second term in 2002. When the informant elaborates on his idea he latently criticizes the teaching methods of the ATs (Ex15.8Line190) and indicates (like other informants) his underlying opinion of how technology should be approached and instructed (see Section 15.4.1). Whether or not the method of instructing computers en block during the teacher's holiday will be successful is difficult to say. Investigating field notes on the course held by the ATs shows that besides latencies people's attendance was varying a lot, although the number of people was increasing day-by-day. To attend the course was ones own choice and as a matter of fact, football games, or wedding plans, were prioritized to the course. Some could not attend as they had arranged other advanced training, although the course dates had been settled two months in advance, with the agreement of all teachers. Others attended although they had cancelled first. This indicates, among other occurrences[12], a polychronous orientation, in which planning is seen as an unrealistic behavior and neglected. This appears to be contradicting, as a school is bound to strict timelines, which represents also elements of a monochronous oriented local environment[13]. Anyhow, both attitudes emerge and flow into each other[14]. This crops up also during the ordinary teaching day. People agreed on lessons in the afternoon, but did not show up. The interviews of this study were made during a teaching period and when the interviewees were asked if they wanted to give feedback, they usually claimed not to have time due to their duties (workload, looking after children), but several times they suddenly appeared in the computer lab and wanted to have their interview. They had "*found some time*" to do the interview. I interpret this as some kind of mixture of both attitudes and applying a thick description passage to the computer course. This shows that the attendants' punctuality increased day by day. When the ATs insisted on the participants timeliness and explained that they were going to start teaching half an hour after the scheduled

---

[12]During the course several attendants were several hours late.

[13]The example of the intended punishment of the girl who was running late, underlines this. Of course, these elements are an effect of taking over the British (Missionary) school system and its formal way of education.

[14]This emerges also in the way public transport is realized. On the one hand there are strict timetables when buses are leaving (monochronous), but on the other hand it happens that if the bus is full, it leaves several hours in advance (polychronous). It can also be that the bus does not leav until it is full. So it is less a matter of the official timeline, but more a matter of rentability for the owner.

time, it seemed as if the participants took over this monochronous attitude. Thus, the ATs wielded power and enforced their cultural expectations on how a computer course should be performed. Day-by-day this perception of time was negotiated. Finally, the participants had given up, obviously, the ATs tried to start the course regardless of whether people were missing or not which made them appear rather punctual. It would be interesting how the situation would have been if the attendees had to contribute a little fee (to increase attendance) for the course and if the course was assigned with an official certificate. Of course, both aspects are more assigned to technological oriented cultures than to indigenous ones. However, it would be interesting to find out if regular and punctual attendance would have been adapted faster. By that the instruction of computers could have been performed according in a way participants had expected - see Section 15.4.1.

Ways of Usage

General. As shown in previous sections, staff members are using computers for private things. This is regardless if they are teaching or non-teaching staff. If they are allowed to use computers in general, they use them also for private things. Although this is an exception in the below example (Ex15.9); this fact is not mentioned by the interviewee himself/herself. Either it is "admitted" after requesting for it or latently made clear by subsequent contradictions. Even in this example she immediately relativizes her private usage, as she emphasizes the official usage as well. This depicts her uncertainty regarding the private usage of computers. It shows also the preferred private usage, to write messages/letters to friends. This form of communication is related to the benefit of printing (which is discussed below in Section 15.2.1), but the underlying principle remains as follows: If people do not find the time to visit someone, they write a short letter[15], staple it and hand it over to someone who is supposed to give it to you. The staple is supposed to attest that the message is unread. This makes it possible for no personal encounter and is rather reliable and cheaper than sending via post service which costs $ 0.50 per letter. If one types the letter instead of writing it by hand, the process of communicating itself is not changed but the form of representation. A letter is a form of self-representation and by using "neat"[16] computer work one ascribes this value to himself/herself.

An example of how teachers can be pushed to use computers privately was discussed on Page 163. Apparently teachers can be motivated to use computers: one informant reports that he had to hand in computer work to absolve his course at the university. This measurement "forces" students to enhance their computer skills, although only those who are employed in a school with computer access or those who can afford to outsource the typing work will pass the course. Of course, the latter one will not develop further computer skills. People put a lot of effort into advanced training, which emphasizes the appreciation of formal education. However, the personal benefit

---

[15] An exemplary letter could be as follows: "Dear Mr. Ruhanga, I have a transport for you to Kasese tomorrow morning. Be at my place at 7 a.m.. Yours Sincerely, Mrs. Akumaswhahese"

[16] See below in Section 15.2.1 for this discussion.

from and motivation for such courses could not be enlightened and to attend courses seems more to be a matter of life for the holidays than any real noticeable professional profit.

Among the usages for private purposes is also the creation of different types of cards: "[...] *things I like, these congratulations, these ahm these ahm, good luck wishes birthday cards like that, putting all that, wedding cards like that,*[...]" - In10Line214-216. For all these occasions cards are created and distributed, which is once more bound to the action of printing (Section 15.2.1). The extent of this application is not transparent, it was at least mentioned in the interviews and accepted enthusiastically in the computer course of the teacher's holidays, when the creation of invitations was trained.

Investigating the ways of usage shows that profession related tasks are found in terms of making marksheets[17], creating exam sheets, and writing official letters. As discussed above, teachers tend to outsource their tasks to the secretaries. As the following examples (Ex15.9) show, some informants claim that, if needed they could do the preparation of exams by themselves. The knowledge for these tasks seems to be conveyed successfully and can be recalled if needed.

92 **R:** hm *[*.........*]* So how are you using a computer now? For what purposes?
93 **H:** Maybe when I have *[*.*]* exams to set and I've handed them over to the secretaries very
94 late, and they seem so busy and they can't do it, I can come and do it, I usually come
95 and do it myself and making marksheets from whenever I want and for some personal
96 issues like writing letters to my friends sometimes I feel I should not write them, write to
97 them in pen and I come in, put it on, and then send it *(*very soft, not audible*)*
98 **R:** Mhm, so, for private, just writing letters for example
99 **H:** hm and for the school.

Example 15.9: *Private usage of computers*

Other professional usages were not named by the teachers themselves but by pupils, when they claim that they are sometimes in the computer classroom for English lessons and write letters - Ex13.3. Obviously, the computer cannot be integrated into other lessons yet. So I conclude that such skills of the most teachers are, due to lack of practice at best stagnating. Games are also appreciated to be used in leisure time, which is discussed in detail in Section 15.3. The extent of how much games are played among adults is unknown.

Internet. In accordance to above introduced difficulties of grasping concrete benefits of the Internet it seems as if these competencies are still unused[18]. On Page 123 I have presented how the students were instructed and what situations emerged. During the analysis, indications emphasized that no real benefit can actually be retrieved from the Internet, although high expectations are set upon it. According to one informant they used the Internet once after the ATs departure with Mr. Rukinga,

---

[17] After each test, the ranking for each tested subject (which is based on the average score reached) is hung up in the classroom. The prior results and the total accredited points are presented. According to informants this aims to encourage competition.

[18] I derive from the stance that there are members of staff who can send e-mail messages if they want to. Training was done extensively and scripts could be applied, at least when the AT left in 2003.

but did not use it again[19]. Another expense factor might be the connection fee, which might be considered as wasted money, if one does not know how to benefit from the Internet. Furthermore, the necessity to use the Internet connection to communicate to the Projekt=Uganda is not that urgent as communicating via e-mail has proved to be reliable via Mr. Dekanya. Apparently, the staff is not aware that there is the possibility in the Headmistress's office[20] to dial in, when one states:

1294   **R:** I will help you. Yes because I think with e−mail. Now at the headmistresses office, you
1295   can send e−mails again.
1296   **A:** Yes.
1297   **R:** And even I told Jackie how to send e−mails from the Voice of Kigezi.
1298   **A:** Is it?
1299   **R:** Yes.
1300   **A:** That's good.
1301   **R:** So, she can send an e−mail to me or she can teach you as well as about that.
1302   **A:** The e−mail?
1303   **R:** Mm. Yes. How to send an e−mail. And ...
1304   **A:** From here you can send an e−mail?
1305   **R:** From the headmistresses office.
1306   **A:** Hey!

Example 15.10: *Sending e-mails from the Headmistress's office*

If the Headmistress concealed this fact consciously or simply regarded it as not important enough to mention to her staff is unsolved. Anyhow, in comparison to a mobile phone, going to the Headmistress's office, switching on the computer, typing the message, dialing in and sending it might be considered as circumstantial as well. Investigations retrieved that the social relationships of the teachers are mainly placed within Uganda. Most of them are found in rural areas, and whereas Internet coverage is low, the mobile network proved to be quite good and reliable in 2003. It seems as if whenever there is a need to communicate the mobile phone is used. In 2003 about ten out of fourteen teachers had one[21]. Even when pre-paid credits are rare, at least they were passive users and they were observed several times talking with friends and relatives. If one has low credits he/she "beeps" the one he/she desires to talk with. This means to ring once and hanging up. Thus, the other end becomes in charge of establishing the connection and paying for it[22]. For example: After the AT's return in 2003 he was beeped a few times by some members of staff, but did not receive an e-mail. Doing so, communication is consigned to the other one, who seems to be wealthier. Thereby, this one becomes more powerful as it is up to him/her to establish the communication. Social inequality manifests and is reconstructed by this usage of technology and money becomes even more important in the social relationship. It is

---

[19] This seems to be related to the dismissal of Mr. Rukinga.

[20] The Internet sharing module does not work any longer and due to privacy issue the LAN from the Headmistress's office to the computer lab got disrupted.

[21] During lunch, mobile phones were presented by putting them on the lunch table, which might be considered as a way of representing ones status, but is beyond the scope of this work.

[22] Notice: This emerges also in the process of getting to know each other. Women beep men and if they want to get in contact with her, they call. Informal talks found that this is interpreted as a sign of investment as well. However, this can be seen as a way how roles reconstruct and strengthen themselves: The man is supposed to show that he can take care of the woman.

not within the scope of this research to discuss the way mobile phones are integrated into Kabale, but mobile phones are evidently preferred to the Internet. This seems reasonable as talking coincides more to the local tradition of communicating.

15.2   Effects and Experiences of a Computer

A concept of a computer is always assigned to different types of (un)expected effects and experiences it causes. Besides discussed indirect expectations of the users like the gain of status through the appliance of computers, there manifests also straight expectations to its effects, like printing. Such expectations towards physical and meta-physical characteristics are mutually reconstructing each other and influence the whole concept of a computer. Some of these effects and experiences might be considered as obvious, but others are not. This causes me to list experiences and effects which are named in the course of the interviews and which are straightly related to the physical object[23]. Whereas some are perceived as (personal) benefits (15.2.1), others are connoted negatively and conduce to a reduced usage of computers.

15.2.1   (Personal) Benefits

The following sections deal with the retrieved personal benefits of the users of the computer intervention at the primary school in Kabale.

Storing and Reusing Information

The analysis has demonstrated the perceived benefit of reusing stored information[24]. This applies not only on written official letters (Ex15.12) and the work of the secretaries (*"it keeps our work safely"* - In2Line144), but also on storing music as the following informant states:

136   **R:** no, no sorry, I mean, but you could also listen to music when you are using a radio.
137   **P:** yea, but also here you can, [*.*] you can store the best music you want to listen too all
138   the time and Radio its just
139   **R:** Mhm.
140   **P:** something like that

Example 15.11: *Store music on computer as a benefit*

The informant clearly points out the permanent availability of music as a perceived benefit: Songs can be played whenever the user wants to. This advantage gains in importance, when the usage of listening to music as a measure to *"keep one busy"* arises - see Section 15.3.

---

[23] The indirect expectations like to increase ones social status have been discussed previously.

[24] The emergence of this benefit is no real surprise, although during the first intervention the functionality of saving led to some problems, as it was not reasonable to the participants. It was difficult to understand why one has to keep on clicking the "SAVE"-button during writing a letter and not only once the user has begun using the corresponding program. It seems as if the computer was over-estimated in his functionality.

33 **R:** no, no sorry, I mean, but you could also listen to music when you are using a radio.
34 **P:** yea, but also here you can, [*.*] you can store the best music you want to listen too all
35   the time and Radio its just
36 **R:** Mhm.
37 **P:** something like that

Example 15.12: *Storing information*

Although, the way of using computers to store written documents is common (Ex15.12), it is not utilized to its maxim yet, as this benefit is weakened by the fear that the information might be lost because of the screens' blowing up (Ex14.11). Even if the informant knew that the information would not be lost if the screen blew up, it seems an understandable reaction to still do important things manually in order to avoid any problem.

Printing Information

The preferred usage of typing is inseparably bound to the possibility to print and is considered as a basic functionality of a computer. This emerges not only from an image which is correlated to the machine like the concept of a computer, which implies to *"produce work"*, but also from the fact that the printers were often used during the period of participant observation. The example below (Ex15.13) shows how an informant would start computer teaching. The text passage leaves the impression that if, after the theoretical instruction about *"the work it does"* (Ex15.13Line214) and after the following presentation of the most important parts, the informant turns to the machine's core functionality: printing.

211 **B:** I start by telling them what a computer is. I define a computer,
212 **R:** Mhm.
213 **B:** the work it does, hm.. then from there I can tell them some parts main parts,
214 **R:** hm,
215 **B:** where to begin from and on and on
216 **R:** then on, what next.
217 **B:** I go to the details when I feel that the class has gained more
218 **R:** Mhm.
219 **B:** of the computer knowledge then I say if they will produce like what, say printing and [*.
220   *] doing wo, getting work printed on the [*.*] screen
221 **R:** hm,
222 **B:** then getting out, out of the printer
223 **R:** hm
224 **B:** yea, so I begin from the parts, the work it does, then may be I start teaching them how
225   to use the computer
226 **R:** hm
227 **B:** and in case they have acquired the knowledge they, I say to them to print work

Example 15.13: *Printing as the core functionality of a computer*

Implicitly, the *"work it [itmo a computer] does"* (Ex15.13Line225) is focused on printing. From the choice of words it is also interesting that the informants emphasizes in Ex15.13Line221-223 that the computer *"prints"* e-mails both on the screen and on the printer. A similar thing emerges in Ex15.14, which actually deals with the presentation of how an informant understands the process of sending e-mails. However, as a part of the process of sending e-mails he finishes his narration with the act of printing the e-mail on paper. Once more the informant underlines to print the e-mail on *paper* (Ex15.14Line125). To my understanding the WYSIWYG[25]-like representation on the screen leads to a perception as if the words were already printed on the screen. This explains field notes of the participant observations, which shows that participants are asking how they can get rid of the underlined "misspelled[26]" words, as they do not want to print the lines. Especially, in the time of initial uses this form of representation seems to be confusing. This argument is supported by the fact that the secretaries who, in reference to the [TD|p.1], called the benefit of a computer into question. They criticized why one cannot print[27] without a printer, even though it has already been printed on the screen.

122   **B:** You just switch it on, it just put's itself on, it can imagine you are talking to someone,
123   like ahm let me send you an e−mail and it's arising in his computer
124   **R:** Mhm.
125   **B:** And then he takes it out, prints it on paper then.

Example 15.14: *Printing as the core functionality of a computer*

The computer's preferred functionality as an "advanced typewriter" has emerged during the discussion of the analogies as well. The tremendous importance of printing to the participants is underlined by the fact that although papers are quite expensive, even small letters (e.g. with only "how are you?" on it) are printed. It seems as if the participants have a strong longing for producing their virtual work from the computer to paper in order to achieve practical and haptic results. From the perspective of a possessive oriented approach this can be seen as an attempt to adopt the virtual image as one's own. To give you an example: One informant responds to the question what she wants to print: "*To draw a picture on a paper then it gives it the, this other, the other machine that prints it and I take as mine*" - In15Line120, which indicates a gain of power over the technology. Another reason for this longing could be the self-ascription of representational values to oneself. Showing off computer-made work means to have a real evidence of having computer knowledge, which indicates again an increase of power. Although, no informant talks about this explicitly. Anyhow, the desire to print is strong enough that people neglect to consider printing as a factor of financial expenses[28].

---

[25]WYSIWYG is the acronym for "What You See Is What You Get". A program applying this principle presents objects like texts or pictures exactly they way they are printed.

[26]I put "misspelled" under inverted commas as in fact some words might not be misspelled but not stored in the programs' dictionaries. Therefore, they are marked as unknown or wrong.

[27]Due to the fact that the printers arrived late, the secretaries could type some work, but not print. As mentioned above the secretaries immediately started to type applications, but were disappointed when they realized that they could not print their work.

[28]The school administration is aware of the involved costs and aims to cut down paper expenditures for duplication tasks like exams. The school has purchased a stencil printer to create the exams blueprint. The (neater looking computer made) blueprint is cyclostyled, which is cheaper than printing the exams on the Laser Printer. However, during the daily usage it is no problem to print if a pupil asks to do so and teachers bring their own papers if they need to print for private issues, but costs for toner are not charged, although this is the main part of costs.

However, asking for the reason why the informants like printing, it turns out that it is they way the work looks: *"what should I say about the computers.. [*...*] in fact I like their neat work so when you produce work it's like not that of a type writer"* - In7Line138.

455 **J:** Yea *[*...*]* and it makes the work neat! *[*.*]* And nice
456 **R:** Why?
457 **J:** it, it looks attractive, the work done by the computer,
458 **R:** ah, yea,
459 **J:** if you compare it with that *[*.*]* written using a pen, that of the computer is attractive.

Example 15.15: *The work done by a computer looks attractive*

73 **J:** The work and is not tiresome when you are using on computer, it producing the work so
74   quickly as you need it. And the operations produced this was also neat. So you feel you're
75   actually it is some time you deal with. You are actually it, if you, If you, If you use it
76   properly than it produces actually the information.
77 **R:** Mm. *[*...*]* Can you help me. What, what you mean by neat?
78 **J:** The work on you look on it, they way it is presented.
79 **R:** Ah, yeah, yeah! So, yeah.
80 **J:** The way it is presented is not like the one for a, a, a typewriter it produces or the
81   handwritten work.

Example 15.16: *Neater than handwritten work and the work of a typewriter*

220 **A:** That's the first thing I like on a computer, then they make work neater.
221 **R:** Mm.
222 **A:** It is to used, to use it make to do some work for you, if you come and print this it is
223   clear.
224 **R:** Mm.
225 **A:** for everyone to, to study or to read.

Example 15.17: *Makes work neater*

Concluding from above examples printing is preferred to hand-writing and to work created by a type-writer as it looks nicer (Ex15.15, Ex15.16) and is legible to everyone (Ex15.17). This qualitative advantage is utilized for the creation of congratulation cards as well. They imported clip arts which makes it different to other people's. Besides, such cards are becoming only useful if one prints it, as otherwise they cannot be distributed.

Makes Work Easy

Above listed usages imply the duplication of documents, which is another perceived benefit. The informant of the below listed example calls this advantage as something that "makes work easier" - Ex15.18Line335. This interview shows that as a consequence to the possibility of duplication the saving of time manifests as a further issue (see Section 15.3).

332 **Rn:** you can print in two seconds, I minute; it's finished your work
333 **R:** Yea
334 **Rn:** Yea, *[*.*]* even photocopying, other than first going and writing again with your hands,
335   it makes work more easier and quick
336 **R:** mm
337 **Rn:** mm, that's why it's good and it saves, it saves time

Example 15.18: *Makes work easy and saves time*

Whereas the above discussed example refers to a plausible perception why a computer "makes work easy", the one below shows that sometimes this characteristic is more related to the image of a computer than to in person experienced benefits:

175 **R:** Hm. What would you do if you were a computer teacher? What do you think is the most
176   important, the necessary things to teach? And, yeah.
177 **P:** Hm. *[*...*]* Calculations, drawing tables, timetables and it would make, it would make
178   the child, children understand or make their work easier. Like if they not understand in
179   mathematics some in class they could come and try it on the computer and it would give
180   them the correct answer or help them how to calculate it and it would improve *[*..*]*
181   their skills, different skills like art work. They may be able to draw computer pictures
182   using their hands, if they know how to use with the mouse, we we poor trained *(*??*)* on
183   the computer. Yes. *(*background noises*)*
184 **R:** Hm. So you wouldn't teach writing letters?
185 **P:** I would. Maybe, if, may... if there is Internet it.
186 **R:** Hm.
187 **P:** I'd first teach them, I make them understand. And so that it will make it easy for them
188   to send emails. Yeah.
189 **R:** Hm. But can you help me, what, what do you mean by making work easier?
190 **P:** Hm. *[*...*]* they said it's so, it's so maybe calculation that we don't under..., understand
191   and it can help us. It can help us maybe print, print some exams, when we don't ha, or
192   maybe when a, a, typewriter is out of use or collapsed, it ok it does not function, well.
193 **R:** Hm.
194 **P:** And so the computer would help us and it would hel..., it, it also keeps files. Different
195   information like keeping it maybe, like in a box when maybe at any time it would be
196   stolen or, *[*...*]* or burnt up in fire. It makes work easier.

Example 15.19: *Makes work easy*

I have discussed the connotation of several personal benefits of a computer. Among these we figured out that users consider computers as something which can teach and help to improve ones skills, which has been indicated in Section 14.2.2. A quite illustrative example on an ascribed personal benefit to a computer is the above example, which clearly demonstrates how the computer is perceived and what function is ascribed to it. The 13-year-old child from P7 is asked about the necessary subjects to teach if she were a computer teacher. Consequently, she shifts into the role of a teacher and apparently repeats learnt knowledge and after some concrete examples (calculations, drawing tables) she turns to benefits. One is that the computer can help pupils to teach themselves how to calculate things (Ex15.19Line180) and the other one that it helps to improve other skills like art work. The latter one seems to be referring to a personal experience, whereas the other ones do not. At a first glance the improvement of art skills might look strange, but put into context with the results of participant observation, she apparently refers to the functionality of UNDO, which enables a user to "rub"[29] mistakes. Paper is precious and not naturally at hand. Therefore, it is common to write small things onto ones thenar and rub it away when the information is no longer needed. The value of paper and pencil is also emphasized by the fact that the prize for the best performing pupils of the term was a pencil and a book. These benefits are presented similarly to the one of Ex13.3, in which the pupil presents the benefit that a computer displays spelling mistakes. Such educative characteristics of a computer might be in fact perceived as an assistance, but the universal way how the benefit of making work easier is presented in above example asks for closer inspection.

Even the interviewer asks the informant what she means by "making work easier" in Ex15.19Line189. In her following answer the girl admits that her knowledge is based on sayings of someone else ("*they said*" - Ex15.19Line190) and repeats her previous stated benefits in the manner of a proper pupil. First a computer helps to do calculation which we (there is no definite reading for this pronoun and so this option remains open, as it can refer to any user, pupil, teacher, etc.) do not understand and "*can help us*" (to understand these calculations). The next one refers to the action of printing and the third benefit why a computer makes work easier is that a computer can be seen as a replacement or further option if the typewriter breaks down. The latter one underlines the preferred usage of a computer as a typewriter, but whereas most of the time computers replace typewriters this is just the case if a "*typewriter is out of use or collapsed*" - Ex15.19Line192. Investigating about who the "*they said*" - actor from Ex15.19Line190 might be, we make a find in an interview with a teacher in section Ex15.20Line575-582, where it turns out that in P5 there is a chapter in SST, where computers are presented as modern technology which makes work easier. As the girl is from P7 she might have studied this lesson on computer and talking about benefits of a computer makes it plausible for her to draw upon this knowledge.

575 **C:** and this Bill /∗..∗/ I was teaching, I had just left the college I was teaching about making
576   work easier
577 **R:** mmhm
578 **C:** in P.5 it was SST lesson
579 **R:** mmhm
580 **C:** Then we found a topic were there was modern technology
581 **R:** mmhm
582 **C:** then the word computers came in, how computer makes work easier

Example 15.20: *We learn in SST that a computer makes work easier*

Hence, the assistance by a computer in daily life is something which becomes not only experienced in the course of life, but is also conveyed to those who have not utilized the technology to such an extent. By taking over the experience of others and retelling those it becomes ascribed to the concept. Thereby, the constant flow of concepts, and attitudes manifest, which is culture.

---

[29]A typical statement was: "*Excuse me, I want to rub, where can I do it.*"

15.2.2   Negative Related Experiences

Health Problems are Revealed

An issue concerning the physical appearance of computers and its actual effect on human people are health problems. Several informants complain about eye problems, which they did not expect. Informants even ask the interviewer why he had not made them aware to undertake preventive steps to avoid these consequences: "*At least that's something you never told me when you left, before you left, which I didn't know.*" - In9Line411. This latent criticism is justifiable, but the ATs did not take into account teaching such issues. The eye problems are described as follows:

76 **R:** Mhm. And what do you dislike about them?
77 **H:** Hm? Decide?
78 **R:** dislike, yes
79 **H:** dislike..(*seems to finally understand*).. it's like I don't know if it was my personal
80   experience only but there is a way I would get there for a long time and I would feel
81   weak, as in my head, my eyes and all that
82 **R:** hm
83 **H:** So it's like most of the times I wouldn't want to spend there a long time and you, I get
84   there sit for sometime then when you, I would stand I would somehow feel dizzy like the
85   eyes are, cannot, can't focus on some of the things as quickly as that and that's one of
86   the main reasons that, that actually made me withdraw for sometime but part of it

Example 15.21: *Eye problems*

Some of the named eye problems might result from the reduced adjusting of the focus when one is staring at a screen, or from fatigue. Eyes get stuck in computer focal length and it takes a while to adjust. Informants report this as "*my eyes was [..] not clear*" - In2Line64. Following example depicts the dialogue of the interviewer with an elderly teacher who presents her eye problems as a dislike of computers.

113 **R:** And is there something you dislike?
114 **C:** about them?
115 **R:** Yea
116 **C:** yea
117 **R:** (*laughs*), ...what?
118 **C:** my eyes have a problem
119 **R:** mmm
120 **C:** Okay
121 **R:** Yea
122 **C:** So when I don't use glasses I get pain
123 **R:** mmmhm
124 **C:** Yea so whenever I am on computer I make sure I go with my glasses
125 **R:** mmmhm
126 **C:** mm, but my, my eyes get problems beginning with last year not because of the
127   computers, they started l, last year, they pain me, when I look on the screen
128 **R:** mm
129 **C:** I don't want to look at that screen for a long time
130 **R:** mmm

131 **C:** mm, but if it were not the problem of my eyes they are very ea, good and easy to
132   produce work

Example 15.22: Eyes as a problem

This dialogue (Ex15.22) shows that eye problems might not be factually caused by
the computers, but emerge from the usage of computers.

379 **H:** Maybe what I may say dis, /*..*/ dislike dis, dislike as such I heard: it has an eye effect
380 **R:** Mhm
381 **H:** if you expose your eye to the screen so much
382 **R:** Mhm
383 **H:** Mhm, that's why I feel it, it my now affect to ones eyesight, but it, it has never
384   happened to me.
385 **R:** Mhm
386 **H:** because I don't usually expose it too much, eyes so much.

Example 15.23: Public opinion about computer affecting eyes

Above example reveals that there is a public opinion on the affects of computers
on eyes, which means the user should restrain from too much computer work. Although, not only problems related to eyes are experienced by computer work, but also
headaches due to lack of water or as an effect of not yet identified poor eyesight.

937 **H:** And sometimes my head get strained.
938 **R:** Mhm.
939 **H:** Because I was not taking like water or what to refresh me. And how the kind of get a
940   little bit bored? And so, doing some, /*..*/ know some work on games at least would now
941   relieve me.

Example 15.24: Headache as an effect of computer work

Obviously, people talk about these effects and inform each other how to cope
with the situation, ("*Because I was not taking like water or what to refresh me.*" -
Ex15.24Line939; "*you like to drink more*" - In2Line152) these effects are ascribed to
computers and its usage. However, as a matter of fact these negative experiences
prevent people from an extensive usage (Ex15.21Line86, Ex15.22Line129).

Reliability

Another negative experience is the possibility of damaging computer parts. As depicted by Ex11.2 and Ex11.3, the encounter with a computer is related to the fear of
damaging it and its implied sanctions. When the informant from Ex11.2Line80 claims
that she was afraid of "spoiling" the technology it both means that she is aware of
her capability of ruining the technology and that she searches for a tactic or a solution
to overcome her fears. To my understanding there might be at least two reasons why
children are afraid of damaging computers: to have to pay money and/or to be corporally punished -Ex11.2Line80, Ex11.3Line589. The fear of paying for the damage is
also evident among adults, when one informant states:

56 **H:** so you say if we will be on break if I am tired I just put it off so I, how will I put it off
57   now, if I leave it one and it goes bad and it is faulty and the da (*??*) it's actually it
58   goes wrong, so how much will I pay, I don't know how much, I don't have money for it

Example 15.25: Adults are scared of having to pay for damages

If the manifest fear in this intervention is bigger than in other interventions because of the breaking screens cannot be stated. Apparently, the experience of the screens blowing up has impacted a negative connotation to computers. The experience that the screens were damaged by high voltage is an important "dislike" among the interviewees and emerges in emotional answers like "*So something I hate on computers is when power brings about such a problem*" - In7Line215 or "*Ahaa, the way they blow is what I hate* [...]" - In5Line145. The unusual sharp words[30] underline the embarrassment of the informants. This intense emotions can be also understood as an form of expressing powerlessness. Several express their fear, e.g. Ex15.5, or Ex14.11. The fear emerges also latently, when one informant complains that the break down implies to be at the mercy of those who have technical skills - Ex15.26. The next section will describe the corresponding prevalent opinion how technology shall be approached. .

175 **A:** What I dislike? Maybe maintenance.
176 **R:** Mhm.
177 **A:** When they break down and /*.*/ you don't know what to do, even those who come in /*
178 ..*/ you s, s, sort of see that /*.*/ they maybe it's using try, try and error not knowing
179 what and you say: "Ah, this is coming again" and, pardon, instead of, you know, solving
180 problems, ..
181 **R:** Mhm.
182 **A:** .. otherwise it would be good.

Example 15.26: *Trial and error as a tactic to solve problems*

However, the negative experiences of screens blowing up are not isolated. Ex11.2 shows that the possible infection of a computer with viruses is promoted. Apparently, the informants have a vague concept of a virus, as following example shows: "[...] *and at times its hard for, its hard for us, at times we get scared maybe we have done something wrong or there is a fault* [...] *or maybe there is a virus or anything*" - In19Line81-82. When one girl asks the interviewer how viruses infect a computer, she apparently explicates her virus experience and says:

145 **Y:** a new selection were I need no seat/*??*/, ha. /*..*/ But how exactly do the viruses go
146 there?
147 **R:** Yeah, when you have a discette for example and the virus is on it. When you start the
148 program, you start a game it can copy itself on the harddisk. There are some viruses,
149 which go that easy.
150 **Y:** Ok, I see. Do you know, as they lied us. There is this coloured boy around here.
151 **R:** Mhm.
152 **Y:** He had come with the computer with one of our teachers and they were *(*laughing, deep*
153 *breath*) that speak him came to tell us that his hair is standing on end *(*laughing, deep*
154 *breathing*) that he had a shock from the computer.
155 **R:** Oh.
156 **Y:** We all run to the, to this lab to see what had happened, we found the boy no more, he
157 started laughing. *(*laughing all the time, very quick*)

Example 15.27: *Mixing up human and computer virus*

This explication, although sometimes not very solid English (Ex15.27Line153), presents the existing analogy to a human virus. The girl's anecdote shows that some-one ("*they*" in Ex15.27Line150) had played a trick on someone and claimed that one

---

[30] The word "hate" occurs four times in the whole transcript of 270 pages!

person had been hit by a virus. This virus had been transferred by a computer
(Ex15.27Line154). By the way, the mentioned "*colored boy*" is once more the pupil's
top-gamer who played the trick on his friends. From this story, the people's general
awareness of computer viruses becomes evident, but it's uncertainty concerning its
diffusion and whether it can affect human beings, too. The fooling of friends shows up
the possibility of misleading friends by extra knowledge and represents an act of power.
When the boy "*starts laughing*" (Ex15.27Line154) at the end of the girl's explication,
the situation is emotionally relieved, but if explanations have followed to reveal the
context remains unclear. This is of no importance for us, as the core message of the
girl's anecdote is based on the undefined concept of viruses, which is not solved but
remains blurred. It might not necessarily do harm to human beings, but can affect
computers. Their functionality and when a computer is potentially endangered seems
to be concealed.

In addition to above depicted unreliabilities of a computer some respondents criticize the hanging up of programs and claim:

"*Mhm, at times they have bad habits, [...], they stop working, at times they are
confusing, some way they cannot be understand.*" - In2Line112-113.

"*Sometimes they can get spoiled when you are playing something like a game and
then the thing get stuck*" - In14Line128-129.

The above mentioned problem of getting stuck is mainly mentioned by pupils
who play games in DOS-mode, which has proved to be very instable. Whereas the
first quote (In2Line112-113) draws upon the imagination that the computer aims at
signifying something which she cannot understand, the second one is based on the
concept that the thing is stuck and not the user him/herself. These are two different
concepts of the computer: The first concept guides the user. The user attributes the
mistake to him/herself. The second one denotes the "*thing*"-like characteristic and
implicitly ascribes the fault to the computer and not to the user. In context to the
informant's individual capabilities of using a computer, the statements do not surprise,
because whereas the first statement was made by an average user, the second one is
made by the "top-gamer" of the pupils. Apparently, he has achieved a certain level of
computer knowledge, which allows him to deconstruct the image of a computer.

All in all, these four different ways of experienced unreliabilities (caused damages
by users, power, viruses, and hanging up of programs) do not to influence the user
to refrain from using a computer, whereas experienced physical problems do. The
unpredictable unreliability makes the computer more ominous and more difficult to
control. This makes a computer more powerful and also the people who know how to
control them, too.

15.3   Time

Another benefit which emerges in the above example (Ex15.18) is the one of saving
time through the usage of a computer. This draws upon the saving of time through
duplication. Besides, participants have further ideas how else a computer can be
utilized to save time. One is that typing a letter can help to save time. So far this is

preconditioned by the capability "*if you know how to type quickly on a computer*" - Ex15.28Line452.

449 **R:** What does it mean to make—work easier?
450 **J:** I think if you know how to use a computer hm? And you may need to, to write like a
451 letter, you can use a computer others than using a pen because you may delay using a
452 pen while you know, if you know how to type quickly on a computer, I may take, I may
453 type 3 letters when you are still doing one using a pen. That's why, that's how I may
454 explain that point of being, of making the work easier *[*...*]*
455 **R:** I see—easier because you don't have to type it again.
456 **J:** Yea *[*...*]* and it makes the work neat! *[*.*]* And nice
457 **R:** Why?
458 **J:** it, it looks attractive, the work done by the computer,
459 **R:** ah, yea,
460 **J:** if you compare it with that *[*.*]* written using a pen, that of the computer is attractive.

Example 15.28: *Save time when you know to type quickly*

Some informants perceive this situation to be similar when they claim that "*[...] if the typing, my work, I can type it, all the speed now still disturbs me, speed just to know this case is here, this case is there I am not yet speed, speedy [...] I would be happy to get quicker*" - In10Line289-291. Thus, currently the benefit of saving time concentrates more on the act of printing, than on the typing process. Weighing the factors of printing and typing against each other to retrieve and objective results is impossible because the time needed for typing letters heavily depends on an individual's typing capability.

Another idea how time can be saved emerges by the usage of a scanner[31]. This device allows the teachers to save time, notably because of their experiences in drawing on a computer for exams, it has made them discard the option to draw using the computer. Furthermore, the informant of Ex15.29 notes that they are still using their free hands - see Ex15.29Line90. They prepare the written parts of the exams on the computer, leave space for graphics and draw them manually on the blueprint afterwards. By facilitating the scanner the teachers expect to integrate complex science diagrams like the structure of the ear: "*because of this scanning, the scanner, we are able now to process our science diagrams.*" - In10Line299, which has been identified as a need - Ex15.29Line86.

80 **A:** But then, what was difficult, was say it to draw those pictures. I can not even now I can
81 not do it probably.
82 **R:** Why *[*.*]* or w, where is the problem?
83 **A:** We are using a pencil and a rubber.
84 **R:** Mm.
85 **A:** And the *[*..*]* actually it was hard for us to move the pencil or the piece of paper.
86 Actually it is, it is still a problem, by the way. When you are typing your exams and
87 there is a picture to draw. It is very difficult easy we don't usually draw it using a
88 computer.
89 **R:** Mm.
90 **A:** We, we use our free hands on the stencil, which is a big problem.
91 **R:** Mm.

---

[31] It was brought by the interviewer in 2003, but had not yet been utilized for school purposes when the interviews were made.

92 **A:** And I, I think there is a way doing it faster on the computer.
93 **R:** Yeah, now with the scanner we can do it.

Example 15.29: *Scanning saves time*

If it is in fact a matter of saving time or more the general capability of getting complex graphics into a computer cannot be revealed. Indeed, it is much more important to focus on WHY the wish to save time emerges as a desire and how the application of computers contributes to it. According to my conviction someone wants to save time, if he/she perceives an activity more molesting than another one which he/she prefers to do instead. This encourages the user to be quicker with the less preferred one. This prioritizing implies that synchronous activities fade into the background and get timely ordered in queues[32]. Concluding from the view that the investigated environment builds not merely on polychronous, but also on some monochronous elements I conclude that the wish to save time poses a fortification of a monochronous time-understanding. As depicted above monochronous elements have been transferred by organizational changes as school systems were previously. This monochronous time perception seems to be strengthened by a computer. This is also enabled by the fact that a computer's characteristic is capturing the user's undivided attention. When someone is using a computer other activities are neglected, e.g. people forget to drink, or even to wink. Furthermore, interactive components of programs require the user to (re-)act in a processual way, which is also a sign of a chronologically orientated concept. In general, these elements are mainly perceived as sequential and not in parallel. This leads to the monochronous habit of waiting for the computer to do what it was commanded to do, but not to act in parallel. Multi-tasking is a future reality. I have shown in Ex11.3 that the operating principle of a computer is in fact perceived and represented as a processual one. Thereby, the computer and its immanent characteristics support the monochronous tendencies to chronologically structure the course of daily life

This structuring function of a computer emerges also by the application of favored usages like drawing, listening to music and playing games:

These activities are perceived as relaxing, entertaining, and occupying and are utilized by adults and pupils to keep one busy, when one is bored:

---

[32]This concept of structuring daily life was introduced by Saint Benedict who regulated the daily life in his convents. His well-known doctrine: "ora et labora" contrived a strict segregation of clerical and mundane activities. This form of living shaped especially Christian orientated cultures. Of course, countries with a long Christian tradition have internalized this system more than those who got proselytized like Kigezi one hundred years ago. With a new time system the whole way of living is changed and promotes rather monochronous elements than polychronous. By measurements like the establishement of a school system or installing a Lord's Day, missionaries transferred this new time system to other cultures. Previously, there were no structure of weeks in the area of Kigezi. In these days time measurements in Kigezi were based on lunar cycles and people celebrated events like weddings independently from a calender. The day was not structured by a clock, but started at sunrise and ended at sundown [Karwemera, 2003]. Even nowadays some people orientate themselves on the time-system of the day. It can happen if one asks the time, the answer can easily be "10", although the local person actually means 5 p.m. from my point of view. In this case the respondent refers to the begin of the day at 7 a.m. (This is when the sun rises) and indicates the 10th hour of the day.

" [...] *games you enjoy and play when you are bored you can get cheered up when you are angry or what*" - In21Line41.

"[...] *if you have a computer at home, you can't get bored* [...] *it's there you can relax after sometime, if you don't have work to do, you can go on the computer, listen to music*" - In22Line310-313.

"*Especially when it is raining it's boring, a computer would be interesting to cheer you up*" - In26Line476.

"*I can keep my boredom. Or either [* .*] listening to some music.*" - In9Line981.

On a manifest level a computer is something which cheers you up, when you are bored or angry. Ex15.30 presents it even more explicitly when the girl claims:

136 **R:** In your leisure time?
137 **P:** Playing, playing games.
138 **R:** What kind?
139 **P:** Prince of Persia, Super Mario, Dynablaster, and it, it keeps, it keeps us busy. It keeps us
140   busy, when we are doing that thing. So, it is better to pick up busy than doing that and
141   maybe sitting down there and doing nothing. So it is better to, to use, to play games on
142   the computer or write letter or see different things about computer, draw pictures,
143   different things like ...
144 **R:** Hm.
145 **P:** ... at leisure time.

Example 15.30: *Doing anything on a computer to overcome boredom*

Consequently, to overcome the boredom, which she describes as "*sitting down there and doing nothing*" (Ex15.30Line142), is achieved by an indiscriminated usage of a computer. Obviously, just doing anything on a computer: be it listening to music, drawing pictures, or playing games helps to do so. Such activities are obviously not associated with doing work on a computer, but with enjoying the time on it and actually "killing time". Differences how a computer is used, namely to use a computer to do work or to use a computer to shorten time and to entertain oneself become manifest. By the processual characteristic of a computer these usages have to be placed in a chronological order. Thus, they contribute to a timeline orientated structure of the day. Work and pleasure are not only timely separated, but also by the way a computer is used. The informant of Ex15.31 points out the difference of MSWord and games as well. Whereas typing is connoted to work (although for some informants it is fun, because they treat typing as a game[33]), playing games, drawing and listening to music are activities of entertainment.

241 **L:** Yeah, I would play games with them.
242 **R:** Why?
243 **L:** I play games, because they are interesting to them. I think they are very interesting to
244   kids. Because I think everytime you come here children want to games, games, games,
245   they don't want to go to be entertained by Microsoft Word or other things.
246 **R:** Mhm.

---

[33] The pupils were enjoying the coloring and changing the fonts, and it was perceived more like a game than serious office work. Among adults the typing of letters was not mentioned as a method of "*keeping ones boredom*".

247 **L:** I think that's what they like about games, because they are so interesting and the music
248   they have. Yeah.

Example 15.31: *Children want games not Microsoft Word*

However there seems to be a further aspect causing the application of games by
adults. Adults emphasize the possibility to release ones energy by playing games, when
one is stuck - Ex15.32.

439 **A:** Mhm. Then computer maybe relaxes ones mind, you are tired of this other complications
440   and then you say "Let me now switch on games" and you relax your mind.
441 **R:** So you mean to *[*..*]* just to keep busy in your leisure time? ???
442 **A:** Aha *(*ack*)[*.*]* or relaxing you mind, at times you say "now I am tired of this" it
443   stucks in the mind then ..
444 **R:** Mhm.
445 **A:** ... sort of relaxes it ..
446 **R:** Mhm.
447 **A:** .. in a different way.

Example 15.32: *Games relax you when you are stuck*

Ex15.33 testifies how games are utilized to *"create"* the informants interest in
computers again:

934 **H:** Because I was not taking like water or what to refresh me. And how the kind of get a
935   little bit bored? And so, doing some, *[*..*]* know some work on games at least would now
936   relieve me.
937 **R:** Mhm.
938 **H:** And create me, creating me some interest again.
939 **R:** Yeah.
940 **H:** So, *[*.*]* it was my request to learn games but I don't think there was a programme for
941   me to learn games. It was not there. It was my own interest,
942 **R:** Mhm.
943 **H:** hoping it would help me. To bring my, to pick my interest again.
944 **R:** Mhm.
945 **H:** Which would run. By concentrating on one thing and sometimes not coming out the way
946   you want. Because I would trying to learn, I would try to learn something, but it wouldn
947   't come the way I want.
948 **R:** Yeah.
949 **H:** and I would get bored. So, the only thing to keep boredom would be how to shift to an
950   other what?
951 **R:** Yeah.
952 **H:** Another programme like games and some arts.

Example 15.33: *Games as an escape of problems*

Both examples (Ex15.32, Ex15.33) show games as a possibility when the users are
over-extended. When they are stuck, they do not know how to reach their goals on
the computer. As a consequence they escape from the problem and shift to other
programs. As they are unable to cope with their work; due to their perception; their
attempt to control the technology gets blocked by the computer, they feel powerless.
By playing games they win (at least partly) back the control over the computer and
take heart to confront themselves once more with the previous problem. An interplay

of power between computer and user happens[34]. Obviously, playing games or drawing pictures is more promising to control the device. One informant approves that when he notes that drawing was easier to him than typing - Ex15.34Line73:

71   **B:** Do you remember the drawings?
72   **R:** hm,
73   **B:** yea, that was what was easier for me than typing looking for theses other letters and
74   turning the, the /*.*/ whatever, it was sort of difficult.
75   **R:** hm,

Example 15.34: Drawing is easier than typing

Previous discussions and the above example (Ex15.34) makes it evident that some applications and topics concerning computers are easier to tackle and to be obtained. The extent up to which technological knowledge (in our case computer skills) can be transferred depends not only on the teacher-pupil relationship and facilities, but also on the latent expectations towards a technology and different standpoints on how the technology should be approached. Whereas firstly one has been issued with a prior discussion (Section 14.3) latter ones will be the focus of the following sections.

15.4   Technological Knowledge Transfer - Teaching

It will not be the aim of this study (and not of the following sections) to discuss proper and *"true" (itmo* proper*)* teaching methodologies, but definitely part of the analysis will be to show how the ATs' teaching was *perceived* by the participants and which attitudes they have towards it. This will be followed by a discussion about the perceived problems out of an ATs perspective and how these can be resolved. In addition to that, I will point out major problems which came up while approaching the technology, contributing to the revealed intercultural phenomena, since any human-computer interaction is an intercultural one. The following sections will point out retrieved differences from the ATs to LCTs and the local expectations on how one shall approach the technology. At the end of this section I will discuss the implications of technological knowledge and why, according to my understanding, it appears difficult to **encourage** its distribution.

15.4.1   Perceived Way of Teaching

In general, the applied way of teaching was perceived as a practical orientated one, although the ATs were mainly focusing on theoretical lessons within the first three weeks (the screens were delivered after the intervention's start). Several teacher-informants approve of the "discovery" method and that *"things"* were trained practically and discovered by the participants themselves - Ex15.35, Ex15.36. According to the informant of Ex15.35 this was opposite to other school computer teachers who seem to be too impatient - Ex15.35Line296. Obviously, the teaching way of the ATs is perceived as the better solution: "[...] [discovery method] *is even encouraged in colleges and*

---

[34] Actually, it is more a conflict between the user and himself/herself.

schools [...] so yours is better [...]"[35].

311 **R:** so what do you think its a difference then between the teaching we did and other
312   teachers are doing now?
313 **C:** aahm, the teaching you did was
314 **R:** mmhm
315 **C:** more of a discovery method
316 **R:** mmhm
317 **C:** do you get me?
318 **R:** Yea
319 **C:** while these one we got after yours are a bit of spoon feeding do you understand?
320 **R:** mmhm
321 **C:** what it means by spoon feeding
322 **R:** Yea
323 **C:** when we didn't get the point quickly they would do it for us quickly so that they finish
324   the, and go to

Example 15.35: ATs applied discovery method

323 **I:** One the method was good, because you were using discovering method [*..*] whereby you
324   would direct yourself do this and that and you'd let us find out ourselves.

Example 15.36: Discovery method as the perceived teaching method

In fact the ATs aimed at applying a student centred way of teaching, which apparently runs counter the popular way of teaching dominated by an authoritarian style[36]. This teaching disparity is perceived as a cultural break, which was caused by the ATs. Although, as shown above, positively connoted. However, also critics on the ATs' way of teaching emerge. It seems as if those, who attended less considered the ATs' teaching as unsystematically, unthematically, and too fast. One informant refers this to the fact that the ATs had prepared a syllabus. According to the informant's point of view this enforces a progress of teaching, which is at the participants' expense[37]. When teachers proceed too fast, the students cannot follow their instructions and they do not get the subject matters. Unfortunately, the specification of the location (*"here"* Ex15.39Line695) could not be defined. Apparently the negative effect of a syllabus is restricted to a certain area. At least it seems as if the school is also within the area which is indicated with *"here"*. Anyway, the criticism on the syllabus seems also correlated with testing, which is, in reference to an informant, inevitably connected to teaching (*"in teaching there must be testing"* - In1Line701).

686 **R:** Because we had agreed that all should attend at least three times a week,

---

[35] The justification of the participant's judgement is underlined by her reference to institutions of formal education. This underlines the status of these institutions.

[36] Teachers follow the excathedra method in which active cooperation with the pupils is neglected. From the ATs' perspective this was quite difficult, especially in the beginning phase. When the pupils were asked a question, pupils raised their hands, but could never give an answer. This left the impression that it was unusual for the children actually give an answer to a question. This might also be correlated with the fear to being punished in case of error. Apparently, the sign of raising ones hand is more an acknowledgement of being attentive. The teachers' question seems more to be a proforma one, than a really intended question.

[37] The informant states: *"So, that forced you [*.*] to teach certain topics in a given period to make sure within this period you finish this peri, this topic."* - In9Line593.

687 **A:** mhm.
688 **R:** .. but some simple didn't show up, because it was their leisure time.
689 **A:** So maybe those are the ones who were saying, *[∗..∗]* we are not having things
690   thematically. Systematically. Do you have anything like, like a syllabus.
691 **R:** What we wanted to do?
692 **A:** Yes.
693 **R:** Yes of course.
694 **A:** Mhm. Then I think that's it *(∗low voice∗)*, because when you have a syllabus you know
695   you are going to deal with this then that makes it a problem *[∗..∗]* here.
696 **R:** Mhm.
697 **A:** But as I said, giving something like *[∗.∗]* a test is not bad.

Example 15.37: *Syllabus is criticized and tests wanted*

The expectation to be tested is also evident in Ex15.38. This statement of a teacher
also shows that he completely took over the role of a student, as he discards the option
to ask a teacher to have exams and talks from the perspective of a pupil. Once more,
the strong hierarchical teacher-pupil relationship becomes manifest and is obviously a
reason why the teacher did not ask the ATs to test them.

733 **R:** Do you think we should have made exams with you?
734 **A:** Yes. Short exams. *(∗laughing∗)* Not eeh *[∗..∗]* Short exams!
735 **R:** Why didn't you tell this to us?
736 **A:** Ah. *[∗...∗]*
737 **R:** *(∗small laugh∗)* It just comes into my mind.
738 **A:** It can, it is hard by the way and even the pupils when we are teaching they can't tell us
739   that exams. No. *[∗..∗]* They can't. They can't ask us to give them exams. I don't know
740   why it is it like that even the way we were studying at Makere, you can't ask for an exam
741   . An exam is. you can't ask.
742 **R:** *(∗both laughing∗)* of course you can.
743 **A:** Asking, ah, asking someone to test you.
744 **R:** Yeah!
745 **A:** Ah, it is hard.
746 **R:** Why? *(∗silent∗)*
747 **A:** Because the children never want to be tested.
748 **R:** But you are no child. *(∗both laughing∗)* Your are not a child.
749 **A:** *(∗laughing∗)* You get my point?
750 **R:** Yeah.
751 **A:** Yes.

Example 15.38: *Expectation to be tested*

The below example (Ex15.39) clearly depicts that the expectation to be tested
is set, not only because to do justice to the educational system, which builds upon
explicit testing, but also as some kind of self-control to the participants to make sure
that they had achieved certain teaching goals. Obviously, this form of feedback would
have been desired by the participants in order to get a form of feedback.

717 **A:** But that's why, you, you are here. It's the *[∗.∗]* No, ok. You not got the extent of saying:
718   " You have failed, but when he sees, eh, eh, of the work we have covered I can't even
719   answer half of what as been asked then, maybe he says: better pull up. Or I better put in
720   more effort. Much more than I have put in.

721 **R:** Yeah, but I can tell you what, where the reason I think is, because you all, you are so
722   different learners, ..

Example 15.39: Tests as a form of feedback

According to the ATs' statement in Ex15.39Line721 this was contradicting to the
different technical skills of the students and their inconstant attendance. Therefore, the
ATs had focused more on an individual orientated training and implicit testing. Files
on each participants were kept and their covered subject matters noted. Furthermore,
the ATs did not want to put too much pressure on the grown-up participants, as
they realized their heavy workload. The ATs considered the participants' attendance
as their leisure activity and did not want to discourage their efforts through explicit
tests. Therefore, the ATs integrated previously trained topics in the following lessons
by asking the participants to do tasks practically on their own. To give you an example:
while in the one lesson the plugging in of the parts was trained, the participants
had to plug in them on their own, when they wanted to switch on the computer
next time. Obviously, this disparity could have been resolved if the ATs had known
about the local preference of being tested. However, it seems as if the way of implicit
testing is not perceived as a testing per se. The wish to be explicitly tested and "*to
cover the computer thematically*" is regarded as an attempt to gain control over the
technology - see Section 15.4.4. Only by passing tests on thematically ordered sections,
the participant can climb the ladder of acquired computer knowledge. Finally, this
progress climaxes in a certificate which is the proof of computer knowledge.

15.4.2   Language Difficulties during Computer Teaching

During the process of teaching the ATs encountered several misunderstandings. Most
of them are related to Cushner's theme of language. According to the informants
the misunderstanding related to language were not based on the ATs' or participants'
language skills. Of course, pronunciation issues and the local dialect of speaking
English were perceived in the beginning as "*a bit of the problem but not very much
because afterwards we understood we get used*" - In3Line94. Another informant states
that: "*So the English is O.K. It was just, those dialect problems.*" - In9Line587. Some
words are locally more common than others, but after a few weeks such difficulties
almost vanished or were explicated and negotiated.

In fact it seems to have been rather the problem of using appropriate terms and
analogies to convey the knowledge than of basic language skills. To give an example:
When the ATs were teaching the new computer administrator her task was to install
the scanner software. She had the description right in front of her, but could not
manage as she was confused by the installing guides instruction to "insert the disc".
This actually referred to the CD-ROM, which contained the drivers of the scanner.
According to the informant she associated the "harddisc" with "disc" and did not
know how to proceed - Ex15.40.

627 **J:** I thought that they were talking about the disc that is inside /*..*/ so I thought that they
628   were telling me that I should use the wire to connect the, the, I should insert in the wire
629   then it goes to the hard disk inside, (*high speed*)
630 **R:** hm,

631 **J:** and I failed to understand that they, they are talking about the CD, the one of the
632 scanner
633 **R:** hm,
634 **J:** I thought that it is the wire that will do, that will do that, that it will be connected to
635 the hard disk inside the computer.

Example 15.40: *Computer terminology is not clear*

A further indication for the difficulties which can be found when the ATs were presenting signs of the computer. They called them according to their own made association, although these were not clear to the new users. As Figure 15.1 shows the "arrow" which indicates the dropdown characteristic of a combobox. This sign was perceived by the students as a triangle, as its shape is in fact more similiar to a triangle than to an arrow. For the ATs this represented an arrow, which indicates that one can click there - a symbol.

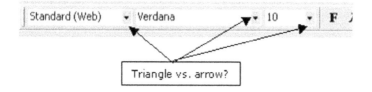

Figure 15.1: Participants do not perceive the triangle as an arrow.

However, sometimes the misunderstandings are related to more complex problems. All participants speak a different mother tongue than English, most of them Rukiga. When the ATs presented the parts of a computer and reached the "tower", one pupil asked why it looks like a "towel". This is due to the fact that in Rukiga there is no letter "r". Difficulties distinguishing these differences correctly, emerged also in daily life. Some people might say "Thank you vely much" or "Little Litz" instead of "Little Ritz". To do justice to an optimal transfer of technology this asks for computer teachers who know the historical context as well as local language particularities.

Another issue which arose both during participant observation and in the course of the interviews is that the English terms of computer programs run counter the local English. Terms like "*copy*", "*edit*", or "*click*" were considered as self-explanatory by the ATs, but should have been explained from the very beginning as they were difficult to understand. To give an example: One informant claims that for example the participants got confused, when the ATs referred to a "*blinking cursor*". Some of them had heard about the cursor before, but got confused when hearing about a "*blinking cursor*". The ATs had used this explication in order to avoid misunderstandings. As a consequence to such difficulties the same informant asks to define most of the applied terms in detail - In9Line766f. Of course, explicating as much as possible is desirable, but takes time and difficult to narrow down. To cope with this situation the informant suggests that if the ATs stayed longer it would be possible to do so. However, to my opinion one can increase his/her explications ad libitum, can break explanations down to a very high granularity, but according to my experience by commanding any single step does not increase the understanding of a computer. The concepts of a computer are too complex to control it. They method of explaining any term and vocabulary, this would lead to a capability to reproduce processes, but not to the understanding of underlying concepts. Only if ways of usages are experienced by oneself and by trial and error during this experience, these processes are internalized. Does and don'ts are recognized and an own way to solve problems can be realized. However, this demands to approach the technology in an exploring manner instead by the way with the aim to control it - see Section 15.4.4.

Note: Beside above mentioned problems there were some coordinative problems like (double-)clicking or typing. These problems got less soon, especially among those students who practiced a lot, which underlines once more the success of trial and error and the importance of training practically. From a perceptional perspective issues like the differentiation between active and passive window was difficult. Visualizing concepts like framing or shading to "indicate" a button or menuitem takes getting used to (e.g. the START-button was not perceived as a "button"). These problems are not extraordinary and according to expert-interviews and to my own experiences in Austria these emerge in any encounter of new computer users. If the teacher is aware of these problems, it is up to his empathy to make the new user aware of the underlying visual presentation concepts of computers. For further teachings this aspects shall not be neglected. However, it is not the aim of this research to discuss usability problems of human-computer-interaction.

Success of Practical Teaching

Measuring the success of teaching is difficult especially since the ATs avoided to make exams and enforced learning by doing and tested implicitly. At a meta-level, how the success of teaching was perceived, it shows that the success of learning is bigger when the students have fun. Apparently doing things practically is more fun to the trainees as one teacher states::

"I don't know why whether it's because it is practically, interesting or what, but they actually love it better than this theory lessons when you go to them and talk and give an exercise and all that." - In8Line304.

This is supported by the statement of a matron, who notes that the acquisition of her knowledge was drawn upon the practical work she had done on the computer:

*"[...] the knowledge I acquired from it. I just got it trough [*.....*] through the practical work."* - In6Line399.

Using computers practically shows promise and is consistent with the used time on a computer. But using a computer also demands understanding certain concepts. For example: To understand that it is not the machine which "tells" the user what to do next, but a causal effect to the user's previous inputs. However, these concepts were perceived differently. It cannot be figured out which user has the biggest capacities as their expected benefit of a computer varied. Even motivation to dedicate oneself to computers varies a lot, but no general tendencies could be revealed. Potential reasons are too manifold (e.g. educational level, time of usage, cleverness, information processing capacity, etc.) to decide on one factor, and due to the doubtful statistical data no serious predications can be made.

Consequences of Acquired Knowledge

The analysis seems to approve the considerations which dealt with the local brain drain in the theoretical part of this thesis. It has shown that people with most advanced computer skills were not motivated to transfer their knowledge at the school. This would have been important for achieving a multiplier effect as to disseminate computer knowledge en gros. However, the two students[38] seemed as unmotivated to do so, which is interpreted as an attempt to preserve their increased hierarchical position. The increase in social status encouraged them keep their knowledge in private, (as computer knowledge is highly connoted). As a consequence, both students which acquired "deep" computer knowledge, and could perform rather complex computer operations (networking, recovering file images) were not available to the school in 2003. During the intervention's first phase, they were no official members of the staff, but they were instructed with the agreement of helping the school in consequence. Apparently, after the ATs departure, they abandoned the school. According to my information, they looked for better jobs in more populated areas and bigger towns, which indicates a local brain drain. Even though, the school was willing to pay them extra money, they denied to support the school. This lead to the disruption of the LAN and apparently the end of the Internet usage.

The school's consequence of educating socially integrated teachers instead of external consultants seems to be promising. These new computer teachers might provide the school with long-term computer knowledge. Of course, if this becomes true should be investigated in a few years from now on.

15.4.3 Differences in Computer Lessons

Obviously, the above described way of teaching seems to be different to the one of the LCTs. The following example clearly depicts how the differences between the ATs and the LCTs are perceived. Some of them are not directly related to the way of teaching, but show that factors influence each other.

101   **R:** When you are remembering the computer lessons of Oswald and me, what did you
102   especially like and what were you specially interested in?
103   **H:** ah, /*.*/, I actually wanted to, by that time I had wanted to, because I had some slight
104   knowledge on the computer before you came so I had wanted to broaden that, to make
105   sure I learn more especially the progra, the different programmes, to make sure that I
106   mastered the little I knew on the different programmes like actually like Microsoft Word,
107   I only knew the typing part of it and the saving part of it so when you finally came I
108   learnt the printing part of it, the drawing diagrams and all that and I hadn't been
109   introduced to especially there were, this /*..*/, what is it, /*..*/ this graphics, mh, so I
110   finally, I wanted to, to have more knowledge on that which you actually had done by the
111   time you left and it lists me a lot by the time you left.
112   **R:** What do you think is the difference between the lessons you got in town and /*..*/ of
113   ours?

---

[38] In addition to Mr. Rukinga, one 25-year-old son of a teacher was intensively instructed during his college holidays. Both were external, but actually with a personal relationship to the school.

114 **H:** The main difference is that mainly because those people were paid, I, I had to pay for
115   those lessons and these were free anyway so that makes a difference on its own. And
116   another thing is that because there were few computers and there were many of us, so we
117   would be allocated a certain period of time so if your time would be over even if you
118   hadn't got what you wanted to get you would leave and someone else gets in but I think
119   here I was free to ask whatever I wanted and use it for the time I wanted to use it and
120   being around with you. But that was like a school, you go there, time goes you go but
121   here we would be around with you anytime we feel like coming and asking even when it
122   was not time for lessons we would be able to do it so I think it was also a good part of it
123   and a difference /∗.∗/ and it's like there were some facilities there that we would not use
124   like the printing /∗..∗/ they would, actually the printing would be done theoretically they
125   would tell you do this command this and put like this and you would end eld (∗??∗) not
126   even see where the printer is working from because the printer would be somewhere in a
127   room, they would just be connected so they would simply tell you when you are going to
128   print you do this like this like this theoretically but with this you would say "follow the
129   following procedures and then print", you print on your own and see what you have done
130   has produced something practically. /∗.∗/ So I think it was also a difference.
131 **R:** Good.. so that, that you see the real result of printing
132 **H:** hm
133 **R:** Yea /∗..∗/ so you think that's a bit different? Ah, /∗.∗/ what would you do if you had to
134   teach computers? How would you start, what do you think is the most important, what
135   is necessary to know about computers?

Example 15.41: *Differences in Teaching Computers*

When the interviewer requests the informant to differ between the lessons she
attended in town, and the ones of his group, he distincts himself from the other group
and introduces new actors, teachers located "in town" - Ex15.41Line112. This evokes
a quite narrative response and results in five differences:

- **Paid lessons vs. free lessons:**
  One main difference lies in the fact that the lessons held by the interviewer and
  his colleagues were free, whereas the courses she was attending were charged. *"I
  had to pay for those lessons and yours were free"* - Ex15.41Line115.

- **A high student computer ratio vs. a one person per computer ratio:**
  The lessons in town are significantly marked by a high student computer ratio
  as there are only "few computers" but many students (*"there were many of us"*
  (Ex15.41Line116)) which are *"allocated a certain period of time"* (Ex15.41Line117)
  in order to use a computer. Apparently, the demand to learn computers is higher
  than the offer.

- **Allocated time to user - free user time:**
  As a consequence it is not possible to make sure of achieving certain educational
  objectives, deduced from the fact that when time has passed other students have
  the right to use the computers. In opposition to that the informant has, at
  the place where the interview is taken, free access to the computers, which is
  depicted by the phrase *"use it for the time I wanted to use it"* - Ex15.41Line119.

- **Theoretical vs. practical teaching:**
  A difference she mentions about the way of teaching and not about the socio-

financial constraints is the one between theoretical and practical learning of
processes. She depicts this difference by the example of printing. Whereas "the
printing would be done theoretically" (Ex15.41Line124) by the teachers in town
and not practically as it was done with "this": "you print on your own and
see what you have done has produced something practically" (Ex15.41Line129-
130). Latently, this implies that there is an advantage of turning a theoretical
process into a practical one and finishing. This enhances the understanding of the
whole process and makes results concrete, which are evidently needed in order to
understand the subject matter. The printer represents the interface to the real
world. She states on the one hand that she learnt the "printing part" during
lessons given by ATs, but on the other hand that the other teachers in town
("*they*" in Ex15.41Line122) had taught her how to print theoretically previously.
Thus, it becomes obvious that she had not covered printing by the theoretical
instructions and relates the concept of learning to a holistic understanding of
certain processes or knowledge areas.

- **School like lesson vs. non-school like lessons:**

  When she compares the other institutions with schools (Ex15.41Line120) with
  fixed lessons and points this fact out as a difference, she did not perceive the
  lessons held by the not-in-town teachers as school adequate lessons. It makes
  it possible to conceive her prevalent image of school, which apparently consists
  of fix scheduled lessons (Ex15.41Line117, Ex15.41Line120). According to her, a
  school is also a place that is left after the lessons are over. Even asking is limited
  to the timeframe of lessons - Ex15.41Line121 - and questions are supposed to
  be appropriate ("*here I was free to ask whatever I wanted*" - Ex15.41Line119),
  which indicates that during the lessons in town she was not free to ask. This
  could be identified to a lack of time, or as the underlying opinion that questions
  during lessons are disruptive. Another reading would be that she did not dare
  to ask, as she considered herself having too little computer knowledge, and did
  not want to make a fool of herself. This difference is mentioned in a narrative
  string about time, which indicates that as allocated time was too little it was
  uncommon to ask. Most likely all other reading versions support the decision
  not to ask as well.

Once more, these differences show that access to computers is not only influenced
by the knowledge on how to use it (way of teaching), but also by other (re)sources;
namely, the time one can spend on it and, the money one has to pay to use it. These
factors are shaped by the amount of available resources and the need to access them. Of
course, the interviewee presents the computer as a resource which should be available
and accessible in an unrestricted way, like at the place the interview is taken.

Whereas most differences are based on socioeconomical constraints the one of the
way of teaching (theoretical vs. practical) might be (also[39]) a social one. Above examples refer to the LCTs which are placed in town. The following example (Ex15.42)

---

[39] In reference to the above discussed issue that the transferring of knowledge is prevented by those
who have gained in status, I want to indicate the possiblity that the chosen method of a theoretical
computer instruction is consciously chosen. Doing so, they prevent learners from acquiring deeper
computer knowledge. However, this is an reading option, but since these computer teachers where not

discusses the way of teaching of Mr. Rukinga and how it was perceived by informants. It becomes evident that he followed a strict disconnection of theoretical and practical subject matters. Apparently, this is also applied by the new computer administrator. According to the informant, the children had computer lessons without any practical usage, which is completely different to the way the ATs were teaching (Ex15.42Line166).

154 **R:** Mhm. So what do you think was the difference between his teaching and ours.
155 **N:** I think *[*...*]* your teaching included most of the things would be done in a very short
156   time and children would understand much better, but in his lessons, he almost took
157   about the whole term, whole term on Microsoft.
158 **X:** Where do I press to change levels?
159 **N:** Microsoft
160 **R:** What?
161 **X:** Where do I press to change levels?
162 **R:** Shift L
163 **N:** Ah. *[*....*]* when, when he was here, he almost took a term to finish Microsoft Word,
164   that's what we have, I think we have never learned Microsoft Excel.
165 **R:** *(*small laughing*)*
166 **N:** And the lessons were very much boring. He would come, sometimes he wouldn't do
167   anything practical, he would just come and talk.
168 **R:** Oh.
169 **N:** And, and then after we, we go out, *[*.*]* and so the teach, the children, on the end came
170   most about, because they didn't have a chance to try out sometimes. *[*.*]* Mhm. So I
171   think, your, the time of teaching was better than his.
172 **R:** And the times of teaching now?
173 **N:** It's also, I don, now it's , *[*..*]* it's, *[*..*]* it's m, much better than that of, of Mr.
174   Rukinga, *[*.*]* because now, now we do some, sometimes in fact most children now like it,
175   because we do *[*..*]* when we do something and we finish it *[*.*]* quickly, we are allowed,
176   allowed to play games, but in his lessons, there was nothing like playing games, playing
177   games would come like at the end of the term when you, when you have nothing to do
178   and you just, just read and scrawl the school. That's when he would play games. But as
179   for now, we, we play games sometimes, sometimes even the whole lesson, and now I think
180   things are getting more serious, because they are also using books, we have to write,
181   note down what we, what we have learned, the steps and, and the way of how to do it.
182   When you are using the computer, how to do different things. Because I think the
183   teacher found out that we f, forget so, we forget easily so they introduced the type of the
184   system of *[*.*]* using books

Example 15.42: *Teaching Microsoft Word the whole term*

Mr. Rukinga's way of teaching is obviously not only coined by a disparity of theory and practice, but also by a stronger focus on one topic during the whole term.

153 **B:** While You people you were on various topics
154 **R:** Mhm.
155 **B:** So *[*..*]* he mainly concentrated on Microsoft Word. *(*arguing P1 ones in background,*
156   *because school finished*)*)
157 **R:** So did you like that?
158 **B:** *(*seems like NO*)*
159 **R:** Why?

interviewed it cannot be taken for granted.

160 **B:** Because it is, for a long time, because we did it for the whole term,
161 **R:** Mhm.
162 **B:** So, you know if you do something for long one particular for long, you'll soon leave it.

Example 15.43: *Teaching Microsoft Word the whole term*

Although, not only the way of teaching is seen as a difference to the ATs, but also the underlying computer skills of the teachers. As indicated in the previous discussion about certificates and the ownership of computers, it seems as if teaching computers is more related to the willingness to teach and the fact to have unlimited access to a computer than of having actually relevant skills. Several informants claim that *"you know sometimes you can be taught by someone who is not well knowing"* - In10Line333 and *"In fact, when you left there was a big problem. [...] And the we didn't, what we had learnt it seems we didn't add on [...], [Mr. Rukinga], he also needed your knowledge."* - In27Line440f. The issue arises if teachers focus on certain topics for a longer period, as they do not know what to teach else. It might also be that these teachers do not know how to transfer their knowledge with their available teaching methods. This speaks for the validity of the above argument that teachers limit their teaching on certain issues and proceed too slow in the view of some informants.

Another difference becoming manifest is that the LCTs are perceived as *"spoon feeding"* - see Ex15.35Line319 or Ex15.7Line92-94. One informants claims that *"[...] those with more knowledge do it for you instead of teaching"* - In3Line322. Instead of explaining the way how to solve the problem of the user, these teachers seem to be impatient and do it for them. In addition to an individual impatience this can be due to several reasons, as both the limited teaching skills and the latent wish to preserve ones expert status might be. It is a matter of fact that by showing someone his/her skills is a presentation of power and as long as one can uphold this position, he/she feels more powerful.

Concluding from the above statements and thoughts this argues for a focus on "train the trainers", to increase their capabilities of applicable teaching methodologies and their technical skills. However, in combination with previously discussed rejection to transfer computer knowledge, it might come to a deadlock, which cannot be easily resolved. If trainers are trained, but (a) stick to limited subject matters, (b) spoon feed their students, or (c) focus their teaching on a theoretical approach, in order to prevent other people from gaining a different status by acquired computer knowledge, situation will not change. Indeed, to transfer knowledge implies a loss of power. Despite all these considerations sound good, evidences are vague. This claims for a further study investigating technological knowledge transfer initiated by LCTs. However, the statements given by the informants indicate that the current situation seems not as a satisfying one (for the participants).

15.4.4   Expectations How Technology Shall Be Approached

One reason which might contribute to the above presented situation is as follows: Concluding from above discussions and examples (Ex15.42, Ex14.18) I state that the expectation of approaching a computer is to control it. This emerges in several ways and not only latently by wishes to cover things thematically, but also explicitly like the following statements manifest:

"*Ok, after I have taught all about Microsoft Word that's when you go to [* .*] to Micros, Microsoft Excel.*" - In3Line205.

"*Me what I think is the what is the good for someone is to cover everything each thing, under each command.* " - In13Line207.

The following example (Ex15.44) shows that above critics on the unthematical way of AT teaching also draws upon the opinion to control the technology. Indeed, it underlines above argument, as if the thematical ordering of topics enables the control of the technology.

634   **A:** Conservatively *(* ?*)* following each other, because the, th that was the, the the general
635   complaint, ..
636   **R:** Mhm. What?
637   **A:** .. the general complaint of having *[*.*]* this today and tomorrow this, *(*laughing*)* before
638   the other one is mastered.

Example 15.44: *Changing topics before one is mastered*

As shown in the previous sections, the practical way of teaching was considered as an appropriate one. However, the current way how the LCTs are teaching seems not to have changed fundamentally. Pupils still have to write down command processes in their books. Another male informant stresses that this theoretical approach is not only applied for programs, but also for the components of the computer - Ex15.45.

275   **H:** They would teach them right from the keyboard from the mouse even every detail of the
276   mouse the left click small time to learn a whole keyboard it would take a whole month

Example 15.45: *First to control the keyboard*

He criticizes the current way of teaching and promotes the change of topics to maintain the eagerness to learn computers - Ex15.46.

288   **H:** the difference is that *[*..*]* instead of starting from scratch. *[*...*]* it's better for someone
289   who is eager like I was, and who would wish to know much more in a short time, so that
290   that I can develop the rest on my own *[*.*]* to start with the basics of a computer *[*..*]*
291   learning how to use like the basic programs like Excel, the Word and how to easily use
292   the mouse and the keyboard than teaching me theoretically having notes, reading notes
293   every day, without hav having a chance of having a whole day at the what
294   **R:** mhm
295   **H:** *(*knocking on laptop*)* at the instrument itself.
296   **R:** mhm.
297   **H:** because here in Uganda you just first learn that first some bloody theory, you are every
298   learning the theory by blackboard taking notes *[*.*]* for almost a month with no a chance
299   *[*..*]* of handling *[*.*]*the equiment itself. *[*..*]* they start the mouse

Example 15.46: *The focus is on theoretical lessons*

Apparently, his opinion runs counter the widespread opinion to master "*whole programs*" first. It would be interesting, if he promotes these insights to his colleagues. Unfortunately, this could not be revealed.

Another indication why the prevalent approach of learning computers is to control the technology becomes apparent in Ex15.26. In this example method of trial and error, an expression of exploring technology on ones own is criticized. The example reveals that the one who calls in the technician expects him/her to control the technology, but not checking around certain aspects. According to this informant trial and error is correlated to not knowing which is contradicting to his /her familiar true way how problems have to be sorted out. In fact, one is completely dependent on skilled personal, which means powerless, and if one does not know how to cope with the situation by oneself, you shift to a meta-level and judge upon the way one deals with a problem.

All these statements strongly argue for the control of the technology, which is definitely impossible to achieve, as a computer is too complex to be controlled by one person. Therefore, it becomes the task for the cultural broker to deconstruct this latent conviction and encourage users to explore the technology. However, factors like the fear of spoiling the computer, difficult retrievable expert advices, high information retrieval costs (Internet), and lack of time have to be daffed aside. Otherwise teaching activities like the one of the ATs becomes more a drop in the ocean.

Chapter 16

Discussion

16.1 Research Question #1

16.1.1 What are the social functions of computers within the geographical area of Kabale (Uganda) and (how) is a computer integrated into a participants' everyday life?

On the foregoing pages, I have clearly stated that the participants of this case study are constantly using the computer lab, which I consider as an integration into the daily life. However, the extent to which a computer is integrated into an individual participant's daily life differs from participant to participant and one cannot make generalizations. What one can say in general, though, is that the extent to which a computer is integrated depends on the way the technology is accessed. For this case study, this means that whereas matrons are excluded from computer access completely, the secretaries use their computer as an "*advanced typewriter*". Among teachers, lack of time seems to be the most decisive reason for why the integration into their daily life is limited. So far, the daily computer usage of teachers is rather restricted and involves, at best, the typing of letters, the creation of marksheets and exams, or the reproduction of work. Apart from a lack of time, other factors like the spoon-feeding by other computer teachers or computer literate people, and the fact that the usage for private issues is limited by the SMC might contribute to this decent integration. So far the prevalent computer knowledge is mainly applied for administrative tasks. Computer-aided instruction is far off reality and limited to typing letters in English. This is similar to experiences made in the SAIL project, where it was claimed that "*When ICT was introduced in schools, it was thought that it would revolutionize the manner of teaching, yet substantial progress is still required to actually integrate ICT into classroom practice*" [Caruana-Dingli, 2005]. Interestingly, all the above mentioned findings - the time problem and the limited kinds of usage- are close to those of Roger Harris's in North Central Borneo (Page 38), Helen Scheeper's Computer-Ndaba experience in South Africa (Page 40), and Nel & Wilkinson's experiences in South Africa (Page 43), which indicates that these difficulties are not culturally determined, but, perhaps, inscribed in the basic concept of a computer. I have also presented other reasons concerning why some teachers retain from using computers too much: Experienced health problems - see Page 213 - and the technology's unreliability (breaking screens, hanging up) lead to a limited usage as well.

However, another circumstance which leads to a limited usage seems to be the respective cultural expectation concerning how the technology should be approached, which is discussed in Q3. To my understanding, it is rather important to note the fact that the participants do not dare to explore the technology on their own. They aim at controlling the technology first and, therefore, become dependent on other local computer experts who, obviously, have problems in passing on their computer knowledge. Whether this is due to their lack of teaching methodologies, their technical skills or, simply, their unwillingness to share knowledge in order to avoid a loss of power[1] is an unsolved question (see Q3). The loss of power seems to asks for a discussion of the social meaning of computers, which is expressed by assumptions, expectations towards and knowledge about computers. I have, in the previous chapters, described that the expectations derived from having access to a computer are high, as improved living circumstances are expected - be it through an increased income or through time saved. A computer makes it possible to perform other, new activities: The benefit gained from a computer seems to be a better life. This "better life" as presented in the interviews is not only characterized by obvious pecuniary advantages but also by an increased social status. Simply by having computer knowledge, a user expects to gain power. This is supported by the fact that the image of a computer is projected. I have exhaustively discussed that a computer itself is introduced as something very powerful which can for example speak all languages, allows to control jets and can even look into the future - see Chapter 14. Different ICT initiatives contribute to this blurring and through each isolated initiative the image of a computer becomes more difficult to understand and less graspable. This is emphasized by the fact that LCTs seem to teach in a formal authoritarian way and apparently, do not convey computer knowledge in concepts, but in processes. All these factors contribute to the presentation of computers in a rather complex way. As most people have little opportunity of "experiencing" a computer on their own (due to costs for usage, maintenance, and access - see Chapter 15) various characteristics and rumors are packed into the one miraculous, but haptic perceivable object: "the computer". Consequently, the local knowledge about computers remains within a small group of some "clever", "developed" and powerful people who know about the facts behind the concept of the computer. Thus, someone who is computer literate aims to gain profit from this situation. This is supported by the fact that the income in local urban area is higher than in the rural one and the biggest economical profit is precast for regions which are supplied with electricity (see Page 116). In fact it seems highly promising to utilize one's own computer knowledge to boost one's income.

As I have shown in the results of the analysis, one social function of computers seems to be a time-structuring one. I have discussed that a computer carries monochronous time values and seems to promote a structuring of the day. With the wish to acquire certificates, the demand for explicit testing, and the prevalent way of approaching and using the technology a focus on formal education is reconstructed. However,

---

[1] This is similar to an observation made during the Computer-Ndaba experience, when teachers concentrated more on a personal advantage in terms of job security - see Page 41. If one does not explain properly, people will become dependent on one's knowledge and his/her job is secured.

the advised way how ICTs should be integrated into the daily life, is an informal one: Through its practical application in daily life and not through specific courses. Computer lessons should not be dedicated lessons, but integrated into the daily life which depends on the self-confidence of the teacher and their knowledge, if they dare to use computers for computer-aided instruction.

Another reason why the integration of computers into the participants' daily lives is still limited is that benefits that are expected to be received through the usage of the Internet are not really experienceable. I have shown that costs and time efforts involved in accessing and performing the (only) known application of communicating via the Internet are in fact higher than communicating via a mobile phone. Asking for the key to the computer lab or saving money to be able to buy a computer by oneself requires a much higher effort than buying a mobile phone. In addition to that the mobile phone works temporarily independent from electricity (most of the time the network still works even though there is no power) and can also be presented as a status symbol. To my understanding the triumphal procession (see Figure 7.3 for a comparison of Internet users and cellular subscribers) of mobile phones hampers the integration of computers. Only if real benefits can be gained from using the Internet (such as shopping possibilities, best price information, or local contents might be) and cost efficient infrastructure becomes available, computers will again reach a certain level of competitiveness. At least these are characteristics which are perceived benefits in technological orientated cultures. To save time, create revenue and save money.

16.1.2 Does the fact of "having computer access" influence power relationships among pupils, staff, non-teaching staff and other schools as well as between them?

Previous discussion has shown that having computer access does influence power relationships among the emerging social actors. This is not only because the school as a whole expects an increased status as a school with an implied attractiveness and higher income, but because an increased status of the individual is also expected. Thus, the school will be able to maintain the lab easier than other schools. As a consequence, the gulf between different schools is widening, which leads to a digital/electrical divide (among schools). I have shown that, already, certain schools do not have sufficient financial means to pay for the used electricity. How should such schools cope with the maintenance costs of a computer lab if they do not attract richer families? Evidently, computer interventions increase the inequality between the schools. Even at the school investigated here, not all school members are treated equally. The SMC decides on the access to the computer lab (see Q2) and some are excluded. Generally, an increase in personal social status is expected if someone has access to computers, and in correlation to the high image of a computer it seems reasonable that this increase of status will take effect.

Whereas hierarchical positions are reconstructed among the teaching staff ("key owner") the position of the computer teacher/administrator becomes more powerful. As a consequence (and as it will be discussed in Q2), the SMC has to undertake certain measure in order to be in charge of the situation again. The computer is also instrumentalized to present and reconstruct ones own hierarchical position, when teachers pass off their work to the bursar. Obviously, computers are used to change the previous hierarchy: even the bursar aims to do so and attempts to establish herself as superior to the secretaries. When the headmistress declines the bursar's efforts, her hierarchical position is weakened and while, consequently, that of the headmistress is restored and strengthened. Another example of how computers are instrumentalized to preserve power is the one where an informant tells the story about how she outwitted her maiden. She audio-recorded the girl's history in advance and plays it to the girl without telling her that she has recorded it. The girl thinks that the computer is very clever ("kalimagezi") and her landlady becomes empowered herself since she is the one who uses the machine

Among pupils, this instrumentalization did not emerge at first sight, because no rivalry was mentioned and, during the interviews, the group presented itself as homogenous, which was signalled in the use of the pronoun "we". However, at home some children are allowed to use computers more extensively than others who might not be allowed to use a computer at all. This might also lead to a gap between computer knowledgeable pupils and those who are not allowed to use computers. An indication of this might be the occurrence where the "colored boy" and the teacher played a trick on the other pupils. This shows that the pupil was able to collaborate with a (powerful) teacher through the application of computer knowledge. Only because the pupil knew that a computer virus cannot affect a human being, the trick could be played. Furthermore, this indicates that computer knowledge might be misused - see Page 215. Another aspect in which the access to computers can influence hierarchical relationships is that of children increasing their power in comparison to the teachers through computer knowledge. As they are more used to studying and have at least some leisure time during their holidays, they have more time to practice. If their parents have the time to teach them, they will soon be ahead of their teachers. Apparently, this seems to be the case already, e.g. when one informant claims: *"For the children [...] because, there are some who are beyond even our knowledge"* - In1Line58. Taking this change into account, this could imply a big change of the whole educational system in which children are usually guided by the elder ones - Page 98. Nevertheless, too less indications could be figured out to discuss this issue in detail.

16.1.3   How is computer knowledge represented and posed?

To get a grasp on the existing computer knowledge proved to be difficult, but apparently a certificate or the ownership of a computer is enough for a person to be considered as computer knowledgeable. I have shown that the fact of being computer literate increases one's status and is very much linked to the expectation of increasing one's income. Indeed, it seems as if the ownership of a computer makes it possible to start up a business and teach computers. During the interviews, the informants present their computer knowledge by slipping in technical terms and by mentioning, for example, underlying concepts such as the "hardware and software concept". During the interviews, some informants try to save their face by applying verbal concepts and attempt to name things properly (Microsoft, harddisk, mouse, name of games), although they sometimes apply words in the wrong context. This underlines the wish to present oneself as computer literate which becomes manifest when one is talking to a computer teacher.

Another verbal concept which is used to represent computer knowledge is the Internet, which is a topic that is discussed over and over during the interviews. However, I have revealed that, in fact, users have hardly any experiences with the Internet themselves. Informants present the benefits they expect from using the Internet, but - as I have shown - these are rather vague than concrete ones (Page 189). Apparently, it is sufficient to be able to talk about the Internet and its possibilities. Sometimes, computer knowledge is also posed by listing rather generalized facts about the technology's characteristics. Computers are considered to be an asset because, "they help you in the future", "help you to earn money", or "entertain you". Often, benefits are presented without the background of a real personal experience, although many participants have such experiences which were discussed in Section 15.2. This underlines the local importance of computers and the positive connotation that goes with a computer. It also indicates the expectation of the increased status of a computer literate person, since only if someone hopes to gain status he/she will pretend to know more about certain topics.

16.2   Research Question #2

In which way does power manifest itself in computer teaching? Which signs of social function can be spotted among various social actors? I.e. a computer teacher or a teacher at a primary school.

As shown in the previous discussion power becomes manifest in different ways in the computer teaching. Computer teaching is a social interaction and any social interaction involves power. However, this way is not limited to the mere process of computer teaching. Another way in which power manifests itself is the way in which physical access to computers is arranged. The physical access to computers of the investigated school is determined by the access to a key to the computer lab (see Page 147) as has been discussed in depth in Chapter 13. By looking at the way in which access to the physical key is constructed the social functions of various social actors can be revealed and investigated. I have depicted (described how) that some social actors (matrons, cooks, etc.) are excluded from access to computers as they cannot actually ask for the key. In comparison, others (e.g. the teaching staff) have more power in terms of accessing computers because they can ask the headmistress or the computer teacher for the key to the computer lab. The teaching staff is allowed to use a computer at any time. Thus, the teaching staff is definitely hierarchically placed above the group of matrons who are denied access. Teachers ask for the key and use computers whenever they want to. Reasons for the matrons' exclusion are diverse and have been discussed. What seems most important to me is that this exclusion shows the possible instrumentalization of computers to prevent other social actors from gaining benefit from computer knowledge. Benefits are definitely expected and even though these are sometimes vague at least the increased status by the capability of being computer knowledgeable, whatever it might be, is expected and preserved.

SMC. Decisions on whether someone is allowed to learn computers at the school's computer lab and whether someone gets the chance to gain benefit from it, are made by the powerful SMC, who can be considered as the virtual owner of the computer lab. As the management body of the school, this board executes power. However, this power emerges not only in terms of physical access through e.g. the introduction of a computer fee for the pupils, but also when the headmistress (an entity of the SMC) encourages her subordinates to study computers. To give an example: Observing the fact that the purser took lessons after she was requested by the headmistress, it becomes obvious that the headmistress has the power to influence the staff when it comes to studying computers. Apart from that, the SMC also decides on the level of usage. Thus, the staff is not supposed to use computers for private purposes. However, in this case, it becomes apparent that the acquired computer knowledge can lever out this usage constraint, e.g. when teachers use their computers privately and protect them with passwords. Such people become empowered in opposition to the SMC. Nevertheless, the regular private usage is still within the power field of the SMC. As parents are supposed to make a financial maintenance contribution, the SMC also controls the children's' access to the computer lab.

Government. Another entity which seems to be in the background at first, but which actually takes up a very powerful social function is the Ugandan government through its curriculum. By limiting the time of computer access through the prescribed curriculum and the mode of nationwide qualifying examinations for secondary schools, computer lessons are cut down in favor of other lessons. So they indirectly prevent the teachers to ask for the key. This sounds reasonable, as the pupil's achievements are compared nationwide to all other schools. The better a pupil scores, the better the secondary school he/she is allowed to enter. Also, the better the school scores in average, the higher the reputation of the school becomes. According to the informants, an increased status of the school leads to an increased attractiveness with students which implies a better financial situation of the school. Since the government has turned out to be an unreliable employer, the teacher's salaries are also paid out of these budgets. The time constraining function of Uganda's educational policy indicated above could be avoided, if the teachers had adequate computer skills and suitable software applications to integrate computers into the everyday teaching. As the analysis has shown, this is far from becoming a reality, as the extent to which computers are integrated into a teacher's everyday life is limited to writing letters, preparing exams, and creating marksheets.

Parents. Parents control computer access at a pupil's home and indirectly also at school. If they cannot afford or do not want to pay for the extra computer fee, the pupil's are not allowed to attend the computer lessons. Thereby, parents have more power than pupils, which emerges in the way in which computers are utilized at home, too. I have shown that the pupils are (obviously) hardly allowed to play games or to use computers for entertainment, although this would be their preferred usage. The expectation linked to a computer of earning money through being computer knowledgeable and, most likely, high acquisition and maintenance costs shape the image of a computer. Apparently, a computer is mainly used for serious and "business orientated" activities such as writing letters or creating business cards. Consequently, the focus in the usage of computers is on further formal educational purposes. This underlines the rootedness of the colonial way of teaching. Children hardly get chances to explore the device on their own and are not even encouraged to do so. According to the examples mentioned above, the strong hierarchic position of teachers and parents and the children's fear to be punished could also be reasons for the pupils' fear to explore technology on their own (if they get the time). Although there are exceptions, parents take up the role of teachers and guide their children in their way of using computers. So far, it seems as if the pupils' computer teachers (at school and at home) continue the (colonial) formal way of approaching technology: to control it. This is different to the way in which the ATs were teaching and represents an intercultural encounter which is discussed in the third research question (Section 16.3).

Computer Teachers. An interesting issue which arises in this context is the fact that the headmistress (and consequently the SMC) does obviously not influence the way in which computer knowledge is transferred.

Although there are indications that the practical way of teaching is regarded as a preferred[2] and more effective one (see Page 227), the SMC does not interfere into the computer teaching process itself. This becomes even more striking when, during the interviews, both the headmistress and the deputy teacher (as entities of the SMC) evaluated the practical way of teaching as positively. Be it the teaching of the ATs or LCTs, none of these groups were criticized for their way of teaching, although they applied rather different ways of teaching. Whereas, the ATs focused more on student-centred methods, the LCTs drew upon the common formal educational way of teaching as I have discussed in Section 8.2 and presented in Chapter 15. One might claim that Mr. Rukinga was dismissed from the school because of his way of teaching, but this claim can be neglected as he was actually dismissed because he did not show up at the school any longer. The reason why the SMC does not interfere might not be because the SMC does not want to control the computer staff, but rather that the SMC has no methods and methodological instruments to validate the quality of teaching. Therefore it is difficult to control them.

Thus, it seems as if the fact of "being a computer teacher" provides enough status, trustfulness, and power to be immune to criticism. To give an example: On the one hand, the headmistress criticizes the new English teacher for "spoiling" the other teachers, which lead to a hampered usage of computers in their daily lives, but on the other hand she does not intervene. Obviously, she has not even admonished the English teacher to change her way of teaching, but asks her to instruct the purser in computers. Perceived "methodological teaching problems" like the one of a syllabus (see Page 222) were not discussed in public and just "accidentally" mentioned during the interviews. However, this might also be due to the fact that the ATs were partly seen in the role of the donor or, at least, as representatives of the donor. One reason might also lie in the more powerful position of the ATs. In reference to Kipnis, who has explored that people who hold power become isolated from criticism because no one likes to bring bad news to power holders (Kipnis in [Cushner & Brislin, 1995], p.315), this seems plausible. And I have shown that computer experts with a technologically orientated cultural background have, in fact, a high status. The reason why the participants did not dare to ask for an exam might also be the strong teacher-student

---

[2]This emerges not only in the course of this case study (Page 229), but also in [Latu & Young, 2004] who presents students who demand more practical examples.

relationship, which is prevalent (see Page 222). The students did not criticize the
computer teachers, as they perceived themselves in the role of a pupil. It is definitely
not the case that the teachers are not aware of other teaching methods, which are obviously already promoted by other institutions. Apparently, there exists an awareness of
the possibility of teaching ICTs in a student-centred way among teachers. As discussed
above, one teacher remarks *"so yours was more of a discovery method [...] which is
even encouraged in colleges and schools [...] so yours was better"* - In7Line317-321,
which indicates that the informant is aware of an actual preference of student-centred
methods in comparison to authoritarian way of teaching. I have shown that most
literature encourages one to apply practical methods in teaching ICTs - see Page 95.
However, apart from criticizing them for their computer skills, no criticism arose in
the course of the computer teachers' instructions by other teachers regarding the way
of teaching - Page 230. Criticism on the way of teaching is (only) voiced by pupils
who claimed that LCTs where teaching the same topics for the whole term (see Page
229)[3].

To my understanding, these observations demand a discussion of two issues: On
the one hand, referring to the thoughts mentioned above, it seems as if currently
the actual *qualification of computer teachers is rather poor.* It might be due to their
technical skills and practicable teaching methodologies that they cannot go beyond
basic lessons. On the other hand, *the position of a computer teacher emerges as a
rather powerful one,* as his/her teaching cannot be controlled by the SMC or other
colleagues any longer. This powerful position is enforced as computers are a highly
appreciated in the sociocultural context. As I have discussed previously, the SMC has
no means of assuring the qualification of teachers and, therefore, of their teaching.
This means that the local hierarchical structure has been changed by the computer
intervention. One way to control the quality of teaching (and consequently renegotiate
the previous relations of power[4]) could be to introduce a computer expert to the SMC.
This might be difficult to achieve, and so it seems reasonable to look for objectified
and standardized measurements to approve the teachers' qualification.

Drawing upon the assumption that the fact of being a computer teacher vests
people with power, it appears to be important to investigate how a computer teacher
becomes entitled to be one. So far, it seems as if the sheer ownership of a computer or
the claim of knowing how to use computers is sufficient as a qualification for teaching.
A standardized computer certificate which provides the SMC with a quality protection
of a certain amount of computer knowledge among computer teachers, might be useful for the SMC to cope with this situation of qualification uncertainty. Of course, I
realize that this represents a rather ethnocentristic perspective and contributes to the
prolonging and reconstruction of the importance of formal education. It is also contradicting to the proclaimed return to indigenous education (see Page 8.2.4), which is
promoted in case ICTs are supposed to be adapted by a non-technologically orientated
culture. However, I have shown that formal education seems to be deeply integrated
into the people's everyday life and especially in the regional context of Kabale. The

---

[3]Unfortunately, such a way of teaching seems to be no exception within this case study. Turner
shows similarities when she realises that the same lessons had been repeatedly taught to the same
student. ([Turner, 2005], p.10).

[4]The discruption of the LAN shows that the power is negotiated as indicated in Page 25.

indigenous form of education seems to be neglected completely (which is argued with the underlying monistic-vitalistic world-view) and formal education is in fact more promising to them as jobs in urban areas are more likely to provide people with more income. By the acquisition of formal qualification, people expect to retrieve a higher income - see Page 114. In correlation to the discussed local brain drain, this also underlines the focus on formal education and, consequently, the wish to acquire computer certificates.

As shown in Section 7.4.3 there is a variety of different ICT initiatives in Uganda. The country's role as a model student and the big interest of other donor countries to make Uganda a "best practice" for development cooperation might be counterproductive to a transparent integration of ICTs. Several separate and uncoordinated initiatives might contribute to the proliferation of different certificates. Once more it becomes the task of the Ugandan Government to achieve transparency in the issuing of certificates. Assuming that one nationwide certificate can be realized there is still a disparity between the dominant focus on formal and the neglect of indigenous education, which inhibits in fact the preferred method of approaching computers: learning by doing. However, any attempt to pin this "different" way of teaching on the available teachers would represent just another form of cultural imperialism. To avoid a way of methodological prescription it is to my conviction that it becomes necessary to point out the contradiction between the prevalent way of approaching technology and the obviously preferred one. Educational institutes have to make it a subject of discussion. In this case the concept of a cultural broker sounds promising. This is not only, because such a person is aware of cultural differences (which includes different ways how to approach technology), but also historical and sociocultural aspects of both cultures to enable intercultural interaction. Even if the computer teacher has the same local cultural background, it is important to combine the different ways how to approach technology. I have already shown that **any** human computer interaction is an intercultural encounter and so any computer teacher has to become a cultural broker, which is manifold encouraged in literature (see Page 14, Page 34f, Page 45).

16.3   Research Question #3

What kind of intercultural phenomena can be observed? When and how is referred to different (cultural) systems?

As discussed above, it is not the aim of this case study to analyze intercultural phenomena at the interface of computer (programs) and users. Difficulties in the usage of computers are obviously no peculiarity of a local cultural context and are shared phenomena among computer beginners across different cultural contexts. To give an example from the Ugandan case study: in order to perform a calculation in Excel, the students were inclined to put the equal sign after the calculation instead of a demanded prefix to initiate the start of the equation[5]. However, such difficulties are no sociocultural and regional deviances, but also occur regularly with other computer beginners like, for example, those in some Austrian computer teaching institutions[6].

---

[5] Formulas like "=A1+A2=" or "A1+A2=" were common instead of the correct way of "=A1+A2".

[6] This is based on the experiences of an interviewed Austrian computer teacher, who has been

Therefore, I have focused and will focus more on intercultural phenomena which involve the teachers' and students' cultural background. Thus, issues like the discussed terminology-related problems or the way in which the technology is approached are subject of this study.

16.3.1   Communication and language use

Some previously presented intercultural phenomena can be assigned to the theme *communication and language use*, which is the *"most obvious problem that must be overcome in the crossing of cultural boundaries"* ([Cushner & Brislin, 1995], p.40). Language problems, such as presented from Page 224 onwards, are placed within this theme which has been identified by Cushner and Brislin [ibi.]. I have shown that computer related terms seem to be self-explaining at a first sight, but must not be taken for granted. I have presented an example for the fact that symbols are not necessarily perceived by a computer learner in the way that a computer-literate teacher might expect. Thus, by verbal interaction, the meanings of symbols have to be explained and misunderstandings resolved. This leads to a more student centred way of teaching since it encourages students to ask for clarification. Obviously, the larger the common language base is, the better a teacher can explain underlying concepts of terms and symbols. However, misunderstandings related to pronunciation ("*r*" vs. "*l*" - Page 225) and the proper selection of words ("*save*", "*edit*", "*triangle*" vs. "*arrow*" - Page 225) occur, as none of the participants speaks English as his/her mother tongue. The approach to focus on a broad common language base is not new: similar results have been retrieved by [Nel & Wilkinson, 2001] and [Latu, 2006] who underline the importance of negotiated, agreed and shared computer terminologies. Both studies have revealed basic language problems of students, which emerge especially during the application of student centred teaching methods of teaching in which the teacher focuses on verbal interactions. As long as a teacher draws upon the formal authoritarian way of teaching, these problems can be concealed. However, in verbal interaction, English deficiencies become apparent both among the teachers and the students. Consequently, teaching in mother tongue would be a preferred scenario as misunderstandings related to communication could be minimized. When some informants emphasize that in the investigated case study only minor dialect problems arose, and that only in the beginning (Page 224), the instruction of computers in Rukiga sounds like a promising example of how to avoid some of these "*minor dialect problems*". However, as discussed in Chapter 8, this would be in contradiction to the current Ugandan education policy which focuses on English as the lingua franca for education - Also, as the investigated school is a boarding school, some students might have a different mother tongue than Rukiga, which shows that the situation could actually becomes even more complex to deal with. Consequently, it seems to be unavoidable to maintain teaching in English and to negotiate as many terms as possible. Of course, there will always be terms which the teacher will omit to explain as he/she will believe that these can be taken for granted. Thus, it becomes important that the students dare to ask for a solid explanation. Especially within the field of computers, technologically orien-

---

teaching computers to beginners for more than 20 years now. He reports similiar problems for computer newbies.

tated terms have to be negotiated and discussed. As computer terms are based on a different sociocultural context, any computer teacher has to become a cultural broker (see Page 34). I go along with Cushner and Brislin who indicate that it becomes a major responsibility of intercultural trainers and educators to help individuals develop appropriate vocabularies from the very beginning. This will allow them to "*discuss freely the encounters they experience, thus enabling them to resolve problems before they get out of hand*" ([Cushner & Brislin, 1995], p.7).

16.3.2   Approaching the technology and learning styles

Another intercultural phenomenon observed during the project was that the practical, explorative way of approaching computers was untypical for the people who were taught and instructed in the basics of computing. I have shown that the local students are used to an authoritarian way of teaching which is characterized more by formalism than by exploration. The local cultural expectation towards control finds its expression in a non-explorative way of teaching, which suggests that the external instructors were causing a culturally misbehaviour. If informants claim that they are afraid of "spoiling" the technology this means that they are aware of their capability of ruining the technology and that they search for a tactic to overcome their fears. Commonly, the solution is to completely rely on the teacher's commands and reproduce them step by step. Even one year after the first contact, they are able to reconstruct their initial computer usage (see Page 129). Unfortunately, this method of reconstructing makes it difficult to learn new ways of using the computer. Apparently the fear of "spoiling things" is greater than the eagerness to explore the technology. To my knowledge, there are at least two reasons why the students are afraid of damaging computers: to might have to pay money and/or could be punished. In order to avoid any damages which might be cause through its misuse[7] they try to control[8] technology first. This constitutes an important difference to the Austrian youth who approaches computers just in the opposite way [Steinhardt, 1994]. The strong internalization of the authoritarian way of teaching indicates that the promoted turn back to former indigenous education (see Page 103) might become difficult and can only be achieved through political changes in educational institutions. This goes along with the demand for an educational reform which I have indicated in Section 8.1 and which is being promoted by the UNESCO as a major point of interest in [UNESCO, 2002]. If mainly student centred education is promoted for the introduction of ICTs (see Page 96f) this automatically implies terms like "*collaboration*", "*consultation*", and "*negotiation*". These pillars of student centred education describe quite exactly the characteristics of a "*cultural facilitator*" [Latu & Young, 2004] and "*cultural broker*" (see Page 34). Charles Ess also recognizes this necessity when he states "*to become such cultural hybrids requires, in part precisely our becoming aware of the diverse cultural values and communicative preferences*" ([Ess & Sudweeks, 2001], p.260). Thus, it becomes important that computer teachers do not only have technical skills but can also draw

---

[7]Anyway, the breaking of screens and the fact that computer lessons cannot be part of the official curriculum enforces the pupils' fear. It prevents them from accessing and studying computers.

[8]This is also indicated by Latu, [Latu & Young, 2004], who presents the fear of users that the computer might "do things" they themselves had not intended to do (see Page 45).

upon intercultural and methodological ones. I have discussed that this requires an awareness of one's own culture and world-view. This acceptance of different world-views (which, evidently, exist) should not be left to chance, but has to be promoted explicitly. Otherwise, one ends up shifting from one world view to another. If this happens the integration of ICTs will imply just another form of cultural imperialism. This does not mean that one teaching method is supposed to be replaced by a new or better one. By offering additional teaching methods a computer teacher can choose from more options (than he/she could do before) and is still able to decide on his/her own whether he/she wants to change his/her teaching style. Culture is change.

Aiming at assigning this phenomenon to intercultural categories leads us to Cushner & Brislin's theme of learning styles which is outlined as follows: "*Even though people desire change and improvement the styles in which people learn best may differ from culture to culture. People involved in change efforts (e.g. teachers [...]) may find that information presented in ways attractive and efficient to them [...] may not lead to desired outcomes*" ([Cushner & Brislin, 1995], p.42). In this case study, the situation seems to be widely similar, but different with regard to certain aspects. In fact, the students appreciated the ATs practical and explorative way of teaching and criticism from pupils are mainly focused on the way of teaching of the LCTs which they considered as too theoretical. Pupils criticized the LCTs' focus on writing down things instead of experiencing them on their own. The practical, active way of teaching of the ATs was apparently attractive to all actors but in order to achieve a better integration of computers in daily life, it might be necessary to combine this way of teaching with intermediate tests as the implicit testing by the ATs was not perceived as a testing at all. The fear of spoiling the technology is still prevalent but "checkpoints" might enable the development of enough self confidence to explore the technology and might reduce the fear of approaching it. By introducing such, the feeling of the capability in controlling the technology could be strengthened and the willingness to take "risks" could be increased. This is supported by the studies of Scheepers, who figured out that it is necessary to place checkpoints in the course to make sure what and how well the participants understand (see Page 41). She notes that the goals should be set within the reach of the participants in order for them to become self-confident enough to be able to achieve them. Nel and Wilkinson also support this method in their seven points plan and postulate regular assessments (see Page 43) [Nel & Wilkinson, 2001].

Whereas the ATs concentrated too much on student centred teaching methods and did not perceive the local cultural expectation of the participants to be tested, the LCTs ignored that their way of teaching ICTs seemed not to be attractive and efficient to their students. This underlines a demand for cultural brokers even in the case of ICTs being instructed by LCTs, because this represents an intercultural encounter as well. Already in 1990, Niels Bjorn-Andersen considers training as the biggest problem: "*Training is often too little, too late, and too technical. Furthermore it is often aimed at the wrong people and is based on an inadequate model of learning. No doubt a very large part of the problems with Information Technology and LICs (Least Industrialized Countries) are due specifically to the poor quality of training*" [Bjorn-Andersen, 1990]. Unfortunately, it seems as if - about 15 years later - the situation has not changed much. Cushner's demand that "*teachers must themselves learn to be flexible so that they can redesign their lesson plans to take advantage of the ways their students typically learn new material in their everyday lives* " ([Cushner & Brislin, 1995], p.197) emphasizes the necessity of more adequately qualified and well educated teachers who are able to shift between different world-views and are willing to negotiate and collaborate with their students. In the following section, I will discuss what the concept of a cultural broker could have been in this case study and where the risks and chances are.

16.3.3   Cultural Broker

Besides the broad common language base discussed above, the concept of a cultural broker promotes the transfer of technological knowledge (see Section 3.1.3). This implies an awareness and readiness of teachers to negotiate cultural conflicts, as computers represent a subculture of their own. As indicated by the NARST (see Page 34f), a crucial capability of a cultural broker is to make the different underlying world views a subject of discussion. The pointing out of differences in these world views (local culture vs. computer) is necessary. I have shown that some participants have already understood the different underlying concepts (see Page 176) on their own, whereas others have not. If one informant presents a computer as something which can look into the future (Ex14.21) he cannot be said to have internalized the cause-effect concept of a computer yet. On the other hand, I have presented that some students - obviously those who have used computers intensively - have realized that the user is the one who tells the computer what to do. To make sure that all students understand this cause-effect mechanism and are, consequently, more encouraged to explore computers, a cultural broker can clear this barrier by discussing these mechanisms and underlying cultural concepts. An effective discussion can be achieved by comparing a computer's mechanism to indigenous scientific and technological principles. This, then, goes along with the concept of Jegede who defines this as a task of the cultural broker - see Page 34. By involving such indigenous scientific and technological principles, the indigenous way of approaching science is not acculturated and (cultural) assimilation might be prevented, which I have indicated in Page 36.

In this case study, there were at least two different teacher groups who could/should have functioned as cultural brokers: the ATs and LCTs. It is difficult to decide which group complies best with the requirements for becoming a cultural broker. We know that both teacher groups might have an increased status through their computer knowledge.

The ATs have deep computer knowledge and willingly apply the more effective way of teaching in a student centred and practical way. However, they completely lack knowledge of local culture, forms of English dialect, customs, attitudes, and cultural expectations. For a certain period of time, they might cope with certain issues through intercultural trainings and by reading books, but they won't be able to reach the level of a LCT. For example, this ignorance emerges, when the ATs were not aware that they were supposed to issue certificates, although this is of tremendous importance to the local population. Certificates enable a gain of social status which runs counter Hofstede who claims that the gain of status through certificates is a characteristic of individualistic cultures (East Africa is supposed to be a collectivistic one) [Hofstede, 1991]. The experience of this case study shows that culture is nothing which can be taken for granted, but is constantly changing.

For LCTs, it seems to be easier to become cultural brokers as technical skills and teaching methodologies can be acquired with less effort. However, turning these quality ensuring measures into reality demands a willingness to give away power and reduce ones status as a teacher. Formal education plays an important role in Uganda and to work as a teacher implies an increased social status. This is not only expressed in the way teachers treat other people, but also by their earnings[9]. I have presented that the income of teachers is much higher than that of an average worker. Teachers, then, are fully aware of their formal qualifications which make them feel superior to less qualified people (Page 103). In correlation with the high expectations which are set towards a computer, it seems as if the crucial point in this context might be the readiness of a computer teacher to share his/her computer knowledge and to "donate" some of his/her ascribed status. This means that, if teachers really wanted their students to develop a better understanding of computers, they would have to give up their elite status as a computer teacher (which is evidently even more appreciated than being a computer knowledgeable person). By changing the teaching paradigm to student centred methods, the students would become less dependent on the teacher's recited lessons and could generate more computer knowledge on their own. However, the teachers' own limited computer skills might prevent them from teaching in a less formal way, because perhaps they still have internalized the attitude that technology has to be controlled. So this becomes, once more, a deadlock situation. On the one hand, capacity building efforts to increase the technical skills among the computer teachers seem to be necessary. On the other hand, those people who train the trainers might not be motivated to do so, because they might loose power. I believe that once a critical mass of computer knowledgeable people has been assembled, a snowball-effect can be achieved. However, this implies a general loss of status among computer literate people and it seems as if something like this can only be achieved through political interventions. Only if large numbers of people are computer knowledgeable, one might be motivated to share and transfer one's knowledge, as no further loss of power must be feared. At first, however, infrastructure related issues like power, or computer access have to be resolved. Otherwise, computers will rather contribute to a digital divide than to a more balanced society, since only those who have regular access to computers get the opportunity to exploit its benefits (although it seems as if these are currently limited to the reproduction of (neat!) work, storing information, and the increased social status).

Apparently, a "perfect cultural broker" for this case study would have been a person, who had been aware of the different world views, "*regionally used teaching methods/habits*", "*ways of sanctions*", use and status of "*local language*", the social role of "*involved institutions*", and knowledge on historical and cultural pecularities. This person would have had also a profound computer knowledge and the willingness to take into account a loss of power by transferring the knowledge. "Deep" computer knowledge is therefore important as it ensures several different ways of explaining and solving the students' computer problems.

---

[9]I derive from the stance that one method of determining social status involves a comparison of the height of income.

16.4   Research Question #4

In what way does the multifunctional role as an interviewer, teacher, colleague and donor influence the progression of interviews? Should and can donors evaluate such interventions on their own?

The analysis has shown that the Austrian teacher embodies a personal union consisting of four roles: instructor, interviewer, donor and evaluator. However, the roles of instructor and evaluator are difficult to separate. All in all, I can say that during the interviews, manifest roles of the actors are constantly changing, mainly the one of the interviewer and the one of the teacher. Obviously, members of staff have internalized a strong teacher-student relationship and if members of staff change roles they perceive the interviewer as a teacher rather than as an evaluator. The role of a colleague did not emerge at all and the one of a donor hardly ever, although, admittedly, this role became manifest. Multiple roles as a researcher are problematic and have to be treated carefully during the whole process of research and evaluation. I have shown that both the interviewer and the interviewee slip into different roles several times, which demands an extra effort during the research process in order to avoid any role confusion. Especially during the interviews, it becomes evident that the interviewer aims at achieving a domination-free sphere in order to retrieve narrative statements, which, per se, underlines his dominant position. This problematic situation could not be avoided a priori in the context of this case study. However, the application of objective hermeneutics in combination with the problem centred interview proved to be reliable in carving out such latent changes of roles and led to a solid reconstruction of the informants' thinking. Premature assumptions and conclusions were identified and discarded. Whenever reading options emerged which indicated that an informant aimed at saving his/her face, all reading versions remained open. In the course of the interviews one reading version by the other was closed by the principle of sequentiality. Indeed, there were some statements in which the informant attempted to save their face. These attempts were revealed in the progress of the interview. To give an example: one informant claimed that she used a computer in everyday life. However, during the progress of the interview, it turned out that she did not even know how to start a program. Also, she contradicted herself and, taking into account the notes of the participant observation, it became apparent that she definitely tried to present a "deeper" computer knowledge than she had actually acquired. Another one would be Ex11.8, in which the informant tried to avoid the question about the private usage of a computer.

Naturally, an external researcher would not have caused that many role confusions (i.e. interviewer, instructor and donor), but this kind of research would imply an evaluation by an outsider and this is to be considered even more problematic. If the case study had been done by an external researcher it would demand a difficult phase of establishing trustfulness. Also, this would lead to a more severe distortion of the results: Context knowledge would be missing, because the researcher would always remain as a formal guest and not have access to the internal environment of the school. This is approved by Turner's experiences [Turner, 2005] depicted in Section 3.1[10]. Results

---

[10] Unfortunately, I cannot compare this research questions to other available case studies as it is

show that the insights, which were retrieved by the methodological combination of problem-centred interviews and objective hermeneutics, are strongly connected and validated by other meta-documents like participant observation or thick description. Concerning the role confusion that occurs when the researcher is also a teacher it is to my conviction that criticism on the way of teaching and the differences to local teaching methods can be carved out best, if the interviewer is a teacher himself/herself[11]. A fact which becomes manifest in the interviews is that shared experiences during the teaching, help to refer to specific situations to illustrate problematic and memorable situations. Thus, the interviews get less formal, and a relationship of trust is established. Interviewer and interviewee can refer to the same context, if they want to point out certain issues. Some exemplarily statements taken from an interview illustrate this:

"*I remember (*small laugh*) the other time when you were saying "hold", "hold"*" - In10Line20.

"*Don't you remember the other day when we lifted them, when you were fixing them at your place?*" - In5Line34.

When informants refer to shared experiences like above mentioned ones, this enables one to explore the participants' lifeworlds and to evaluate to what extent a computer is integrated into their daily lives. As mentioned above, the ATs are considered to be the donors or representatives of the donor as well. From my point of view, it is better to risk a possible confusion of roles than to have the study carried out by an external researcher, although evaluating an "own" intervention is always difficult since a personal bias and future plans of the organization might (latently) prevent the revelation of crucial issues. For this research, political motivations seem to be less important as the Projekt=Uganda is no professional NGO, but mainly the effort of one single person. Of course, any activity of an NGO is an expression of a political attitude. However, it might be that bigger organizations are more in need of justifying their activities to governments. Public pressure is higher and they themselves feel more obliged to present positive results to their donators. The collaboration between the ATs and the Projekt=Uganda presented in this case study was just a temporal one, which was settled from the very beginning. As future plans with the Projekt=Uganda have always been far-off reality, any economical or political motivations to fulfill the NGO's expectations can be neglected. Nevertheless, there is the risk of a latent bias and one criticism on qualitative studies is that their results are difficult to track and verifiable. One method which supported the quality assurance of the interpretation was that of context-free discussions. Thereby, no personal developmental issues or politically motivated considerations like "*should these people have a computer*" impacted the text analysis.

Deriving from the stance that evaluating a computer intervention means to find out the extent to which a computer is integrated into the participants' everyday life, above discussion of the research questions Q1 and Q2 has shown that is obviously possible to do so.

---

unknown wheter the respective researchers have been involved in teaching as well.

[11] The interviewer was in fact mainly perceived as a teacher.

Chapter 17

Conclusions

In the following sections I aim at drawing a conclusion from previous analytical and discussional parts. My conclusions are structured according to the critical aspects identified by Nel and Wilkinson [Nel & Wilkinson, 2001]. I aim to show the necessity to go beyond those, namely to explore the matter of power, which plays an important role for the success of ICT interventions.

Nel and Wilkinson have figured out seven critical issues which I have presented in Page 43. In relation to this case study they are considered as follows:

1. The first issue claims for an advanced **awareness** of IT teachers in the nature and extent of the social, economical and cultural problems of students in LDCs, which I understand as the demand for IT teachers to become cultural brokers. Mainly those who deal with literature on science teaching promote the concept of cultural brokers. Any HCI encounter represents an intercultural one and, as I have discussed in Q3, the applied concept of a cultural broker becomes inevitable to avoid that computer related interventions become another sign of cultural hegemony.

2. I have shown that the image of a computer is high and computer knowledge is obviously assigned to high expectations. Together with the fear of damaging the technology these issues are, according to Nel and Wilkinson, supposed to be addressed in a period of **orientation**. When the South African authors claim for an informal discussion of fears, expectations and what a computer can and cannot do, this argues for a computer teacher, who reveals fears and expectations in a trusted environment. Such teachers have to refrain from their current strong hierarchical positions and have to be able to discuss critical issues frankly. Thereby, negative (unexpected) experiences can be avoided and trust into the technology established. A planned intervention always implies cultural changes and by placing some computers to a certain place new situations and demands emerge. To my understanding, it would be good, and someone might call it "*fair*", to make aware of upcoming problems and risks of the technology instead of promoting a panacea. Such problems can be both health related ones and the increased financial effort to maintain a lab. This includes costs for power and physical maintenance, but also costs for personal and specialists.

3. Another issue which emerges both in Nel and Wilkinson and within this study is the obvious demand to change existing **teaching methods**. I have discussed that the teacher centred way of teaching was a point of criticism from the participants towards their experiences with the LCTs. Although [Scheepers & de Villiers, 2000] claim that there were no differences in the examination results between the two teaching methodologies (see Page 39), for this case study the practical, student orientated way of teaching was definitely preferred by the participants (see Section 15.4.1). It seems as if the own exploration of technology and the practical usage of a computer is more helpful to the students than theoretical, abstract studies. Obviously, and in accordance with the major stance in literature, the underlying concepts of a computer are too complex to be learned by heart and are internalized in a better way, if they are experienced throughout practical usage.

4. Difficulties related to **language** are also among these seven points. Instructing computers in one's mother tongue might be the most promising way of avoiding language misunderstandings. Several technical terms have to be explicated anyway (see Section 15.4.1), but even in daily talks misunderstandings become manifest. If someone teaches in his mother tongue he/she is definitely more precise and affluent in his/her explanation of certain terms. However, for this case study two problems arise: The first, a political one, refers to the fact that English is the official language of teaching/education. The second one draws upon the fact that the investigated school is a boarding school and not all pupils are Bakiga. So it seems to be inevitable to stick to teaching in English in certain situations, but nevertheless I want to recommend to reconsider the paradigm of teaching computers in English if possible (e.g. in local training institutions).

5. Physical **access** is the basis for the development and acquisition of computer knowledge. Lack of usage, due to limited time (spare time in case of the teachers) poses a barrier to the integration of computers into the interview participants' everyday life. In opposition to Nel and Wilkinson, difficulties in relation to finance and transportation seem to play a minor role. The computer lab is placed on the compound and the school can maintain the lab by the financial contribution made by the pupils' parents. Discussing the issue *access* means also thinking about the issue that computer access is restricted to social actors. This shows clearly that computers are instrumentalized as means of power. In addition to the expected future development that income increases mainly among those who are connected to electricity and who live in towns, nourishes the assumption that computer interventions promote the digital divide. At least this could be dampened if computers would be accessible to everyone (the public) for free, and not only to some elitarian social actors. However, many smaller interventions are mainly based on personal relationships between donors and "*recipients*", as for example clerical institutions are. This means that interventions are dedicated to certain closed groups. These "*recipients*" have to take up the responsibility to maintain the equipment, but also the power to manage the donation. This happened also in this case study. Unfortunately, the collected data material did

not allow an investigation to find the reason why the school excluded non-school members from the access[1]. If this is in order to prevent damages caused by outsiders and/or to ensure the computer's permanent accessibility to the school remains unsolved. A further (quite reasonable) argument why the school did not make the computer lab publicly accessible seems to be the fear of losing the technology through its break down. For future interventions different physical and environmental circumstance should be taken into account and tested in advance, e.g. by causing voltage peaks. Consequently, reasons for the (non-)access of outsiders can be cut down.

6. Obviously, it is inevitable to assess the success of teaching explicitly. Both studies (Nel and Wilkinson's and the one at hand) have identified that **assessment** techniques need to be revisited. Nel and Wilkinson promote assessment techniques of co-operative learning methods. The Ugandan case study shows that explicit tests should be made to increase the students's self-confidence in approaching the technology. Thereby, the students claim to develop a feeling of having covered instructed topics as a whole. The (latent) implicit testing by the ATs was not perceived as a way of testing at all - at least for this case study. Colonial methods of testing are expected, as the announcement of an exam or the issuing of a certificate, if a topic is covered adequately.

7. The encouraged **feedback** given by students to their teachers in order to improve and adapt their lessons seems to be reasonable. However, I have presented the strong hierarchical teacher-pupil relationship, which is not only manifested by the excathedra teaching method, but also by different forms of punishment, be it corporally or financially. Not only this inequality in terms of power might lead to distorted feedbacks, but also the revealed poor qualification of the computer teachers. But how would a computer teacher react if someone asked him/her to teach something new instead of last year's topics?

All these facts show once more that the whole success of technology transfer remains within the hands of the computer teachers. Whereas Nel and Wilkinson have revealed exactly those issues which are dominant in the investigated intervention as well, they do not discuss aspects of power of computer teachers. I have pointed out that obviously the extent on how successful computer knowledge is transferred to the local environment depends heavily on the capability of the teacher to take up the role of a cultural broker. However, this seems to be difficult as several reasons mark their way of teaching. The most preventive factor of changing their way of teaching and consequently passing on computer knowledge in a more successful way, might be their unwillingness to loose their gained status. In previous sections I have intensively elaborated on the high expectations which are set upon acquired computer knowledge and

---

[1] There were plans by the SMC to create a revenue by offering services to outsiders, but this was not realized up to the time the interviews were taken. In fact, two people were allowed to use the computer lab and attend the course: the wife of the Bishop and his daughter. This exceptional status underlines their high social reputation.

described the high social status of a teacher. So it becomes a task to motivate computer teachers to give off some of their power. In the course of this intervention two different teacher groups emerge: At a first glance one might call them the "ATs" and the "LCTs". But at a closer look they, besides their different technical skills, teaching methodology and cultural background, distinguish themselves by one further fact: the spent time within the cultural environment. Generally spoken, ATs are teachers who participate only temporarily in computer teaching and therefore, are more likely to take into account a loss of power, as it would be just a temporary one. However, such teachers are always ignorant when it comes to the local culture (and language) and are not necessarily aware of appropriate teaching methodologies. It can also not be taken for granted that their technical skills are adequately to teach computers[2]. To my understanding local cultural knowledge takes more time to be acquired than new teaching methodologies and the mandatory intercultural awareness as a cultural broker (between computers and ones own inscribed culture). These teaching methodologies are not supposed to replace other ones, but extend the pool of applicable ways of transferring the technology.

However, LCTs (the second group) are coined by their long term availability and might not be motivated and able to pass on computer knowledge properly, as they are lacking adequate teaching methodologies, have internalized the local way of approaching technologies, might fear a loss of status. This case study has also shown that instead of sharing their computer knowledge, the LCTs from the investigated school aim to utilize their knowledge in places where they can gain an even higher profit. This form of local brain drain runs counter the expected multiplier effect; namely that educated people in small towns will help their countrymen to profit from the technology as well. Even payments by the school could not prevent these people from leaving the place to look for a better-paid job. The chosen way of the school to enable their existing staff to become computer teachers, can also be a successful attempt to refrain their computer teachers from leaving the school. However, this should be kept in mind as a possibility, but has to be watched for a longer period of time if this poses a real opportunity.

Consequently, there is a demand to develop and find benefits for local computer teachers in order to increase their willingness to take up the role of a cultural broker. This asks for a computer teacher who sees himself/herself more in the role as a facilitator, than in the one of a teacher. Reasons how such a role-change can be achieved, might be the focus of further studies.

17.1   Implications for Human Resource Development

For the future of Human Resource Development the creation of cultural awareness for "*external instructors*" to enable a more effective and sustainable technological knowledge transfers is important. If it is inevitable to avoid the assistance of an "*external instructor*", the instructor should be able to take over the role of a cultural broker ([Aikenhead, 2002]) or at least know, and be aware of local different factors such

---

[2]I have discussed in detail the proliferation of certificates which makes it not possible to assure the quality of computer teachers.

as "*used teaching methods/habits*", "*sanctions*", use and status of "*local language*" or social role of "*involved institutions*". A careful selection of instructors is obligating to assure skilled and flexible workers in LDCs, as instructors as well as the computer itself are new social actors and elements in an existing social system and might cause a cultural break due to an inappropriate introduction of technology.

17.2   Methodological Review

The collected data has made possible a solid structural analysis and enabled founded statements. I have described some different case studies on teaching computers in LDCs in Chapter 3 and presented their methodological stances and methods. Although the major methodological tenet promotes (Ethnographic) Action Research (Page 51), the insights revealed within this case study, are definitely comparable with those of Action Research related studies. This shows up that Witzel's programme of problem centred interviews is obviously appropriable also in a different cultural context. Even the combination with objective hermeneutics seems to be advisable to carve out role confusions and to limit the researcher's bias. However, among the discussion groups it would have been advisable to analyze the sequences with members of the local society, as it might be possible that ethnocentristic tendencies might influence the process of analysis. A perfect reconstruction of ones mental world and the reasons why someone is saying something is not possible, but the more one knows about the local culture the better he/she can understand the interviewee. The advantage of my applied methodological stance is that the evaluation takes place from an ex-post position, and does not influence the integration itself. Especially by the temporal shifted period of data collection, the cultural change caused by the intervention can be elucidated. It might be that if the intervention is investigated throughout the period of teaching, the influence of a researcher/donor is too strong to negotiate certain aspects of access. E.g. changes between the investigational periods in terms of access like the matrons experienced might not happen as long as the research is on site. At the same time this methodological perspective becomes also a disadvantage, as suggestions on how to improve the integration and teaching cannot be presented and explored. Therefore, it gets difficult to evolve new solutions for an enhanced integration. Improvements have to be retrieved from texts, but cannot be realized. It has therefore been one aim of this research to present the ongoings of this intervention to show up future risks and chances for other donors. If discussions on aspects of power might be fruitful anyway, since all participants are involved, is doubtful. By making power an issue of discussion the renegotiation starts and as a member of a different cultural background and representative of the donor it might become difficult to avoid a hegemonial position. Consequently, to my understanding, a local researcher with sufficient computer knowledge to explore the field of interest could offer new perspectives on ICTs in LCDs. Of course, as this researcher would be computer knowledgeable, a hierarchical difference is to be expected, but the strong teacher-student relationship might be avoided. By a comparison of both studies intercultural differences within researches and possible different perceptions of computer related problems can be revealed. Such a confrontation enables donors to find out if "donor-influenced" researches can be considered as serious options for evaluating interventions.

An obvious benefit of the applied methodological approach is that this study has revealed almost all of those critical issues, which had been identified by several distinct studies in one case study. It has shown that technology transfer seems to be locally independent, since its results can be compared with those in Asia, South Africa, and Australia - see Chapter 3.2. It seems, as if the predominant position of technological-oriented cultures and the inscribed cultural values of computers enforce similar activities to make technological knowledge transfer successful. This is not only limited to infrastructural measures (e.g. electricity, Internet, etc.), but also to e.g. educational reforms (as the introduction of standardized certificates or application of different teaching methods are) and the change of structuring the daily life (polychronous vs. monochronous values). This study has revealed (for the first time) that the expectations towards gaining social status (and towards computers) seem to be correlated with the wish to become computer knowledgeable and obviously hampers the knowledge transfer. I have shown that power plays an important role in terms of access to computers and for the transfer of computer knowledge. Computers are instrumentalized as means of power and contribute more to a digital divide than they erode poverty.

17.3 Outlook

To broaden the results of this case study and of those which were incorporated and presented, further case studies ought to be done by examining computer (school-) initiatives. This offers the possibility for comparative analysis of the results with a view to existing similarities and differences on how power is reconstructed and negotiated through the integration of computers among social/educational levels e.g. primary pupils in different countries. Thereby, the focus should be on the role of computer teachers.

What kind of benefits can be offered to them (by whom?) to remain locally? Are children who are computer knowledgeable changing the form of education, as they tell their parents what to do? Who can take up such a role? Can the creation of a nationwide computer certification institution pose a useful solution? Further case studies should focus on the underlying and present expectations towards computers and what circumstances are necessary to make computer labs accessible to the public. What kind of interventional structures are supposed to exist, in order to give the chance to all people to use computers?

Finally, to my conviction, it is not a researcher's or donor's task to decide if computer interventions are useful or not. It is more the issue of making aware of possible implications, risks and promising approaches to the underlying circumstances. It is up to ones own ethical reasoning if he/she is willing to support other people in disadvantaged countries, who have the demand (and right) to utilize a technology, which is supposed to improve their living circumstances. Be it financially or mentally.

Bibliography

[Afemann, 2003] Afemann, U. (2003). Kommunikationstechnologien in Lateinamerika. *Widerspruch*, 45, 79–92.

[African, Demographics, 2004] African, Demographics (2004). African demographics. Internet: http://www.africandemographics.com/files/UgandaWeb.htm. Accessed: 22.05.2004.

[Aikenhead, 2002] Aikenhead, G. (2002). Cross-Cultural Science Teaching: Rekindling Traditions for Aboriginal Students. *Canadian Journal of Science, Mathematics and Technology Education*, 2(3), 287–304.

[Aikenhead, Glen, 2001] Aikenhead, Glen (2001). Cross-Cultural Science Teaching: Praxis. A paper presented at the at the annual meeting of the National Association for Research in Science Teaching, St. Louis, March 26-28, 2001. Internet: http://www.usask.ca/education/people/aikenhead/narst01.htm. Accessed: 20.09.2006.

[Allen, 2000] Allen, P. (2000). *Interesting Times: Life in Uganda Under Idi Amin.* Book Guild, Limited.

[Annan, 2003] Annan, K. (2003). Text of the video message by Secretary-General Kofi Annan to the fifth meeting of the UN Information and Communications Technology Task Force in Geneva. Press Release SG/SM/8867, PI/1506. 12.9. 2003.

[Annan, 2005] Annan, K. (2005). Secretary-general's message to the eighth meeting of the un ict task force. Delivered by Mr. Jose Antonio Ocampo, Under-Secretary-General for Economic and Social Affairs.

[Appleton, 1998] Appleton, S. (1998). Changes in Poverty in Uganda, 1992-1996. Centre for the Study of African Economies. University of Oxford.

[Audenhove, 2001] Audenhove, L. v. (2001). *Information Technology in Context. Studies from the perspective of developing countries*, chapter Information and Communication Technology Policy in Africa: A Critical Analysis of Rhetoric and Practice, (pp. 277–290). Ashgate Publishing. Aldershot.

[Avgerou & Madon, 1993] Avgerou, C. & Madon, S. (1993). *Computers and Society - citizenship in the information age*, chapter Development, self-determination and information, (pp. 4–12). Intellect Ltd. Oxford.

[Avgerou & Walsham, 2000] Avgerou, C. & Walsham, G. (2000). *Information technology in context: Implementing systems in the developing world.* Ashgate Publishing. Aldershot.

[Avgerou & Walsham, 2001] Avgerou, C. & Walsham, G. (2001). *Information Technology in Context. Studies from the perspective of developing countries*, chapter Introduction: IT in Developing Countries, (pp. 1–9). Ashgate Publishing. Aldershot.

[Ayodele et al., 2005] Ayodele, T., Cudjoe, F., Nolutshungu, T. A., & Sunwabe, C. K. (2005). African perspectives on aid: Foreign assistance will not pull africa out of poverty. Internet: http://www.cato.org/pubs/edb/edb2.html. Accessed: 29.08.2006.

[Baker et al., 1995] Baker, D., Island, S., & P., T. (1995). The effect of culture on the learning of science in non-western countries: The result of an integrated research review. *International Journal of Science Education*, 17(6), 695–704. Taylor & Francis.

[Baker, 2001] Baker, W. (2001). *Uganda:The Marginalization of Minorities.* An MRG International Report.

[Baxter, 1960] Baxter, P. (1960). *East African Chiefs: A Study of Political Development in Some Uganda and Tanganyika Tribes*, chapter The Kiga, (pp. 278–310). Faber. London. London.

[Berger & Luckmann, 1966] Berger, P. & Luckmann, T. (1966). *Social Construction of Reality: A Treatise on the Sociology of Knowledge.* Anchor Books. New York.

[Bijker, 1995] Bijker, W. E. (1995). Of bicycles, bakelites, and bulbs : toward sociotechnical change. *Cambridge, MA: MIT Press.*

[Bjorn-Andersen, 1990] Bjorn-Andersen, N. (1990). *Information technology in Developing Countries - for Better for worse? Information Technologies in Developing Countries.* North Holland. Amsterdam.

[Blumer, 1954] Blumer, H. (1954). What is wrong with social research. *American Sociological Review*, 14, 3–10.

[BMAA, 2002] BMAA (2002). THE FOREIGN MINISTRY. Internet: http://www.bmaa.gv.at/up-media/98_Tabelle_gesamt2002vorl.pdf. Accessed: 26.09.2003.

[Bolander, 2003] Bolander, K. (2003). Student centred learning. Internet: http://www.uwic.ac.uk/ltsu/student_centred_learning.htm. Accessed: 01.07.2003.

[Bosworth, 1995] Bosworth, J. (1995). *Land and Society in South Kigezi. Uganda.* PhD thesis, Oxford.

[Bourdieu, 1970] Bourdieu, P. (1970). *Zur Soziologie der symbolischen Formen.* Suhrkamp.

[Bourdieu, 1991] Bourdieu, P. (1991). *Language and Symbolic Power*. Polity Press. Cambridge. Cambridge.

[Bourdieu, 1992] Bourdieu, P. (1992). *Die feinen Unterschiede. Kritik der gesellschaftlichen Urteilskraft*. Suhrkamp. Frankfurt/M.

[Burton, 2001] Burton, S. (2001). *Knowledge, Information and Development: An African Perspective*, chapter Development communication: towards a social action perspective, (pp. 215–228). School of Human and Social Studies. University of Natal. Pietermaritzburg.

[Caruana-Dingli, 2005] Caruana-Dingli, M. (2005). Integrating ICT and multicultural aspects within a classroom: the SAIL project. *Intercultural Education*, 16(4), 395–404.

[Checkland & Scholes, 1990] Checkland, P. & Scholes, J. (1990). *Soft systems methodology in action*. John Wiley and Sons, Inc.

[Chemonics International Inc. Washington, 2001] Chemonics International Inc. Washington, D. (2001). *USAID/UGANDA SO 7 Assessment of Strategic Agriculture & Environment Options*. Technical report, USAID/ Uganda.

[CIA, 2006] CIA (2006). Cia-factbook. Internet: https://www.cia.gov/cia/publications/factbook/geos/ug.html. Accessed: 03.09.2006.

[Cisler & Yocam, 2003] Cisler, S. & Yocam, K. (2003). Connect-ED Evaluation (Uganda). USAID.

[Cobern, 1991] Cobern, B. (1991). *World view theory and science education research. NARST Monograph No. 3*. National Association for Research in Science Teaching. Manhattan.

[Cobern & Aikenhead, 1997] Cobern, B. & Aikenhead, G. (1997). Cultural Aspects of Learning Science. In *Paper presented at the 1997 annual meeting fo the NARST, Chicago*.

[Cobern, 1996] Cobern, W. W. (1996). Worldview theory and conceptual change in science education. *Science Education*, 80(5), 579–610.

[Cole, 1959] Cole, K. (1959). *Kenya: Hanging in the Middle Way*. Church Army Press. Oxford.

[Comenius-Institut, 2004] Comenius-Institut (2004). Eckwerte interkulturalität. Internet: http://marvin.sn.schule.de/c̆i/download/bg_lp_eckwerte_interkulturalitaet.pdf. Accessed: 20.10.2004.

[Couprie et al., 2006] Couprie, D., Goodbrand, A., Li, B., & Zhu, D. (2006). Soft Systems Methodology. Internet: http://sern.ucalgary.ca/courses/seng/613/F97/grp4/ssmfinal.html. Accessed: 08.08.2006.

[Cushner & Brislin, 1995] Cushner, K. & Brislin, R. W. (1995). *Intercultural Interactions, A Practical Guide*, volume Cross-Cultural Research and Metodology. Sage Publications. Thousand Oaks, second edition.

[Davis, 1989] Davis, F. (1989). Perceived Usefulness, Perceived Ease of Use, and User Acceptance of Information Technology. *MIS Quarterly. Minnesota*, 13(3), 318–340.

[Demorgon, 1999] Demorgon, J. (1999). *Interkulturelle Erkundungen - Möglichkeiten und Grenzen einer internationalen Pädagogik*. Campus-Verlag. Frankfurt/Main.

[Denzin & Lincoln, 2005] Denzin, N. K. & Lincoln, Y. S., Eds. (2005). *The SAGE Handbook of Qualitative Research, Third Edition*. Sage Publications. Thousand Oaks.

[DFID - Department for International Development, 2001] DFID - Department for International Development (2001). IMFUNDO: Partnership for IT in Education Inception Report. Internet: http://www.imfundo.org. Accessed: 10.07.2006.

[Diamond, 1999] Diamond, J. (1999). *Guns, Germs, and Steel: The Fates of Human Societies*. W. W. Norton & Company.

[Dyson, 2004] Dyson, L. E. (2004). Cultural Issues in the Adoption of Information and Communication Technologies by Indigenous Australians. In F. Sudweeks & E. C. (Eds.), *Proceedings of the Fourth International Conference on Cultural Attitudes towards Technology and Communication (CATaC), Karlstad, Sweden, 27 June-1 July 2004, Murdoch University, Murdoch W.A.* (pp. 58–71).

[Ellinger, 2006] Ellinger, S. (2006). Grounded Theory als methodischer Zugang für Werteforschung in der Lernbehindertenschule. Internet: http://wwwalt.uni-wuerzburg.de/sopaed1/ellinger/habil.pdf. Accessed: 18.10.2006.

[Espeland, 2003] Espeland, R. H. (2003). The relationship between Bakigas and Batoros in Kyenjojo District, Uganda. Master's thesis, Institute of Social Anthropology. Bergen.

[Ess, 2004] Ess, C. (2004). Cross-Cultural Communication Online: How Diverse Cultural Values and Communicative Preferences Shape Users and Uses of Computer-mediated Communication Technologies. Presented 03. 08. 2004, Posner Center Board Room, Carnegie Mellon University.

[Ess & Sudweeks, 2001] Ess, C. & Sudweeks, F. (2001). On the edge: Cultural barriers and catalysts to IT diffusion among remote and marginalized communities. *new media & society*, 3, 259–269.

[Freedman, 1976a] Freedman, J. (1976a). Joking, affinity and the exchange of ritual services among the Kiga of Northern Rwanda. An essay on joking relationship theory. *Man, New Series*, 12(1), 154–165.

[Freedman, 1976b] Freedman, J. (1976b). *Principles of relationship in Rwandan Kiga society*. PhD thesis, Xerox Univ. Microfilms. Princeton.

[Gadamer, 1960] Gadamer, H. (1960). *Wahrheit und Methode: Grundzüge einer philosophischen Hermeneutik.* Mohr. Tübingen.

[Geertz, 1973] Geertz, C. (1973). *The Interpretation of Cultures,* chapter Thick Description: Toward an interpretive theory of culture, (pp. 3–30). Basic Books. New York.

[Geertz, 1983] Geertz, C. (1983). *Dichte Beschreibung. Beiträge zum Verstehen kultureller Systeme.* Suhrkamp Verlag. Frankfurt am Main. Frankfurt am Main.

[Gerster & Zimmermann, 2003] Gerster, R. & Zimmermann, S. (2003). *ICTs and Poverty Reduction in Sub-Saharan Africa.* Building Digital Opportunities (BDO) Programme.

[Glaser & Strauss, 1967] Glaser, B. G. & Strauss, A. (1967). *The Discovery of Grounded Theory.* Aldine Transaction.

[Golden, 1978] Golden, D. (1978). Technology Transfer from the Developed to the Less Developed Countries. *J. Moneta Information Technolgy, JCIT.* North Holland. Amsterdam.

[Gomes, 2003] Gomes, B. d. A. F. (2003). *Die Praxis der Entwicklungszusammenarbeit,* chapter Entwicklungszuammenarbeit (EZA), (pp. 13–25). Mandelbaum.

[Gotschi, 2003] Gotschi, E. (2003). Education policies in uganda. Master's thesis, Johannes Kepler University Linz.

[Gürses, 1998] Gürses, H. (1998). Der andere Schauspieler. Bemerkungen zum Kulturbegriff. *polylog - Zeitschrift für interkulturelles Philosophieren,* 2, 62–81.

[Haddad & Draxler, 2000] Haddad, W. D. & Draxler, A., Eds. (2000). *Technologies for Education: Potentials, Parameters and Prospects.* UNESCO and The Academy for Educational Development (AED).

[Haidar, 1997] Haidar, A. (1997). *Report of the Joint Research Project: Effects of Traditional Cosmology on Science Education,* chapter Western Science & Technology Education and the Arab World, (pp. 6–14). MITO. Japan.

[Hall, Edward T., 1976] Hall, Edward T. (1976). *Beyond Culture.* Anchor Books. New York.

[Hall, Stuart, 2002] Hall, Stuart (2002). *Cultural studies.* Argument-Verlag. Hamburg.

[Haqqani, 2003] Haqqani, A. B. (2003). *The Role of Information and Communication Technologies inGlobal Development - Analyses and Policy Recommendations.* UN Information and Communication Technologies Task Force.

[Harris et al., 2001] Harris, R., Bala, P., Songan, P., Khoo Guatlien, E., & Trang, T. (2001). Challenges and Opportunities in Introducing Information and Communication Technologies to the Kelabit Community of North Central Borneo. *new media and society,* 3, 270–295.

[Hawkins, 2002] Hawkins, R. (2002). *The Global Information Technology Report*, chapter Ten Lessons for ICT and Education in the Developing World, (pp. 38–43). Oxford University Press. Oxford.

[Heaton L. and G. Nkunzimana, 2006] Heaton L. and G. Nkunzimana (2006). What makes a technology appropriate or appropriable? In *Proceedings of the Fifth International Conference on Cultural Attitudes towards Technology and Communication (CATaC), Tartu, Estonia, 28 June-1 July 2006, Murdoch University, Murdoch W.A.*

[Heeks, 2000] Heeks, R. (2000). *Lessons for Development from the 'New Economy'*. IDPM. Manchester.

[Heeks, 2002] Heeks, R. (2002). Information Systems and Developing Countries: Failure, Success, and Local Improvisations. *The Information Society*, 18, 101–112.

[Heidenreich, 1994] Heidenreich, G. E. (1994). *Arbeit und Beruf bei Handwerkern der Bakiga in Uganda*. Bayreuth University. Bayreuth. Bayreuth.

[Heinrich-Böll-Stiftung, 2003] Heinrich-Böll-Stiftung (2003). Akteursgruppen des WSIS. Internet: http://www.worldsummit2003.de/de/web/166.htm. Accessed: 29.08.2006.

[Hodge & Kress, 1988] Hodge, R. & Kress, G. (1988). *Social Semiotics*. Polity Press. Cambridge.

[Hofstede, 1980] Hofstede, G. (1980). *Culture's Consequences: International Differences in Work-related Values*. Sage. Beverly Hills.

[Hofstede, 1991] Hofstede, G. (1991). *Cultures and Organizations*. McGraw-Hill. New York.

[Hofstede & Hofstede, 2005] Hofstede, G. & Hofstede, G. (2005). *Cultures and Organizations - Software of the Mind. Intercultural Cooperations and Its Importance for Survival, 2nd. Edition*. McGraw-Hill. New York.

[Hyman & Lowe, 1959] Hyman, L. & Lowe, J. (1959). Kiga-Nkore dictionary by Charles Taylor. Internet: http://www.cbold.ddl.ishlyon.cnrs.fr. Accessed: 24.05.2004.

[Hödl, 2003] Hödl, G. (2003). *Die Praxis der Entwicklungszuammenarbeit*, chapter Die Anfänge - vom Empfänger - zum Geberland, (pp. 28–45). Mandelbaum. Wien.

[IDRC et al., 1997] IDRC, ITU, & UNESCO (1997). Proposal for International Co-operation On Multipurpose Community Telecentre Pilot Projects in Africa. Internet: http://www.itu.int/ITUD/univ_access/reports/telepro2.html. Accessed: 01.02.2007.

[IFIP WG 8.2., 2003] IFIP WG 8.2. (2003). Scope and Aims of IFIP WG 8.2. Internet: http://www.ifipwg82.org/scope.php3. Accessed: 16.09.2003.

[IICD, 2000] IICD (2000). ICT Roundtable Workshop Education Uganda (2000) - Summary Report. Internet: http://www.IICD.org. Accessed: 29.08.2006.

[IMFUNDO PROJECT TEAM, 2000] IMFUNDO PROJECT TEAM (2000). Imfundo Project: Second Report to the Advisory Group Meeting of 14 September 2000. Internet: http://www.dfid.gov.uk/pubs/files/imfundo/AdvisoryGroupPaper-14.pdf. Accessed: 15.08.2006.

[infoDev, 2005a] infoDev (2005a). ICTs and the Education MDGS Briefing Sheet (March 2005). Internet: http://www.infodev.org/en/Publication.140.html. Accessed: 14.06.2006.

[infoDev, 2005b] infoDev (2005b). Knowledge Maps: ICTs in Education. Internet: http://www.infodev.org/files/2907_file_Knowledge_Maps_ICTs_Education_infoDev.pdf. Accessed: 29.08.2006.

[ITU, 2001] ITU (2001). The Internet in an African LDC. Internet: http://www.itu.int/ITU-D/ict/cs/uganda/material/uganda.pdf. Accessed: 06.09.2006.

[Jegede, 1996] Jegede, O. (1996). Culture and citizenship: An African perspective. Inaugural Conference of the Australian Key Centre for Cultural and Media Policy, Australian Key Centre for Cultural Policy, Griffith University, 61.

[Jegede, 1994] Jegede, O. J. (1994). STS Education: International Perspectives on Reform, chapter African cultural perspectives and the teaching of science. Teachers College Press. New York.

[Jegede & Aikenhead, 1999] Jegede, O. J. & Aikenhead, G. (1999). Transcending cultural borders: Implications for science teaching. In Paper presented at pre-conference workshop of annual meeting of the NARST, Boston, MA.

[Kamppuri & Tukianinen, 2004] Kamppuri, M. & Tukianinen, M. (2004). Cultur in Human-Computer Interaction Studies: A survey of ideas and definitions. In Ess, Charles and Sudweeks, F. (Ed.), Proceedings of the Fourth International Conference on Cultural Attitudes towards Technology and Communication (pp. 43-57).

[Karwemera, 2000] Karwemera, F. (2000). LEARN Runyankore - Rukiga - English. Kigezi Printers and Stationers. Kabale.

[Karwemera, 2003] Karwemera, F. (2003). Interview with Festo Karwemera, Bakiga-Expert in Lower Bugongi, Kabale. Recorded on: 30.7. 2003.

[Karwemera, 2006] Karwemera, F. (2006). Biography. Internet: http://www.banyakigezi.org/gen/_area/bios/karwemera.htm. Access: 24.01.2006.

[Kearney, 1984] Kearney, M. (1984). World view. Chandler & Sharp Publishers. Novato.

[Kenny & Qiang, 2000] Kenny, N.-S. & Qiang (2000). ICTs and Poverty. Internet: http://www.worldbank.org/poverty/strategies/chapters/ict/ict0829.pdf. Accessed: 20.11. 2003.

[Kiyaga-Nsubuga, 2005] Kiyaga-Nsubuga, J. (2005). The role of local leadership in improving the delivery of services at community level: Uganda's experience with primary education. Prepared for the Ministerial Conference on Leadership Capacity Building for Decentralised Governance and Poverty Reduction for Sub-Saharan Africa.

[Kozma, 2002] Kozma, R. (2002). ICT and Educational Reform in Developed and Developing Countries. Internet: http://web.udg.es/tiec/orals/c17.pdf. Accessed: 29.08.2006.

[Kress, 1989] Kress, G. (1989). *Linguistic processes in sociocultural practice*. Oxford University Press.

[Kroeber & Kluckhohn, 1952] Kroeber, A. L. & Kluckhohn, C. (1952). *Culture. A Critical Review of Concepts and Definitions*. Peabody Museums. Cambridge.

[Krummacher, 2004] Krummacher, A. (2004). Der Participatory Rural Appraisal (PRA) - Ansatz aus ethnologischer Sicht. *Working Papers Nr. 36, Department of Anthropology and African Studies, Mainz*.

[Kurt, 2004] Kurt, R. (2004). *Hermeneutik. Eine sozialwissenschaftliche Einführung*. UKV. Konstanz.

[Kwitonda, 1995] Kwitonda, A. (1995). *UGANDA - A Century of Existence*, chapter A century of school and education in Uganda, (pp. 220–233). Fountain Publishers. Kampala.

[Latour, 1987] Latour, B. (1987). *Science In Action: How to Follow Scientists and Engineers Through Society*. Harvard University Press, Cambridge Mass., USA.

[Latu, 2006] Latu, S. (2006). Pacific islanders and ict. *Proceedings of the Fifth International Conference on Cultural Attitudes towards Technology and Communication (CATaC), Tartu, Estonia, 28 June - 1 July 2006, Murdoch University, Murdoch W.A.*

[Latu & Young, 2004] Latu, S. & Young, A. (2004). Teaching ICT to Pacific Island Background Students. *Presented at the Sixth Australasian Computing Education Conference (ACE2004), Dunedin, Conferences in Research and Practice in Information Technology, 30*.

[Lenzen, 1991] Lenzen, D. (1991). *Das Fremde*, chapter Multikulturalität als Monokultur. Leske u. Budrich. Opladen.

[Long, 1992] Long, N. (1992). *Battlefields of knowledge*. Routledge. Oxford.

[Long, 2001] Long, N. (2001). *Development Sociology: Actor Perspectives*. Routledge. Oxford.

[Long, 2002] Long, N. (2002). An Actor-oriented Approach to Development Intervention. *Report of the APO Seminar on Rural Life Improvement for Community Development, Japan, 22-26 April 2002*, (pp. 47–61).

[Ludwig, 2003] Ludwig, W. (2003). Genfer UNO-Weltgipfel zur Informationsgesellschaft. Anspruch und Wirklichkeit eines Aushandlungsprozesses. *Widerspruch*, 45, 93–103.

[Luig, 1968] Luig, U. (1968). Preliminary Observations on Kinship, Friendship and Voluntary Associations among the Kiga in Mulago. University of East Africa, Makerere Institute of Social Research. In *Conference Papers. Kampala* (pp. 233–244).

[Lévi-Strauss & Eribon, 1988] Lévi-Strauss, C. & Eribon, D. (1988). *De prés et de loin*. Edition Odille Jacob. Paris.

[Madon, 2000] Madon, S. (2000). The Internet and Socio-economic development: Exploring the interaction. *Information Technology and People*, 13(2), 85–101.

[Mansell & Wehn, 1998] Mansell, R. & Wehn, U., Eds. (1998). *Knowledge Societies: Information Technology for Sustainable Development*. World ICT and Development Report prepared for the United Nations. Oxford University Press. Oxford.

[Mayanja, Meddie, 2003] Mayanja, Meddie (2003). The uganda vsat school-based telecenter end of year status report. Internet: http://info.worldbank.org/etools/docs/library/91628/telecentres/telecentres /docs/updates/ug-statusreport-end02.pdf. Accessed: 20.09.2006.

[Menon & Naidoo, 2003] Menon, M. & Naidoo, V. (2003). COL Experiences in ICT for School Education. Africa Regional Conference in Teacher Training on 'Use of ICT in the Classroom'. Nairobi: 4-6 November 2003. Internet: http://www.schoolnetafrica.net/fileadmin/resources/COL_experiences_in_ICTs _for_Education.pdf. Accessed: 29.08.2006.

[Miller, 2006] Miller, J. (2006). Perspectives and Policies on ICT in Africa. Springer. New York.

[Ministry of Finance, 2001] Ministry of Finance (2001). Uganda Poverty Status Report 2001 Summary. Internet: http://www.finance.go.ug/. Accessed: 10.11.2006.

[Ministry of Finance, Planning and Economic Development, 2001] Ministry of Finance, Planning and Economic Development (2001). Poverty Reduction Strategy Paper (PRSP) - Progress Report 2001.

[Ministry of Works, Housing and Communications et al., 2002] Ministry of Works, Housing and Communications, the President's Office, & National Council for Science and Technology (2002). *National Information and Communication Technology Policy*. Ministry of Works, Housing and Communications and The President's Office and National Council for Science and Technology.

[Mugisha, 2002] Mugisha, S. (2002). Using ICTs in Development: The case of Uganda. Internet: http://www.itcd.net. Accessed 12.10.2006.

[Museveni, 1995] Museveni, Y. (1995). Does Africa matter? *New Perspectives Quarterly*, 12(4), 53–55.

[Mwanahewa, 1995] Mwanahewa, S. A. (1995). *Uganda - A Century of Existence*, chapter From cross-cultural inconsistency to conflict: A logical exposure, (pp. 97–109). Okoth P. Godfrey. Fountain Publishers. Kampala. Uganda.

[Myuganda, 2004] Myuganda   (2004).   Myuganda.co.ug.   Internet: http://myuganda.co.ug. Accessed: 09.01.2004.

[Nath, 2000] Nath, V. (2000). ICT Enabled Knowledge Societies for Human Development.   IFIP Working Group 9.4. Newsletter, Vol. 10. No 2. Internet: http://www.iimahd.ernet.in/egov/ifip/aug2000.htm. Accessed: 05.09.2006.

[Nel & Wilkinson, 2001] Nel, G. & Wilkinson, L. (2001). Where is the "Any key", Sir? Experiences of an African Teacher-To-Be. In *Proceedings of Society for Information Technology and Teacher Education (SITE) International Conference 2001*.

[NEMA, 1997] NEMA, Ed. (1997). *District State of Environment Report, Kabale, 1997*. Internet: http://www.nemaug.org/. Accessed: 10.10.2006.

[NEMA, 2001] NEMA (2001). Uganda State of the Environment Report 2000 Version 2. Kampala, Uganda: National Environment Management Authority, Ministry of Natural Resources, Government of Uganda.

[NEMA, 2004] NEMA   (2004).   Kabale   district.   Internet: http://www.nemaug.org/UPLOADS/KABALE.pdf. Accessed: 09.01.2004.

[NEMA, 2005] NEMA (2005). *District State of the Environment Report 2004 for Kabale*. Internet: http://www.nemaug.org/. Accessed: 10.10.2006.

[Ngologoza, 1967] Ngologoza, P. (1967). *Kigezi and its People*. Fountain Publishers. Kampala.

[Niavarani, 2005] Niavarani, J. (2005). frauenrechte - länderprofil u g a n d a. VIDC. Wien.

[Nocera, 2002] Nocera, J. A. (2002). Ethnography and Hermeneutics in Cybercultural Research Accessing IRC Virtual Communities. *Journal of Computer-Mediated Communication*, 7(2).

[Nocera, 2006] Nocera, J. A. (2006). The politics of technology culture. In *Proceedings of the Fifth International Conference on Cultural Attitudes towards Technology and Communication (CATaC), Tartu, Estonia, 28 June-1 July 2006, Murdoch University, Murdoch W.A.*

[Nulens, Gert, 1997] Nulens, Gert (1997). The policy of the World Bank Studies on Media Information and Telecommunication. *Socio-cultural aspects of information technology in Africa*, 23(2), 15–23.

[Nzita, Richard and Mbaga-Niwampa, 1993] Nzita, Richard and Mbaga-Niwampa (1993). *Peoples and Culturese of Uganda*. Fountain Publishers. Kampala.

[Obrecht, 2005] Obrecht, A. (2005). *Einführung in die Entwicklungssoziologie*, chapter Partizipative Entwicklungsforschung zwischen Humanitärer Hilfe und Entwicklungszusammenarbeit, (pp. 237–265). Mandelbaum. Wien.

[Odedra-Straub, 2002] Odedra-Straub, M. (2002). A Way Forward ... *EJISDC*, 10(1), 1–2.

[Odedra-Straub, 2003] Odedra-Straub, M. (2003). E-mail correspondence. E-mail.

[Odhiambo, 1972] Odhiambo, J. R. (1972). *Science Education in Africa*, chapter Understanding of science: The impact of the African view of nature. Heinemann. London.

[Oevermann, 2002] Oevermann, U. (2002). Klinische Soziologie auf der Basis der Methodologie der objektiven Hermeneutik - Manifest der objektiv hermeneutischen Sozialforschung. Internet: http://www.ihsk.de/publikationen/manifest.pdf. Accessed: 21.07.2006.

[Ogunniyi, 1988] Ogunniyi, M. I. (1988). Adapting Western Science to Traditional African Culture. *Journal of Science Education*, 10(1), 1–9.

[Okee-Obong, 2003] Okee-Obong, J. (2003). Interview with Okee-Obong, Ugandan Sociologist, Center of Austrian African Cooperation. Recorded on: 14.11.2003. Interview.

[Okee-Obong & Langthaler, 2006] Okee-Obong,   J.   B.   &   Langthaler,   H.   (2006).   Migration   und   entwicklung.   Internet: http://www.archiv.gruene.at/planet/planet44/index.php?seite=themen&tid=41774. Accessed: 06.09.2006.

[Okot-Uma, 1992] Okot-Uma, R. (1992). *Social Implications of Computers in Developing Countries*, chapter A Perspective of Contexttual, Operational and Strategy: Problems of informediation in developing countries, (pp. 10–25). McGraw-Hill. New York.

[Okoth, 1995] Okoth, P. G. (1995). *Uganda - A Century of Existence*, chapter Uganda's gepolitical significance since 1894, (pp. 3–35). Fountain Publishers. Kampala.

[Okunoye, 2003] Okunoye, A. O. (2003). *Knowledge Management and Global Diversity: A Framework to Support Organisations in Developing Countries*. PhD thesis, University of Turku, Department of Information Technology.

[Olivier, 1994] Olivier, A. P. S. (1994). An Information Technology Policy towards Development in a Developing Country: The situation in South Africa. *Working Paper Series - University of Pretoria - Department of Informatics*, 19.

[olpc, 2006] olpc (2006). One Laptop per Child. Internet: http://www.laptop.org. Accessed: 20.11.2006.

[Orlikowski & Gash, 1994] Orlikowski, W. & Gash, D. C. (1994). Technological Frames: Making Sense of Information Technology in Organisations. *ACM Transactions on Information Systems*, 12(2), 174–207.

[Parsons, 1951] Parsons, T. (1951). *The social system*. Free Press. New York.

[Pichlhöfer, 2000] Pichlhöfer, H. (2000). *Discourse-semiotic analysis of advertisements in Tanzania, Mozambique, and Austria with respect to intercultural communication*. PhD thesis, Vienna University.

[Ploeg, 1989] Ploeg, J. D. v. d. (1989). *Encounters at the Interface: A Perspective on Social Discontinuities in Rural Development*, chapter Knowledge Systems, Metaphor and Interface: The Case of Potatoes in the Peruvian Highlands. Wageningen Agricultural University. Wageningen.

[Plooy & Roode, 1999] Plooy, N. d. & Roode, J. (1999). The Social Context of Implementation and Use of Information Technology. *Working Paper Series - University of Pretoria - Department of Informatics*, 44.

[PMZ, 2006] PMZ (2006). Kritik am 100-Dollar-Laptop lässt Negroponte kalt. Internet: http://www.heise.de/newsticker/meldung/71714.

[Polenz, 1981] Polenz, P. (1981). *Wissenschaftssprache*, chapter Über die Jargonisierung von Wissenschaftssprache und wider die Deagentivierung., (pp. 85–110). Fink. München.

[Postma, 2001] Postma, L. (2001). A Theoretical Argumentation and Evaluation of South African Learners' Orientation towards and Perceptions of the Empowering Use of Information: A Calculated Prediction of Computerized Learning for the Marginalized. *new media & society*, 3, 313–326.

[Postman, 1992] Postman, N. (1992). *Technopoly: The surrender of culture to technology*. Vintage Books. New York.

[Press, 1997] Press, L. (1997). A framework to characterise the global diffusion of the Internet. Notes for presentation at INFO'97, Havana. Internet: http://som.csudh.edu/fac/lpress. Accessed: 01.02.2007.

[Proenza, 2003] Proenza, F. (2003). *Connected for Development - Information Kiosks and Sustainability*, chapter A Public Sector Support Strategy for Telecenter Development: Emerging Lessons from Latin America and Caribbean, (pp. 9–14). Department of Economic and Social Affairs.

[Pryor & Ampiah, 2003] Pryor, J. & Ampiah, J. G. (2003). *Understandings of Education in an African Village: the Impact of Information and Communication Technologies. Report on DFID Research Project Ed2000-88.* DFID.

[Rapoport, 1970] Rapoport, R. (1970). Three dilemmas in action research. *Human Relations*, 23(6), 499–513.

[Sahay & Avgerou, 2002] Sahay, S. & Avgerou, C. (2002). Introducing the Special Issue on Information and Communication Technologies in Developing Countries. *The Information Society*, 18, 73–76.

[Sahay et al., 1994] Sahay, S., Palit, M., & Robey, D. (1994). A Relativist Approach to Studying the Social Construction of Information Technology. *European Journal of Information Systems*, 3(4), 248–258.

[Samli, 1991] Samli, A. (1991). Information technology transfer to developing countries - Is it really taking place? Internet: http://www.odedra-straub.com/publications/7-Informationfer.pdf. Accessed: 14.07.06.

[Scheepers, 1999] Scheepers, H. (1999). The Computer-Ndaba experience: Introducing IT in a rural community in South Africa. Presented on the The Twenty Second IRIS Conference, Keuruu, Finland.

[Scheepers & de Villiers, 2000] Scheepers, H. & de Villiers, C. (2000). Teaching of a computer literacy course in South Africa: A case study using traditional and co-operative learning. *Information Technology for Development*, 9(3-4), 175–187.

[Scheepers & Mathiassen, 1998] Scheepers, H. & Mathiassen, L. (1998). Out of Scandinavia - Facing social risks in IT development in South Africa. *Journal of Global Information Management*, 8(2), 36–49.

[Schiff & Ozden, 2005] Schiff, M. & Ozden, C., Eds. (2005). *International Migration, Remittances & the Brain Drain.* Palgrave Macmillan. New York.

[Schneeberger & Sedlacek, 2002] Schneeberger, R. & Sedlacek, O. (2002). Project Report - ProjektÜganda.

[SchoolNet Africa et al., 2003] SchoolNet Africa et al. (2003). Workshop Report - ICTs in African Schools Workshop.

[Schütz & Luckmann, 1973] Schütz, A. & Luckmann, T. (1973). *The Structures of the Lifeworld.* Evanston.

[Scriven, 1967] Scriven, M. (1967). *The Methodology of Evaluation.* Rand McNally. Chicago.

[Selinger & Gibson, 2004] Selinger, M. & Gibson, I. (2004). Cultural Relevance and Technology Use: Ensuring the Transformational Power of Learning Technologies in Culturally Defined Learning Environments. In *World Conference on Educational Multimedia, Hypermedia and Telecommunications (EDMEDIA) 2004 Lugano, Switzerland* (pp. 5310–5317).

[Sen, 1999] Sen, A. (1999). *Development as Freedom*. Anchor. New York.

[Slay, 2002] Slay, J. (2002). Human Activity Systems: A Theoretical Framework for Designing Learning for Multicultural Settings. *Educational Technology and Society*, 5(1).

[Snyman & Hulbert, 2004] Snyman, M. & Hulbert, D. (2004). Implementing ICT Centres for Development in South Africa: Can Cultural Differences Be Overcome? . In Sudweeks, F. and Ess, Charles (Ed.), *Proceedings of the Fourth International Conference on Cultural Attitues towards Technology and Communication. Karlstad, Sweden, 27 June-1 July 2004* (pp. 626–630).: Western Australia: Murdoch University Murdoch.

[Soeffner, 2003] Soeffner, H.-G. (2003). *Qualitative Forschung. Ein Handbuch*, chapter Sozialwissenschaftliche Hermeneutik, (pp. 164–175). Rowohlt. Hamburg.

[Ssekamwa & Lugumba, 1971] Ssekamwa, J. C. & Lugumba, S. M. E. (1971). *Development and Administration of Education in Uganda*. Fountain Publishers. Kampala.

[Ssekamwa & Lugumba, 1973] Ssekamwa, J. C. & Lugumba, S. M. E. (1973). *A History of Education in East Africa (Fountain series in education studies)*. Fountain series in education studies. Fountain Publishers. Kampala.

[Steinhardt, 1994] Steinhardt, G. (1994). *Wartezeit*, chapter Der Computer als neues Kulturelement in der Lebenswelt Jugendlicher. Österreichischer StudienVerlag. Innsbruck.

[Stuart Mathison, 2003] Stuart Mathison, F. (2003). ICTs and Human Development in Asia - On Overcoming the "Forever Pilot" Syndrome. In *A discussion paper prepared for the Asia-Australasia Regional Conference of the International Telecommunications Society Perth, Australia, 22-24 June 2003*.

[Sy, 2001] Sy, P. (2001). Barangays of IT - Filipinizing mediated communication and digital power. *new media and society*, 3(3), 296–312.

[Tacchi et al., 2003] Tacchi, J., Slater, D., & Hearn, G. (2003). *Ethnographic Action Research*. UNESCO.

[Taylor, 1969] Taylor, B. K. (1969). *The Western Lacustrine Bantu. (Nyoro, Toro, Nyakore, Kiga, Haya, Zinza)*. IAI. London.

[Taylor, 1985] Taylor, C. (1985). *Nkore-Kiga. Croom Helm descriptive grammars series*. Routledge. London.

[Thomas et al., 2003a] Thomas, A., Kienast, E.-U., & Schroll-Machl, S. (2003a). *Handbuch Interkulturelle Kommunikation und Kooperation Band 1*. Vandenhoeck & Ruprecht. Göttingen.

[Thomas et al., 2003b] Thomas, A., Kienast, E.-U., & Schroll-Machl, S. (2003b). *Handbuch Interkulturelle Kommunikation und Kooperation Band 2*. Vandenhoeck & Ruprecht. Göttingen.

[Thompson, 1981] Thompson, A. R. (1981). *Education and development in Africa : an introduction to the study of the role education may play in national development intended primarily for teachers in training and in service.* Palgrave Macmillan. New York.

[Tipps, 1973] Tipps, D. C. (1973). Modernization theory and the comparative study of societies: A critical perspective. *Comparative Studies in Society and History*, 15(2), 199–226.

[Todaro, 1989] Todaro, M. P. (1989). *Economic Development in the Third World.* Longman. New York.

[Totolo, 2005] Totolo, A. (2005). An Exploration of the Theories that Explain the Failure of Information Technology Adoption in Africa. In *Conference On Information Technology Education - Proceedings of the 6th conference on Information technology education.*

[Trompenaars & Hampden-Turner, 1998] Trompenaars, F. & Hampden-Turner, C. (1998). *Riding The Waves of Culture - Understanding Diversity in Global Business.* McGraw-Hill. New York.

[Trust, 2006] Trust, S. (2006). What is a Simputer? Internet: http://www.simputer.org/simputer/about/. Accessed: 01.02.2007.

[Turner, 2005] Turner, S. V. (2005). The Use of ICT in Secondary Schools in Ghana. In *Proceedings of the 8. International Working Conference of IFIP WG 9.4.*

[Tusubira, 2004] Tusubira, F. F. (2004). Policies and Laws that direct the ICT Sector in Uganda - A Critique. Internet: http://www.fftusubira.com. Accessed: 20.11.2004.

[UBOS, 2002] UBOS (2002). 2002 Uganda Population and Housing Census. Internet: http://www.ubos.org. Accessed: 20.09.2006.

[UBOS, 2003] UBOS (2003). Uganda National Household Survey 2002/2003. Internet: http://www.ubos.org. Accessed: 20.09.2006.

[Uganda Bureau of Statistics, 2004] Uganda Bureau of Statistics (2004). Uganda bureau of statistics. Internet: http://www.ubos.org. Accessed: 09.01.2004.

[UNDP, 1991] UNDP (1991). Human Development Report. Oxford University Press. Oxford.

[UNDP, 2005] UNDP (2005). *Human Development Report 2005.* UNDP.

[UNDP, Thematic Trust Fund, 2001] UNDP, Thematic Trust Fund (2001). Information and Communication Technology (ICT) for Development. Internet: http://www.undp.org/trustfunds/TTFICTe.PDF. Accessed: 16.09.2003.

[UNESCO, 2002] UNESCO (2002). Information and Communication Technologies in Teacher Education: A Planning Guide. Internet: http://unesdoc.unesco.org/images/0012/001295/129533e.pdf. Accessed: 29.08.2002.

[UNESCO, 2003] UNESCO, Ed. (2003). *Measuring and monitoring the information and knowledge societies: a statistical challenge.* UNESCO Publications for the WSIS.

[UNESCO, 2004] UNESCO (2004). Unesco in action education the price of school fees. Internet: http://portal.unesco.org/. Accessed: 20.09.2006.

[Vansina, 1973] Vansina, J. (1973). *Oral Tradition (De la Tradition orale. Engl.) A study in historical methodology. Transl. by H. M. Wright.* Penguin. Harmondsworth.

[Walsham, 2000] Walsham, G. (2000). *Information Technology in Context. Studies from the perspective of developing countries,* chapter IT, Globalisation and Cultural Diversity, (pp. 291–303). Ashgate Publishing. Aldershot.

[Walsham & Sahay, 2005] Walsham, G. & Sahay, S. (2005). Research on Information Systems in Developing Countries: Current Landscape and Future Prospects. *Forthcoming Journal for Information Technology for Development.*

[Welsch, 1995] Welsch, W. (1995). Transkulturalität. Zur veränderten VerfaSStheit heutiger Kulturen. *Zeitschrift für Kulturaustausch. Berlin,* 45(1), 39–44.

[Wernet, 2006] Wernet, A. (2006). *Einführung in die Interpretationstechnik der Objektiven Hermeneutik.* Leske u. Budrich. Opladen.

[Wimmer, 2004] Wimmer, F. M. (2004). Überlegungen zur Frage nach Massstäben kultureller Entwicklung. *Journal für Entwicklungspolitik,* XX(3), 11–45.

[Witzel, 1985] Witzel, A. (1985). Das problemzentrierte Interview. *Qualitative Forschung in der Psychologie,* (pp. 227–253). Bertz. Weinheim und Basel.

[World Bank, 2000] World Bank (2000). Millennium development goals. Internet: http://ddp-ext.worldbank.org/ext/GMIS/gdmis.do?siteId=2&menuId=LNAV01HOME1. Accessed: 01.02.2004.

[WTO, 2004] WTO (2004). Development: Definition. Internet: http://www.wto.org/english/tratop_e/devel_e/d1who_e.htm. Accessed: 20.11. 2004.

[Wulf, 1998] Wulf, C. (1998). *Deutsche Gegenwartspädagogik. Vol. 3,* chapter Bildung als interkulturelle Aufgabe. Schneider. Hohengehren.

[Zwicky & Zwicky, 1982] Zwicky, A. & Zwicky, A. (1982). *Sublanguage. Studies of Language in Restricted Semantic Domains,* chapter Register as a Dimension of Linguistik Variation, (pp. 213–218). de Gruyter. Berlin.

[ÖFSE, 2006] ÖFSE (2006). UGANDA - "PRSP" als Strategie zur Armutsbekämpfung. Internet: http://www.öfse.at/download/PRSP/Uganda-Bericht10.pdf. Accessed: 01.02.2007.

List of Figures

List of Tables

Appendix A

Categories of Text Passages

Actors and their Roles

Title	ExNr	Page
A beginning sequence	11.1	130
Reproducing knowledge	11.2	130
Measurements of sanctions	11.4	134
Computer lessons allow usage at home	11.5	134
Beginning sequence of an interview	11.6	135
Pupils are changing their roles	13.1	147
Short answers	13.2	147
Longer statements	13.3	148
Teacher asks for private usage	13.8	152
Headmistress and new computer administrator	13.9	153
Bursar asks secretaries for help	13.10	154
Parents punishing pupils	13.13	156
Interviewer changes roles	13.18	160
ATs are perceived as donors	13.19	162
ATs are representing the donors	13.20	163
Interviewer refers himself as a colleague	13.21	164
Radio as opinion leader	13.22	164
Radio as an information source	13.23	165
Governments curricula influence computer lessons	13.24	166
Mr. Dekanya emerges as a powerful actor	13.25	166

Access

Title	ExNr	Page
Matrons are exclused from access	13.4	149
Matrons are not allowed to access computers	13.5	150
Unrestricted access for teachers	13.7	151
Parents are granting access	13.11	155
Parents want to see the capability	13.12	155

Power and Status

Title	ExNr	Page
Computer improves status of school	14.39	187
To control things	14.43	190
Computer makes you powerful	14.44	191
Computer improves status of school	14.45	192
Computer improves the status of a pupil	14.46	193
Sending e-mails from the Headmistress's office	15.10	208

Terms

Title	ExNr	Page
Informants explain 'kalimagezi'	14.1	168
Kalimagezi is not commonly used	14.2	169
Explanation of kalimagezi	14.3	169
Something IS kalimagezi	14.4	170
Nicknames are used to get familiar	14.5	171

Analogies

Title	ExNr	Page
Analogies of a computer	14.6	172
A Computer is a machine and does work	14.7	173
Computers promote unemployment	14.8	173
Something that processes things	14.9	173
Screen is like a TV	14.10	173
Information gets lost by blowing screens	14.10	174
Typewriter analogy	14.13	175
Computer-typewriter comparison	14.15	176
Typewriter requires energy	14.16	176
Switch on like a radio	14.17	176
Computer is like a person	14.19	177
Computer is like a person	14.20	177

Concepts

Title	ExNr	Page
Computer behaves strange	14.21	177
How to explain a computer	14.28	181
Informant knows the concept of a computer	14.22	178
Mixing up human and computer virus	15.27	217

Expectations of a computer

Title	ExNr	Page
A computer leads you to the right thing	14.12	174
People say that they are human like	14.23	178
Computer speaks all languages	14.24	179
A computer sets one free	14.25	180
Who has created a computer?	14.26	180
Hear about the Internet	14.27	181
Computer and a mobile phone	14.29	182
Computer as a product of developed countries	14.30	183
Become developed through computers	14.31	184
Globalization	14.32	184

Expected personal benefits

Title	ExNr	Page
Vague perception of benefit of communication	14.33	185
Improve living circumstances	14.34	185
Earn a living	14.35	186
Certificates for the future	14.36	186
Certificates for the future	14.37	187
Informants are expecting certificates	14.38	187
Earn money by producing documents	14.40	188
Get a computer to employ oneself	14.41	189
Own an office and earn money	14.42	189
Owning a computer	14.49	194
Makes work easy	15.19	213
We learn in SST that a computer makes work easier	15.20	214
Save time when you know to type quickly	15.28	219
Scanning saves time	15.29	219

Usage and related experiences

Technological Knowledge Transfer - Teaching

Title	ExNr	Page
Differences in teaching computers	11.3	133
Aim of teaching	11.7	137
Parents are teaching as well	13.14	156
Relatives are teaching as well	13.15	156
A secretary is taking lessons in town	13.16	157
Other computer teachers emerge as actors	13.17	158
Usage at home means to study computer	15.2	198
Computer lessons for teachers	15.8	204
ATs applied discovery method	15.35	224
Discovery method as the perceived teaching method	15.36	224
Syllabus is criticized and tests wanted	15.37	224
Expectation to be tested	15.38	225
Tests as a form of feedback	15.39	225
Computer terminology is not clear	15.40	226
Differences in Teaching Computers	15.41	229
Teaching Microsoft Word the whole term	15.42	232
Teaching Microsoft Word the whole term	15.43	232
Changing topics before one is mastered	15.44	234
First to control the keyboard	15.45	234
The focus is on theoretical lessons	15.46	234

www.ingramcontent.com/pod-product-compliance
Lightning Source LLC
Chambersburg PA
CBHW071544080326
40689CB00061B/1807

* 9 7 8 3 6 3 9 0 6 9 2 6 6 *